A Glimpse into Geometric Representation Theory

CONTEMPORARY MATHEMATICS

804

A Glimpse into Geometric Representation Theory

Virtual AMS Special Session
Combinatorial and Geometric Representation Theory
November 20–21, 2021

Mahir Bilen Can
Jörg Feldvoss
Editors

AMS AMERICAN MATHEMATICAL SOCIETY

2020 *Mathematics Subject Classification.* Primary 14C17, 14M15, 14P20, 58A50, 17B56, 20G10, 05E05, 33D52, 14D24, 20G25.

Library of Congress Cataloging-in-Publication Data

Names: Can, Mahir, editor. | Feldvoss, Jörg, 1960– editor.

Title: A glimpse into geometric representation theory : virtual AMS Special Session on Combinatorial and Geometric Representation Theory, November 20–21, 2021 / Mahir Bilen Can, Jörg Feldvoss, editors.

Description: Providence, Rhode Island : American Mathematical Society, [2024] | Series: Contemporary mathematics, 0271-4132 ; volume 804 | Includes bibliographical references.

Identifiers: LCCN 2023051563 | ISBN 9781470470906 (paperback) | ISBN 9781470476649 (ebook)

Subjects: LCSH: Representations of groups–Congresses. | Representations of algebras–Congresses. | Combinatorial analysis–Congresses. | Geometry–Congresses. | AMS: Algebraic geometry – Cycles and subschemes – Intersection theory, characteristic classes, intersection multiplicities. | Algebraic geometry – Special varieties – Grassmannians, Schubert varieties, flag manifolds. | Algebraic geometry – Real algebraic and real analytic geometry – Nash functions and manifolds. | Global analysis, analysis on manifolds – General theory of differentiable manifolds – Supermanifolds and graded manifolds. | Nonassociative rings and algebras – Lie algebras and Lie superalgebras – Cohomology of Lie (super)algebras. | Group theory and generalizations – Linear algebraic groups and related topics – Cohomology theory. | Combinatorics – Algebraic combinatorics – Symmetric functions and generalizations. | Special functions – Basic hypergeometric functions – Basic orthogonal polynomials and functions associated with root systems (Macdonald polynomials, etc.). | Algebraic geometry – Families, fibrations – Geometric Langlands program: algebro-geometric aspects. | Group theory and generalizations – Linear algebraic groups and related topics – Linear algebraic groups over local fields and their integers.

Classification: LCC QA176 .G55 2024 | DDC 515/.7223–dc23/eng/20240326

LC record available at https://lccn.loc.gov/2023051563

Contemporary Mathematics ISSN: 0271-4132 (print); ISSN: 1098-3627 (online)

DOI: https://doi.org/10.1090/conm/804

∞ The paper used in this book is acid-free and falls within the guidelines
established to ensure permanence and durability.
Visit the AMS home page at `https://www.ams.org/`

10 9 8 7 6 5 4 3 2 1   29 28 27 26 25 24

Contents

Preface vii

From motivic Chern classes of Schubert cells to their Hirzebruch and CSM classes
PAOLO ALUFFI, LEONARDO C. MIHALCEA, JÖRG SCHÜRMANN, and CHANGJIAN SU 1

Locally semialgebraic superspaces and Nash supermanifolds
MAHIR BILEN CAN 53

A survey of support theories for Lie superalgebras and finite supergroup schemes
CHRISTOPHER M. DRUPIESKI and JONATHAN R. KUJAWA 87

Wreath Macdonald polynomials, a survey
DANIEL ORR and MARK SHIMOZONO 123

Meromorphic connections on the projective line with specified local behavior
DANIEL S. SAGE 171

Preface

The **2021 Fall Southeastern Sectional Meeting of the American Mathematical Society** (Meeting ♯ 1173) was held virtually on November 20 and 21, 2021. As part of this conference, a special session, entitled **Combinatorial and Geometric Representation Theory**, was organized by Mahir Bilen Can (Tulane University) and Jörg Feldvoss (University of South Alabama). A closely related session, entitled **Cohomology, Representation Theory, and Lie Theory**, was organized by Christopher Bendel (University of Wisconsin-Stout), Christopher Drupieski (DePaul University), Lauren Grimley (Spring Hill College), Paul Sobaje (Georgia Southern University), and Sean Taylor (Spring Hill College). This volume contains some of the speakers' and organizers' articles on the topics of these two special sessions.

This volume features articles contributed by both the speakers and organizers, focusing on the diverse facets of geometric representation theory, a field that bridges various areas of algebra and geometry. Broadly defined, geometric representation theory aims to elucidate algebraic actions by embedding them within a geometric framework, often associating them with manifolds or varieties. This marriage of algebra and geometry leverages a rich array of tools, including quotients, invariants, cohomology, and more, to uncover the hidden properties of algebraic structures under consideration. Combinatorial techniques frequently play a pivotal role in this process, though the specific methodology employed invariably hinges on the chosen category. In this volume, the authors venture into the exploration of geometric representation theory from diverse viewpoints, offering various insights and applications. The volume contains five articles, each presenting a unique perspective on the intersection of algebra and geometry. The titles and the authors of these manuscripts are as follows:

- *From motivic Chern classes of Schubert cells to their Hirzebruch and CSM classes* by Paolo Aluffi, Leonardo C. Mihalcea, Jörg Schürmann, and Changjian Su,
- *Locally semialgebraic superspaces and Nash supermanifolds* by Mahir Bilen Can,
- *A survey of support theories for Lie superalgebras and finite supergroup schemes* by Christopher M. Drupieski and Jonathan R. Kujawa,
- *Wreath Macdonald polynomials, a survey* by Daniel Orr and Mark Shimozono,
- *Meromorphic connections on the projective line with specified local behavior* by Daniel Sage.

We now briefly summarize the contents of these articles.

When considering a category, should we prioritize our focus on its objects or its morphisms? Naturally, the correct response is 'it depends!' For instance, take the example of a group $(G, \cdot)$, which establishes a category with a single object. In this case, the emphasis should be on the morphisms. On the opposite end of the spectrum, consider the category of positive integers, where there exists an arrow from a to b if a divides b. Within this category, we can narrow our attention to the subcategory comprised of prime numbers. It is evident that this subcategory is remarkably intriguing, yet it does not owe its appeal to the absence of arrows. One category, where both objects and morphisms seem to play equal roles, is the category of Schubert varieties.

In their comprehensive and well-written article, Aluffi, Mihalcea, Schürmann, and Su discuss the Grothendieck motivic group of a Schubert cell $X(w)^\circ$, consisting of equivalence classes of morphisms $[Z \to X(w)^\circ]$ modulo the usual (in the words of the authors) additivity relations. There is a motivic Chern transformation, which is a group homomorphism, between the Grothendieck motivic group and the polynomial ring in one variable with coefficients in the Grothendieck ring of vector bundles on the flag manifold G/B:

$$\mathrm{MC}_y : \mathrm{K}_0(\mathrm{var}/X(w)^\circ) \to \mathrm{K}(G/B)[y],$$

where G is a simply connected semisimple complex Lie group and B is a Borel subgroup with a maximal torus T. In fact, there is a T-equivariant version of this homomorphism, and that is the focus of attention in this article. The equivariant Chern-Schwartz-MacPherson (CSM) class of a Schubert cell is the leading term of the image of the motivic Chern transformation. The primary objective of this article is multifaceted. It seeks to offer an exposition of the CSM classes, including the specializations at $y = -1$ and $y = 0$ resulting from the motivic Chern transformation. Additionally, it endeavors to present the proofs of numerous results that have, until now, been confined to expert discourse within this specialized field. There are meticulously studied examples and engaging discussions centered around open conjectures that connect abstract K-theoretic concepts with objects of combinatorial significance. However, in this work, the authors go much beyond a simple expository article. They use the motivic Chern transformation to define two equivariant variants of the Hirzebruch transformation, which appear naturally in the Grothendieck-Hirzebruch-Riemann-Roch formalism. They find recursive descriptions for the equivariant Hirzebruch classes of Schubert cells, their Poincaré duals, and their Segre versions. Furthermore, they explain the functoriality properties needed to extend their results to Schubert cells in partial flag manifolds.

A Nash manifold is an object of a category that serves as intermediary between the category of smooth manifolds and the category of smooth real algebraic varieties. The underlying topological space of a Nash manifold can be viewed as a smooth semialgebraic space. Examples of Nash manifolds naturally emerge as orbits when a real algebraic group acts on a real algebraic variety. A broad generalization of the category of Nash manifolds is given by the category of locally semialgebraic spaces.

In his article, Can discusses two new directions in supergeometry by introducing locally semialgebraic superspaces along with Nash supermanifolds. It is a well-known result of Batchelor that every smooth supermanifold comes from an appropriate vector bundle. By adapting Batchelor's theorem to his framework, Can shows that all locally semialgebraic superspaces and affine Nash supermanifolds come from appropriate vector bundles in suitable categories. In fact, the main idea behind Can's work can be traced back to work of Blattner and Rawnsley, who discovered that Batchelor's theorem can be obtained from formal properties of sheaves on D-spaces that they introduced. Using this idea, Can develops analogs of D-spaces in the contexts of locally semialgebraic spaces and affine Nash manifolds. While all the necessary tools for constructing the category of locally semialgebraic superspaces were developed by Delfs and Knebusch in the mid-1980's, the toolkit for affine Nash supermanifolds requires more preparation. In his article, Can introduces a natural topology to establish the softness of the structure sheaf for an affine Nash manifold. He shows that under this topology, the higher cohomology of quasi-coherent sheaves on an affine Nash manifold vanishes. He then uses these results to prove that every affine Nash supermanifold comes from an appropriate vector bundle on the affine Nash manifold. The article ends with a brief discussion of the author's notion of an affine Nash supergroup.

Homological and geometric methods have a long tradition in the modular representation theory of finite groups and restricted Lie algebras. More recently, Suslin, Friedlander, Bendel, and Pevtsova have generalized this to finite group schemes which also unifies the two different original strands. In their article Drupieski and Kujawa give an overview of the current knowledge about cohomological support varieties for finite-dimensional Lie superalgebras and finite supergroup schemes. The paper is divided into two parts, where the first part reviews support varieties for Lie superalgebras and supergroup schemes over algebraically closed fields of characteristic zero (or more specifically, over the complex numbers) and the second part discusses support varieties for (restricted) Lie superalgebras and (infinitesimal) supergroup schemes over algebraically closed fields of prime characteristic $\neq 2$. The authors do an excellent job in explaining and motivating this quite technical topic. In the second part they even briefly recall the non-super theory developed in the 1980's, 1990's, and early 2000's. The paper is interspersed with enlightening comments and a plethora of open questions and problems, and because of this, it should give anyone a very good first introduction to these important topics in geometric representation theory.

In the first part the authors mainly report on the joint work of Kujawa with Boe and Nakano on support varieties of classical Lie superalgebras (i.e., finite-dimensional complex Lie superalgebras $\mathfrak{g} = \mathfrak{g}_{\overline{0}} \oplus \mathfrak{g}_{\overline{1}}$ for which there is a connected reductive algebraic group $G_{\overline{0}}$ with Lie algebra $\mathfrak{g}_{\overline{0}}$ whose action on $\mathfrak{g}_{\overline{1}}$ differentiates to the adjoint action of $\mathfrak{g}_{\overline{0}}$ on $\mathfrak{g}_{\overline{1}}$). In this case the support varieties are defined using relative cohomology. The main results in this part are the description of the relative cohomology ring as the ring of invariants for certain finite groups and the identification of the support varieties with so-called rank varieties by using two kinds of detecting sub-superalgebras. As a consequence of this identification, one obtains the very much desired tensor product property, namely, that the support variety of the tensor product of two supermodules is the intersection of the support varieties of the two factors. Moreover, if in addition $\mathfrak{g}$ is basic (i.e., $\mathfrak{g}$ admits a non-degenerate invariant supersymmetric even bilinear form similar to the classical Cartan-Killing form of a semi-simple complex Lie algebra) the dimensions of the support varieties have a combinatorial interpretation in terms of the defect $\operatorname{def}(\mathfrak{g})$ of the Lie superalgebra $\mathfrak{g}$ and the atypicality $\operatorname{atyp}(\lambda)$ of a weight λ for the Cartan subalgebra of the even part $\mathfrak{g}_{\overline{0}}$. Namely, $\operatorname{def}(\mathfrak{g})$ is the dimension of the relative cohomology ring, or equivalently, the dimension of the support variety of the trivial $\mathfrak{g}$-supermodule and atyp(λ) is the dimension of the support variety of the simple $\mathfrak{g}$-supermodule of highest weight λ. Both these numbers have some representation-theoretic significance as ingredients in the Kac-Wakimoto conjecture and its generalization by Kujawa and collaborators. In addition to what is mentioned here, the authors discuss also other aspects of support varieties: for example, the complexity of a supermodule, Duflo-Serganova varieties, Balmer's 'universal' support theory etc. The first part concludes with the authors' results on support varieties for finite supergroup schemes in characteristic zero.

In the slightly longer second part the authors discuss their joint work on support varieties of Lie superalgebras and supergroup schemes in odd characteristics. Here the main result is the identification of the support variety of a $\mathfrak{g}$-supermodule M with the Frobenius twist of the rank variety of M in the odd nullcone $\mathcal{N}_{\mathrm{odd}}(\mathfrak{g})$ of the Lie superalgebra $\mathfrak{g}$. The odd nullcone $\mathcal{N}_{\mathrm{odd}}(\mathfrak{g})$ is the conical subvariety of the affine variety $\mathfrak{g}_{\bar{1}}$ consisting of those odd elements of $\mathfrak{g}$ having a zero supercommutator with itself and the rank varieties are the precise analogs of the Duflo-Serganova varieties which in characteristic zero played a role in describing the complexity of a simple supermodule. Again, a consequence of this identification is the tensor product property. Contrary to group algebras of finite groups or restricted universal enveloping algebras of finite-dimensional restricted Lie algebras, the universal enveloping superalgebra $U(\mathfrak{g})$ of a finite-dimensional modular Lie superalgebra $\mathfrak{g}$ is not finite-dimensional (unless $\mathfrak{g}$ is purely odd). Hence, $U(\mathfrak{g})$ is not-self-injective, and thus projective and injective modules over $U(\mathfrak{g})$ are not the same. But $U(\mathfrak{g})$ is a Gorenstein algebra which guarantees that $U(\mathfrak{g})$-supermodules have finite projective dimension exactly when they have finite injective dimension. This in conjunction with the previous result on the identification of the support variety with a rank variety implies that the vanishing of the support variety of a supermodule detects finite projective dimension. The work on finite supergroup schemes in odd characteristics is much more complicated than in characteristic zero, and consequently, the results are more preliminary and mainly are valid only for *unipotent* supergroup schemes. A recent result of Benson, Iyengar, Krause, and Petsova characterizes the nilpotency of elements in the cohomology ring of a finite unipotent supergroup scheme G and the projectivity of a G-supermodule. As a consequence, the authors obtain for an infinitesimal unipotent supergroup scheme G of height $\leq r$ a universal homeomorphism from the scheme of multiparameter supergroups of height $\leq r$ onto the maximal spectrum of the cohomology ring of G.

The article of Orr and Shimozono is a masterfully written expository account of the recent theory of wreath Macdonald polynomials. Symmetric functions have a history that predates the remarkable achievements of invariant theory in the 18th century. However, one of the most significant advancements in the theory of symmetric functions occurred relatively recently, in the mid-1980's, when Ian Macdonald introduced his famous orthogonal symmetric functions. These functions are defined with coefficients drawn from the field $\mathbb{Q}(q,t)$, where q and t are two formal parameters. In the following decade and a half, Macdonald's positivity conjecture, that is, the coefficient of a Schur function in a Macdonald symmetric function is a polynomial in q and t with non-negative integer coefficients, shaped the whole field. Finally, in 2001, Haiman proved the Garsia-Haiman $n!$-conjecture, settling Macdonald's original positivity conjecture as a byproduct. Haiman's proof contained a profound geometric insight, involving an equivalence between the derived categories of torus-equivariant coherent sheaves on the Hilbert schemes of points and coherent sheaves on the plane that are equivariant under the direct product of a torus and a symmetric group. This development also explained in a combinatorial way the isomorphism

$$\Phi : \bigoplus_{n\geq 0} K^{\mathbb{T}}(\mathrm{Hilb}_n)_{\mathrm{loc}} \xrightarrow{\cong} \Lambda_{\mathbb{Q}(q,t)}, \tag{1}$$

where the source is the direct sum of all localizations of the torus equivariant K-theories of the Hilbert schemes of n points in the plane and the target is the space of symmetric functions with coefficients in $\mathbb{Q}(q,t)$.

Haiman proposed extensions of his results in the setting of Γ-fixed points of the Hilbert scheme of points in the plane, where Γ is a finite subgroup of $\mathrm{SL}(2,\mathbb{C})$. An elegant theory of symmetric functions for the hyperoctahedral groups was on the horizon. Subsequently, Bezrukavnikov, Finkelberg, and Losev proved all geometric theorems that were conjectured by Haiman making the wreath Macdonald symmetric function theory available to researchers. It turns out that, by the work of Wen, the wreath analog of the isomorphism (1) is an intertwining map between the two representations of the quantum toroidal $\mathfrak{gl}_1$ algebra U_1^{tor} (also known as the elliptic Hall algebra). This breakthrough has led to a description of the vertex operators, whose joint eigenbasis is defined by the wreath Macdonald polynomials. Orr and Shimozono masterfully explain this highly technical framework in an accessible manner, keeping in mind their forthcoming papers. In addition to this outstanding exposition, the authors also put forward some of their own conjectures regarding the 'wreath Macdonald P polynomials' and the 'diagonal wreath Macdonald-Kostka polynomials.'

Let S be a finite subset of the complex projective line $\mathbb{P}^1$. Given a representation ρ of the fundamental group of the complement of S in $\mathbb{P}^1$, can we construct a differential equation with rational coefficients and regular singularities, these being precisely the points of S, whose monodromy representation is ρ? This is the classical Hilbert-Riemann problem, celebrated for its enduring complexity and significance. An equally profound problem can be stated in terms of algebraic group actions. The *additive Deligne-Simpson problem* asks for the existence of k-tuples $(A_1,\ldots,A_k) \in \prod_{i=1}^k \mathcal{O}_i$, where $\mathcal{O}_i$ $(i \in \{1,\ldots,k\})$ is an adjoint orbit of a complex reductive group G, such that $\sum_{i=1}^k A_i = 0$, and the A_i's have no common invariant subspace. This problem has an alternative formulation that extends the scope of the original Hilbert-Riemann problem: Is it possible to find a Fuchsian connection with residues precisely at the designated orbits? And if such a connection exists, is it unique? More generally, is there a moduli space of meromorphic connections with this specified local behavior?

These questions have been a major source of progress in various areas of mathematics, including differential equations, algebraic geometry, differential geometry, representation theory and others. In his state-of-the-art exposition, Sage discusses different versions of these problems as the allowed singularities become progressively more complex. Beginning with a review of Crawley-Boevey's solution to the Deligne-Simpson problem, Sage goes on to explain many recent results on this topic. He introduces the concept of unramified irregular singularities and elaborates on the construction of moduli spaces for unramified framable connections, as well as Hiroe's solution to the unramified Deligne-Simpson problem. Sage constructs the moduli space of framable connections, where all ramified singular points are toral, and formulates the toral Deligne Simpson problem. He then specializes in the class of meromorphic connections on the so-called "standard Coxeter torus." These connections, known as Coxeter connections, represent certain generalizations of the Frenkel-Gross and Airy connections. Much of the knowledge in the literature concerning these Coxeter connections were discovered by Kamgarpour and Sage himself. Sage describes the solution to the Deligne-Simpson problem for Coxeter connections, determining when they become rigid under additional assumptions. From this point onwards, Sage's article enters to a new framework, where he discusses a related question: To what extent does specified local behavior determine a global G-connection? He carefully explains the prerequisites for answering this question and formulates several profound conjectures that are certain to shape future studies in this subject.

We are very grateful to the editorial staff of *Contemporary Mathematics*, and in particular, to Christine Thivierge for her patience and help in editing these proceedings. Moreover, we wish to thank the anonymous referees for their insight and the high standards employed in writing their reports. Finally, we would like to take this opportunity to express our appreciation to the participants for making the sessions such a success, to the speakers for their very interesting talks, and especially to the authors for contributing to this volume.

Mahir Bilen Can
Jörg Feldvoss

Contemporary Mathematics
Volume **804**, 2024
https://doi.org/10.1090/conm/804/16110

From motivic Chern classes of Schubert cells to their Hirzebruch and CSM classes

Paolo Aluffi, Leonardo C. Mihalcea, Jörg Schürmann, and Changjian Su

ABSTRACT. The equivariant motivic Chern class of a Schubert cell in a complete flag manifold $X = G/B$ is an element in the equivariant K-theory ring of X to which one adjoins a formal parameter y. In this paper we prove several folklore results about motivic Chern classes, including finding specializations at $y = -1$ and $y = 0$; the coefficient of the top power of y; how to obtain Chern-Schwartz-MacPherson (CSM) classes as leading terms of motivic classes; divisibility properties of the Schubert expansion of motivic Chern classes. We collect several conjectures on the positivity, unimodality, and log concavity of CSM and motivic Chern classes of Schubert cells, including a conjectural positivity of structure constants of the multiplication of Poincaré duals of CSM classes. In addition, we prove a 'star duality' for the motivic Chern classes, showing how they behave under the involution taking a vector bundle to its dual. We use the motivic Chern transformation to define two equivariant variants of the Hirzebruch transformation, which appear naturally in the Grothendieck-Hirzebruch-Riemann-Roch formalism. We utilize the Demazure-Lusztig recursions from the motivic Chern class theory to find similar recursions giving the Hirzebruch classes of Schubert cells, their Poincaré duals, and their Segre versions. We explain the functoriality properties needed to extend the results to partial flag manifolds G/P.

CONTENTS

1. Introduction
2. Preliminaries
3. Operators in equivariant cohomology and K-theory of flag manifolds
4. Equivariant motivic Chern classes
5. The parameter y in motivic Chern classes
6. Equivariant Hirzebruch classes
7. The Chern-Schwartz-MacPherson classes as leading terms
8. Positivity, unimodality, and log-concavity conjectures
9. Star duality
Acknowledgments
References

2020 *Mathematics Subject Classification.* Primary 14C17, 14M15; Secondary 14C40, 20C08, 14N15.

Key words and phrases. motivic Chern class, Hirzebruch class, Schubert cells, Adams operation, Demazure-Lusztig operator, Grothendieck-Hirzebruch-Riemann-Roch, log-concave sequence.

1. Introduction

Let X be a quasi-projective complex algebraic variety and denote by $\mathrm{K}_0(\mathrm{var}/X)$ the Grothendieck motivic group consisting of equivalence classes of morphisms $[f : Z \to X]$ modulo the usual additivity relations. Also, denote by $\mathrm{K}(X)$ the Grothendieck ring of vector bundles on X. The *motivic Chern transformation* defined by Brasselet, Schürmann and Yokura [**BSY10**] is the assignment for every such X of a group homomorphism

$$\mathrm{MC}_y : \mathrm{K}_0(\mathrm{var}/X) \to \mathrm{K}(X)[y],$$

uniquely determined by the fact that it commutes with proper push-forwards and that it satisfies the normalization condition

$$\mathrm{MC}_y[\mathrm{id}_X : X \to X] = \lambda_y(T^*_X) = \sum y^i[\wedge^i T^*X] \in \mathrm{K}(X)[y]$$

if X is nonsingular. If $Z \hookrightarrow X$ is a locally closed subset, the *motivic Chern class* of Z (regarded in X) is defined by

$$\mathrm{MC}_y(Z) := \mathrm{MC}_y[Z \hookrightarrow X] \in \mathrm{K}(X)[y];$$

here y is a formal indeterminate.

If X admits a torus action, there is an equivariant version $\mathrm{MC}^T_y : \mathrm{K}^T_0(\mathrm{var}/X) \to \mathrm{K}_T(X)[y]$ defined in [**FRW21, AMSS24**]. We will work in this context, and omit the superscript T from the notation to increase legibility.

Our main object of study in this paper will be the classes $\mathrm{MC}_y(X(w)^\circ)$, the (torus equivariant) motivic Chern classes of Schubert cells $X(w)^\circ$ in the flag manifolds G/B, for G a complex, semisimple, Lie group, and $B \subseteq G$ a Borel subgroup. By functoriality, these determine the motivic Chern classes in the 'partial' flag manifolds G/P, with $P \supseteq B$ a standard parabolic subgroup.

The motivic classes $\mathrm{MC}_y(X(w)^\circ)$ are closely related to the study of representation theory of the Hecke algebra of G, and through this connection they play a prominent role in several related topics: (K-theoretic) stable envelopes and integrable systems [**RTV15, AMSS24, FRW21**], Whittaker functions from p-adic representation theory [**MS22**], characteristic classes of singular varieties [**FRW17**]. In interesting situations they recover point counting over finite fields [**MS22**] (see also §4.3 below), and are closely related to the study of the intersection (co)homology and the Riemann-Hilbert correspondence for arbitrary complex projective manifolds [**Sch09**]. In Schubert Calculus, the motivic Chern classes, and their (co)homological counterparts, the *Chern-Schwartz-MacPherson (CSM)* classes, provide deformations of the usual Schubert classes, which, provably or conjecturally, satisfy remarkable positivity, unimodality, and log-concavity properties; see §8 below.

Among the main goals of this paper is to gather in a single place several folklore results concerning properties of (torus equivariant) motivic classes $\mathrm{MC}_y(X(w)^\circ)$. These include results on the specializations at $y = -1$ and $y = 0$ of the parameter y; the coefficient of $y^{\dim X(w)}$ in $\mathrm{MC}_y(X(w)^\circ)$; how to recover the CSM classes as the initial terms of the motivic Chern classes; divisibility properties of Schubert expansions. We also state several conjectures and prove a new duality for motivic Chern classes.

Our main new contribution is a treatment of the (torus equivariant) *Hirzebruch transformation* $\mathrm{Td}^T_{y,*}$, and in particular a study of the *Hirzebruch classes* $\mathrm{Td}^T_{y,*}(X(w)^\circ)$ of Schubert cells, as an application of properties of motivic Chern classes. Similarly to the motivic Chern transformation, the Hirzebruch transformation $\mathrm{Td}^T_{y,*} : \mathrm{K}^T_0(\mathrm{var}/X) \to \widehat{H}^T_*(X;\mathbb{Q}[y])$ is a functorial transformation defined uniquely by a normalization property, with values in a completed (Borel-Moore or Chow) homology group; see §6. In the non-equivariant context this transformation was defined in [**BSY10**], and it arises naturally in the context of the Grothendieck-Hirzebruch-Riemann-Roch (GHRR) formalism. The 'unnormalized' variant of this transformation was studied by Weber [**Web16, Web17**]. As in the case of MC, we will omit the superscript T from the notation, since all the classes considered in this paper are equivariant by default.

In this paper we extend the definition of the Hirzebruch transformation to the equivariant context, for arbitrary quasi-projective complex algebraic varieties X with a torus action. As hinted above, there are two variants of the Hirzebruch transformation. The 'unnormalized' variant is defined in Theorem 6.1 as the composition

$$\widetilde{\mathrm{Td}}_{y,*} := \mathrm{td}_* \circ \mathrm{MC}_y : \mathrm{K}^T_0(\mathrm{var}/X) \to \widehat{H}^T_*(X;\mathbb{Q}[y])$$

of the (equivariant) Todd transformation td_* constructed by Edidin and Graham [**EG00**] with the motivic Chern transformation. The 'normalized' version of Definition 6.4 is the composition

$$\mathrm{Td}_{y,*} := \psi^{1+y}_* \circ \widetilde{\mathrm{Td}}_{y,*} : \mathrm{K}^T_0(\mathrm{var}/X) \to \widehat{H}^T_*(X;\mathbb{Q}[y]) \subseteq \widehat{H}^T_*(X;\mathbb{Q}[y,(1+y)^{-1}])$$

of a certain Adams operator with the unnormalized transformation. The Adams operator acts by multiplying by powers of $1+y$ (see §6). A technical subtlety is that *a priori* $\mathrm{Td}_{y,*}$ requires coefficients in $\mathbb{Q}[y,(1+y)^{-1}]$, but it can be shown that $\mathbb{Q}[y]$ suffices. In fact, an important property of $\mathrm{Td}_{y,*}$ is that the specialization at $y=-1$ is well-defined. This specialization recovers the (equivariant) MacPherson's transformation [**Mac74, Ohm06**] $c^T_* : \mathcal{F}^T(X) \to H^T_*(X)$ from the group of (equivariant) constructible functions to homology; see Corollary 6.6 and also [**AMSS17**, §3.2] for a summary of Ohmoto's definition and a discussion of alternative (equivalent) definitions.

We cover general preliminaries in §2 and then focus on the case of flag manifolds. In §3 we recall the definition of the Demazure-Lusztig (DL) operators determining recursively the motivic Chern classes $\mathrm{MC}_y(X(w)^\circ)$ of Schubert cells in flag manifolds. We show (Proposition 3.5) how one can recover the cohomological DL operators as initial terms of the Chern character applied to the K-theoretic DL operators; in other words, how one recovers the action of the degenerate Hecke algebra as a limit of the action of the Hecke algebra. In §4 we investigate several properties of the motivic Chern classes, including a divisibility property by powers of $1+y$ of the coefficients arising in their Schubert expansions (illustrated in Example 4.10), and the fact that the integral of a motivic Chern class of a Schubert cell is equal to the number of points over the finite field $\mathbb{F}_q$ with $q=-y$ elements, see Proposition 4.13 and Remark 4.14. In §5 we study the effect of specializing the parameter y on motivic Chern classes. In a nutshell, $y=-1$ recovers the fixed point classes, and $y=0$ the ideal sheaf classes, see Theorem 5.1.

In §6 we obtain 'Hirzebruch operators' calculating recursively the (equivariant) Hirzebruch classes $\widetilde{\mathrm{Td}}_{y,*}(X(w)^\circ)$ and $\mathrm{Td}_{y,*}(X(w)^\circ)$ (see Theorem 6.11), their Poincaré duals (see Theorem 6.12), and the Segre-Hirzebruch classes (see Theorem 6.13). Our treatment in §6 is particularly extensive, so we summarize some of our results here for the convenience of the reader.

Let P_i be the minimal parabolic group associated to the i-th simple root, and denote by $p_i : G/B \to G/P_i$ the natural projection. The BGG operator ∂_i^H is defined as $(p_i)^*(p_i)_*$. Define the unnormalized and the normalized variants of the Hirzebruch operators

$$\widetilde{\mathcal{T}}_i^{\mathrm{Hir}}, \mathcal{T}_i^{\mathrm{Hir}} : \widehat{H}_T^*(X, \mathbb{Q})[y] \to \widehat{H}_T^*(X, \mathbb{Q})[y]$$

by

$$\widetilde{\mathcal{T}}_i^{\mathrm{Hir}} := \widetilde{\mathrm{Td}}_y(T_{p_i})\partial_i^H - \mathrm{id}; \quad \mathcal{T}_i^{\mathrm{Hir}} := \mathrm{Td}_y(T_{p_i})\partial_i^H - \mathrm{id}\,.$$

We prove in Theorem 6.11 that these operators determine recursively the (un)normalized Hirzebruch classes of Schubert cells. More precisely,

$$\widetilde{\mathrm{Td}}_{y,*}(X(\mathrm{id})^\circ) = \mathrm{Td}_{y,*}(X(\mathrm{id})^\circ) = [X(\mathrm{id})]_T$$

and

$$\widetilde{\mathcal{T}}_i^{\mathrm{Hir}}(\widetilde{\mathrm{Td}}_{y,*}(X(w)^\circ)) = \widetilde{\mathrm{Td}}_{y,*}(X(ws_i)^\circ),$$
$$\mathcal{T}_i^{\mathrm{Hir}}(\mathrm{Td}_{y,*}(X(w)^\circ)) = \mathrm{Td}_{y,*}(X(ws_i)^\circ)$$

for all $w \in W$ and all simple reflections s_i such that $ws_i > w$ in the Bruhat ordering. In Theorem 6.13 we prove analogous statements for the 'Segre-Hirzebruch classes' $\frac{\widetilde{\mathrm{Td}}_{y*}(X(w)^\circ)}{\widetilde{\mathrm{Td}}_y(TX)}$ and $\frac{\mathrm{Td}_{y*}(X(w)^\circ)}{\mathrm{Td}_y(TX)}$. We find the Poincaré duals of these classes, along with the appropriate version of the DL operators which determine them; see Theorem 6.12. Perhaps not surprisingly, the theory we find is essentially equivalent to that of motivic Chern classes. For instance, the Hirzebruch operators are images of the DL operators in K-theory via a Todd transformation. In particular, the Hirzebruch operators satisfy the same relations as the DL operators in K-theory (see Lemma 6.9 and Remark 6.10), implying that they give an action of the Hecke algebra on the equivariant (co)homology of G/B.

In §6.3 we study the specializations at $y = 0$ and $y = -1$ of the Hirzebruch classes and operators, and we recover from this point of view the Todd transformation $\mathrm{td}_*(\mathcal{I}_w^T)$ of the boundary ideal sheaves $\mathcal{I}_w^T$ of Schubert varieties, respectively the CSM classes $c_{\mathrm{SM}}(X(w)^\circ)$ of Schubert cells. Furthermore, we recover the DL operators giving recursions for these classes. To illustrate, the Grothendieck-Hirzebruch-Riemann-Roch implies that

$$\langle \mathrm{td}_*(\mathcal{I}_u^T), \mathrm{ch}(\mathcal{O}^{v,T}) \rangle = \delta_{u,v},$$

where ch denotes the (equivariant) Chern character. The specialization $y = 0$ in Theorem 6.12 gives the operators, listed in Proposition 6.14, which determine the classes $\mathrm{td}_*(\mathcal{I}_u^T)$ and $\mathrm{ch}(\mathcal{O}^{v,T})$ recursively.

In §7 we revisit the procedure giving the degenerate Hecke algebra action as a limit, in order to explain how CSM classes may be computed directly as the leading terms of the motivic Chern classes, in the case of Schubert cells; see Theorem 7.1. The CSM and motivic classes give bases of the (equivariant) cohomology and K theory rings of flag manifolds.

Based on computational evidence, in §8 we discuss several conjectural properties of CSM and motivic classes, concerning positivity and log-concavity of the coefficients of their Schubert expansions and the positivity of their structure constants.

Finally, in §9 we prove a new 'star duality' for motivic Chern classes, for the duality $\star : \mathrm{K}_T(X) \to \mathrm{K}_T(X)$ which sends the class $[E]$ of a vector bundle to $[E^\vee] = [\mathrm{Hom}_{\mathcal{O}_X}(E, \mathcal{O}_X)]$. The duality for the motivic classes, stated in

Theorem 9.1, generalizes a known relation between ideal sheaves and duals of structure sheaves of Schubert varieties proved by Brion [**Bri05**, Prop. 4.3.4].

2. Preliminaries

2.1. Equivariant (co)homology. Let X be a quasi-projective complex algebraic variety. In this paper we will deal with the Borel-Moore homology group $H_*(X)$ of X and the cohomology ring $H^*(X)$, with rational coefficients. As an alternative, one could use Chow (co)homology; there is a homology degree-doubling cycle map from Chow to Borel-Moore, and our constructions are compatible with this map. This map is an isomorphism in some important situations, such as the complex flag manifolds studied later in this note. We refer to [**Ful84**, §19.1] and [**Gin98**, §2.6] for more details about Borel-Moore homology and its relation to the Chow group. In case we speak of (co)dimension we always assume that our spaces are pure dimensional; in addition, by (co)dimension we will mean the *complex* (co)dimension. Any subvariety $Y \subseteq X$ of (complex) dimension k has a fundamental class $[Y] \in H_{2k}(X)$. Whenever X is smooth, we can and will identify the Borel-Moore homology and cohomology via Poincaré duality.

Let T be a torus and let X be a variety with a T-action. Then the equivariant cohomology $H^*_T(X)$ is the ordinary cohomology of the Borel mixing space $X_T := (ET\times X)/T$, where ET is the universal T-bundle and T acts by $t\cdot(e,x) = (et^{-1}, tx)$. The ring $H^*_T(X)$ is an algebra over $H^*_T(\mathrm{pt})$, the polynomial ring $\mathrm{Sym}_{\mathbb{Q}}\,\mathfrak{X}(T) \simeq \mathbb{Q}[t_1,\dots,t_s]$ in the character group $\mathfrak{X}(T)$ (written additively) and with $t_i \in H^2_T(\mathrm{pt})$; see e.g., [**Kum02**, §11.3.5]. One may also define T-equivariant Borel-Moore homology and Chow groups, related by an equivariant cycle map; see e.g., [**EG98**]. Every k-dimensional subvariety $Y \subseteq X$ that is stable under the T action determines an equivariant fundamental class $[Y]_T$ in $H^T_{2k}(X)$.

As in the non-equivariant case, whenever X is smooth, we will identify $H^T_*(X)$ and $H^*_T(X)$. In particular, when $X = \mathrm{pt}$ is a point, the identification sends $a \in H^*_T(\mathrm{pt})$ to $a \cap [\mathrm{pt}]_T$. If X is smooth and proper, then there is an $H^*_T(\mathrm{pt})$-bilinear Poincaré (or intersection) pairing $H^*_T(X)\otimes H^*_T(X) \to H^*_T(\mathrm{pt})$ defined by

$$\langle a, b\rangle = \int_X a\cdot b \quad ,$$

where the integral stands for the push-forward to a point. Equivariant vector bundles have equivariant Chern classes $c_i^T(-)$, such that $c_j^T(E)\cap -$ is an operator $H_i^T(X) \to H^T_{i-2j}(X)$; see [**And12**, §1.3], [**EG98**, §2.4].

We address the reader to [**And12, Knu, AF23**] for background on equivariant cohomology and homology.

2.2. Equivariant K theory. For any T-variety X, the equivariant K theory ring $\mathrm{K}_T(X)$ is the Grothendieck ring generated by symbols $[E]$, where $E \to X$ is a T-equivariant vector bundle, modulo the relations $[E] = [E_1] + [E_2]$ for any short exact sequence $0 \to E_1 \to E \to E_2 \to 0$ of equivariant vector bundles. For any proper morphism $f : X \to Y$ there is a functorial push-forward $f_* : \mathrm{K}_T(X) \to \mathrm{K}_T(Y)$ defined by $f_*([E]) = \sum(-1)^i[R^i f_*(E)]$. The ring $\mathrm{K}_T(X)$ is an algebra over $\mathrm{K}_T(\mathrm{pt}) = R(T)$, the representation ring of T. This may be identified with the Laurent polynomial ring $\mathbb{Z}[e^{\pm t_1},\dots,e^{\pm t_r}]$ where e^{t_i} are characters corresponding to a basis of the character lattice in the Lie algebra of T. We will often denote the tuple $(e^{\pm t_1},\dots,e^{\pm t_r})$ simply by e^t. An introduction to equivariant K theory may be found in [**CG09**, Chapter 5]. The equivariant K ring of X admits a 'vector bundle duality' involution $\star : \mathrm{K}_T(X) \to \mathrm{K}_T(X)$ mapping the class $[E]$ of a vector bundle to the class $[E^\vee]$ of its dual. This is not an involution of $\mathrm{K}_T(\mathrm{pt})$-algebras; it satisfies $(e^\lambda)^\vee = e^{-\lambda}$. Under mild hypotheses (e.g., X projective) there is also a 'Serre duality' involution $\mathcal{D} : \mathrm{K}_T(X) \to \mathrm{K}_T(X)$ inherited from (equivariant) Grothendieck-Serre duality and defined by

$$\mathcal{D}([F]) := [R\,\mathrm{Hom}(F, \omega_X^\bullet)] = [\omega_X^\bullet] \otimes [F]^\vee \in \mathrm{K}_T(X)$$

for $[F] \in \mathrm{K}_T(X)$, where $\omega_X^\bullet \simeq \omega_X[\dim X]$ is the (equivariant) dualizing complex of X. Thus if X is nonsingular, $[\omega_X^\bullet] = (-1)^{\dim X}[\omega_X]$ with ω_X the equivariant canonical bundle of X. Observe the multiplicativity

$$\mathcal{D}([E] \otimes [F]) = \mathcal{D}([E]) \otimes [F]^\vee.$$

In later sections of this paper we will primarily be concerned with flag manifolds $X = G/B$, with T acting on X by left multiplication. In this case X is a smooth projective variety and the ring $\mathrm{K}_T(X)$ is naturally isomorphic to the Grothendieck group $K_0(\mathfrak{coh}^T(\mathcal{O}_X))$ of T-linearized coherent sheaves on X. This follows from the fact that every such coherent sheaf has a finite resolution by T-equivariant vector bundles. There is a $\mathrm{K}_T(\mathrm{pt})$-bilinear pairing

$$\langle -, - \rangle : \mathrm{K}_T(X) \otimes \mathrm{K}_T(X) \to \mathrm{K}_T(\mathrm{pt}) = R(T)$$

$$\langle [E], [F] \rangle := \int_X E \otimes F = \chi(X; E \otimes F),$$

where $\chi(X; E)$ is the (equivariant) Euler characteristic, i.e., the virtual representation

$$\chi(X; E) = \int_X [E] = \sum_i (-1)^i H^i(X; E).$$

Note that

$$\langle \mathcal{D}[E], [F]^\vee \rangle = \int_X \mathcal{D}([E \otimes F]) = \chi(X; E \otimes F)^\vee = (\langle [E], [F] \rangle)^\vee,$$

by equivariant Grothendieck-Serre duality (the corresponding result [**Har77**, Chapter III, Theorem 7.6] also holds equivariantly, e.g. as a very special case of [**LH09**, Part II, Theorem 25.2 and Theorem 28.11]).

Let y be an indeterminate. The *Hirzebruch* λ_y*-class* of an equivariant vector bundle E is the class

$$\lambda_y(E) := \sum_k [\wedge^k E] y^k \in \mathrm{K}_T(X)[y].$$

The λ_y-class is multiplicative, i.e., for any short exact sequence $0 \to E_1 \to E \to E_2 \to 0$ of equivariant vector bundles there is an equality $\lambda_y(E) = \lambda_y(E_1)\lambda_y(E_2)$ in $\mathrm{K}_T(X)[y]$. (As pointed out in [**Nie74**, §1.2], this is part of the λ-ring structure of $\mathrm{K}_T(X)$, cf. [**SGA71**, 6.V.2.4].)

2.3. The Chern character. For a pure-dimensional T-variety X, Edidin and Graham [**EG00**] defined an equivariant Chern character

$$\mathrm{ch}_T : \mathrm{K}_T(X) \to \widehat{H}_*^T(X) := \prod_{i \le \dim X} H_{2i}^T(X)$$

such that:

- If $V \subseteq X$ is a T-invariant subvariety, then $\mathrm{ch}_T[\mathcal{O}_V] = [V]_T$+l.o.t. (lower order terms). (Non-equivariantly, see [**Ful84**, Theorem 18.3(5)].)
- If $\mathcal{L}$ is an equivariant line bundle with first Chern class $c_1^T(\mathcal{L})$, then $\mathrm{ch}_T[\mathcal{L}] = e^{c_1^T(\mathcal{L})} \cap [X]_T$. In particular, $\mathrm{ch}_T(e^\lambda) = e^\lambda \in \hat{H}_*^{\mathbb{C}*}(\mathrm{pt})$ for all characters λ.
- ch_T commutes with pull-backs.
- If X is smooth, then after identifying $H_*^T(X) \simeq H_T^*(X)$ via Poincaré duality, ch_T is a ring homomorphism.

We will generally omit the subscript T in the notation, since the equivariant context is assumed throughout the paper. A fundamental result is the Grothendieck-Hirzebruch-Riemann-Roch (GHRR) theorem. In the equivariant case, this was proved in [**EG00**]. For now we state the following particular form; in §6 below we will need more general versions. Let $f : X \to Y$ be a smooth proper T-equivariant morphism of smooth T-varieties, and let $a \in \mathrm{K}_T(X)$. Then

$$\mathrm{ch}\, f_*(a) = f_*(\mathrm{ch}(a) \cdot \mathrm{Td}(T_f)).$$

where $\mathrm{Td}(T_f)$ is the equivariant Todd class of the relative tangent bundle of f. Recall that if $E \to X$ is an equivariant vector bundle with Chern roots $x_1, \ldots, x_e$, then

$$\mathrm{Td}(E) = \prod_{i=1}^{e} \frac{x_i}{1 - e^{-x_i}} = \prod_{i=1}^{e} (1 + \frac{1}{2} x_i + \ldots);$$

see e.g., [**Ful84**, Example 3.2.4].

3. Operators in equivariant cohomology and K-theory of flag manifolds

The goal of this section is to introduce the Schubert basis in the equivariant K ring of flag manifolds and recall the definition and basic properties of the Demazure and (cohomological and K-theoretic) Demazure-Lusztig (DL) operators acting on the K-theory ring. An important fact which we will use later, and for which we could not find a reference, is that the cohomological DL operators may be recovered from certain initial terms of the K-theoretic ones; cf. Proposition 3.5.

3.1. Schubert data. Let G be a complex semisimple, simply connected, Lie group, and fix a Borel subgroup B with a maximal torus $T \subseteq B$. Let B^- denote the opposite Borel subgroup. Let $W := N_G(T)/T$ be the Weyl group, and $\ell : W \to \mathbb{N}$ the associated length function. Denote by w_0 the longest element in W; then $B^- = w_0 B w_0$. Let also $\Delta := \{\alpha_1, \ldots, \alpha_r\} \subseteq R^+$ denote the set of simple roots included in the set of positive roots for (G, B). Let ρ denote the half sum of the positive roots. The simple reflection for the root $\alpha_i \in \Delta$ is denoted by s_i and the corresponding *minimal* parabolic subgroup is denoted by P_i, containing the Borel subgroup B.

Let $X := G/B$ be the (complete) flag variety. It has a stratification by Schubert cells $X(w)^\circ := BwB/B$ and opposite Schubert cells $Y(w)^\circ := B^- wB/B$. The closures $X(w) := \overline{X(w)^\circ}$ and $Y(w) := \overline{Y(w)^\circ}$ are the Schubert varieties. With these definitions, $\dim_{\mathbb{C}} X(w) = \mathrm{codim}_{\mathbb{C}} Y(w) = \ell(w)$. The Weyl group W admits a partial ordering, called the Bruhat ordering, defined by $u \leq v$ if and only if $X(u) \subseteq X(v)$.

Let $\mathcal{O}_w^T := [\mathcal{O}_{X(w)}] \in \mathrm{K}_T(G/B)$ be the Grothendieck class determined by the structure sheaf of $X(w)$ (a coherent sheaf), and similarly $\mathcal{O}^{w,T} := [\mathcal{O}_{Y(w)}]$. The equivariant K-theory ring has $\mathrm{K}_T(\mathrm{pt})$-bases $\{\mathcal{O}_w^T\}_{w\in W}$ and $\{\mathcal{O}^{w,T}\}_{w\in W}$ for $w \in W$. Let $\partial X(w) := X(w) \smallsetminus X(w)^\circ$ be the boundary of the Schubert variety $X(w)$, and similarly $\partial Y(w)$ the boundary of $Y(w)$. It is known that the dual bases of $\{\mathcal{O}_w^T\}$ and $\{\mathcal{O}^{w,T}\}$ are given by the classes of the ideal sheaves $\mathcal{I}^{w,T} := [\mathcal{O}_{Y(w)}(-\partial Y(w))]$, respectively $\mathcal{I}_w^T := [\mathcal{O}_{X(w)}(-\partial X(w))]$. I.e.,

$$\langle \mathcal{O}_u^T, \mathcal{I}^{v,T}\rangle = \langle \mathcal{O}^{u,T}, \mathcal{I}_v^T\rangle = \delta_{u,v}. \tag{3.1}$$

See e.g., [**Bri05**, Proposition 4.3.2] for the non-equivariant case; the same proof works equivariantly. See also [**GK08**, Proposition 2.1]. It is also shown in [**Bri05**] that

$$\mathcal{O}_w^T = \sum_{v\le w} \mathcal{I}_v^T \quad \text{and} \quad \mathcal{I}_w^T = \sum_{v\le w} (-1)^{\ell(w)-\ell(v)} \mathcal{O}_v^T. \tag{3.2}$$

(Again, Brion's argument also works in the equivariant context.) For any weight (character) λ of T, we denote by $\mathcal{L}_\lambda$ the G-homogeneous line bundle

$$\mathcal{L}_\lambda = G \times^B \mathbb{C}_\lambda.$$

Let P be a standard parabolic subgroup of G, i.e., $B \subseteq P \subseteq G$. Such a subgroup is determined by a subset $\Delta_P \subseteq \Delta$; for instance, $\Delta_B = \emptyset$. Denote by W_P the subgroup of W generated by the simple reflections s_i such that $\alpha_i \in \Delta \smallsetminus \Delta_P$. Let W^P denote the subset of minimal length representatives of W/W_P. By definition, $\ell(wW_P) = \ell(w)$ for $w \in W^P$. Similarly to G/B, the partial flag manifold G/P has finitely B and B^- many orbits – the Schubert cells – indexed by the elements $w \in W^P$:

$$X(wW_P)^\circ = BwP/P; \quad Y(wW_P)^\circ = B^- wP/P;$$

as before $\dim X(wW_P)^\circ = \operatorname{codim} Y(wW_P)^\circ = \ell(wW_P)$. The Schubert varieties $X(wW_P)$, $Y(wW_P)$ in G/P are the closures of the corresponding Schubert cells.

3.2. BGG, Demazure, and Demazure-Lusztig operators. Fix a simple root $\alpha_i \in \Delta$ and denote by $P_i \subseteq G$ the corresponding minimal parabolic subgroup. Consider the fiber diagram:

$$\begin{array}{ccc} FP := G/B \times_{G/P_i} G/B & \xrightarrow{pr_1} & G/B \\ \downarrow{\scriptstyle pr_2} & & \downarrow{\scriptstyle p_i} \\ G/B & \xrightarrow{p_i} & G/P_i \end{array}$$

The **Bernstein-Gelfand-Gelfand** (BGG) operator [**BGG73**] is the $H_T^*(\mathrm{pt})$-linear morphism $\partial_i^H : H_T^i(X) \to H_T^{i+2}(X)$ defined by $\partial_i^H := (p_i)^*(p_i)_*$. The same geometric definition gives the **Demazure** operator $\partial_i : \mathrm{K}_T(X) \to \mathrm{K}_T(X)$ in the (equivariant) K-theory, linear over $\mathrm{K}_T(\mathrm{pt})$; see [**Dem74**]. These operators satisfy

$$\partial_i^H[X(w)]_T = \begin{cases} [X(ws_i)]_T & \text{if } ws_i > w; \\ 0 & \text{otherwise .} \end{cases} \quad \partial_i(\mathcal{O}_w^T) = \begin{cases} \mathcal{O}_{ws_i}^T & \text{if } ws_i > w; \\ \mathcal{O}_w^T & \text{otherwise .} \end{cases} \tag{3.3}$$

From this, one deduces that both operators satisfy the same commutation and braid relations as those for the elements of W. In cohomology, $(\partial_i^H)^2 = 0$, while in K-theory $\partial_i^2 = \partial_i$.

The relative cotangent bundle of the projection p_i is $T^*_{p_i} = \mathcal{L}_{\alpha_i}$. Define the $H^*_T(\mathrm{pt})$-algebra automorphism $s_i : H^*_T(G/B) \to H^*_T(\mathrm{pt})$ by

$$s_i = \mathrm{id} + c_1^T(T^*_{p_i})\partial_i^H = \mathrm{id} + c_1^T(\mathcal{L}_{\alpha_i})\partial_i^H. \tag{3.4}$$

It was proved by Knutson [**Knu**] that this is an automorphism induced by the right Weyl group action on G/B; see [**AM16**] and also [**MNS22**], where both left and right actions are studied. Using this automorphism, the **cohomological Demazure-Lusztig (DL) operators** are $H^*_T(\mathrm{pt})$-linear endomorphism of $H^*_T(G/B)$ defined by

$$\mathcal{T}_i^H = \partial_i^H - s_i; \quad \mathcal{T}_i^{H,\vee} = \partial_i^H + s_i. \tag{3.5}$$

These operators satisfy the same braid and commutation relations as the BGG operators, and, in addition $(\mathcal{T}_i^H)^2 = (\mathcal{T}_i^{H,\vee})^2 = \mathrm{id}$; see [**AM16**, Proposition 4.1]. In other words, these give a twisted representation of the Weyl group W on $H^*_T(G/B)$. This representation was studied earlier by Lascoux, Leclerc and Thibon [**LLT96**], and by Ginzburg [**Gin98**] in relation to the degenerate Hecke algebra. The operators are adjoint to each other, in the sense that for any $a, b \in H^*_T(G/B)$,

$$\langle \mathcal{T}_i^H(a), b\rangle = \langle a, \mathcal{T}_i^{H,\vee}(b)\rangle.$$

It is convenient to consider a homogenized version of this operator. Add a formal variable $\hbar$ of cohomological complex degree 1. Then the homogenized operators are

$$\mathcal{T}_i^{H,\hbar} = \hbar\partial_i^H - s_i; \quad \mathcal{T}_i^{H,\vee,\hbar} = \hbar\partial_i^H + s_i. \tag{3.6}$$

The variable $\hbar$ will arise geometrically from the $\mathbb{C}^*$-action by dilation on $T^*(G/B)$. The restriction of this action to the zero-section $G/B \hookrightarrow T^*(G/B)$ is trivial, and $H^*_{T\times\mathbb{C}^*}(G/B) = H^*_T(G/B)[\hbar]$, where $\hbar$ is interpreted as a generator of $H^2_{\mathbb{C}^*}(\mathrm{pt})$.

We define next the K-theoretic version of the DL operator. Fix an indeterminate y; later, we will set $y = -e^{-\hbar}$. Define the **K-theoretic Demazure-Lusztig (DL) operators**

$$\mathcal{T}_i := \lambda_y(T^*_{p_i})\partial_i - \mathrm{id}; \quad \mathcal{T}_i^\vee := \partial_i\lambda_y(T^*_{p_i}) - \mathrm{id}. \tag{3.7}$$

The operators $\mathcal{T}_i$ and $\mathcal{T}_i^\vee$ are $\mathrm{K}_T(\mathrm{pt})[y]$-module endomorphisms of $\mathrm{K}_T(X)[y]$.

REMARK 3.1. The operator $\mathcal{T}_i^\vee$ was defined by Lusztig [**Lus85**, Eq. (4.2)] in relation to affine Hecke algebras and equivariant K theory of flag varieties. The 'dual' operator $\mathcal{T}_i$ arises naturally in the study of motivic Chern classes of Schubert cells [**AMSS24**] (where the operators are denoted $\mathfrak{R}_i, \mathfrak{R}_i^\vee$). In an algebraic form, $\mathcal{T}_i$ appeared recently in [**BBL15, LLL17, MS22**], in relation to Whittaker functions. The left versions of these operators are studied in [**MNS22**].

As in the cohomological case, the Demazure and Demazure-Lusztig (DL) operators are adjoint to each other and ∂_i is self-adjoint: for any $a, b \in K_T(X)$,

$$\langle \mathcal{T}_i(a), b\rangle = \langle a, \mathcal{T}_i^\vee(b)\rangle \quad \text{and} \quad \langle \partial_i(a), b\rangle = \langle a, \partial_i(b)\rangle.$$

See [**AMSS24**, Lemma 3.3] for a proof.

PROPOSITION 3.2 ([**Lus85**]). *The operators $\mathcal{T}_i$ and $\mathcal{T}_i^\vee$ satisfy the usual commutation and braid relations for the group W. For each simple root $\alpha_i \in \Delta$ the following quadratic formula holds:*

$$(\mathcal{T}_i + \mathrm{id})(\mathcal{T}_i + y) = (\mathcal{T}_i^\vee + \mathrm{id})(\mathcal{T}_i^\vee + y) = 0.$$

An immediate corollary of the quadratic formula is that for $y \neq 0$, the operators $\mathcal{T}_i$ and $\mathcal{T}_i^\vee$ are invertible. In fact,

$$\mathcal{T}_i^{-1} = -\frac{1}{y}\mathcal{T}_i - \frac{1+y}{y}\,\mathrm{id}$$

as operators on $\mathrm{K}_T(X)[y, y^{-1}]$. The same formula holds when $\mathcal{T}_i$ is exchanged with $\mathcal{T}_i^\vee$.

Consider next the localized equivariant K theory ring

$$\mathrm{K}_T(G/B)_{\mathrm{loc}} := \mathrm{K}_T(G/B) \otimes_{\mathrm{K}_T(\mathrm{pt})} \mathrm{Frac}(\mathrm{K}_T(\mathrm{pt}))$$

where Frac denotes the fraction field. The Weyl group elements $w \in W$ are in bijection with the torus fixed points $e_w \in G/B$. Let $\iota_w := [\mathcal{O}_{e_w}] \in \mathrm{K}_T(G/B)_{\mathrm{loc}}$ be the class of the structure sheaf of e_w. By the localization theorem, the classes ι_w form a basis for the localized equivariant K theory ring; we call this basis the *fixed point basis.*

We need the following lemma, whose proof can be found e.g., in [**AMSS24**, Lemma 3.7].

LEMMA 3.3. *The following formulas hold in* $\mathrm{K}_T(G/B)_{\mathrm{loc}}$*:*

(a) *For any weight* λ, $\mathcal{L}_\lambda \cdot \iota_w = e^{w\lambda}\iota_w$*;*

(b) *For any simple root* α_i,

$$\partial_i(\iota_w) = \frac{1}{1-e^{w\alpha_i}}\iota_w + \frac{1}{1-e^{-w\alpha_i}}\iota_{ws_i};$$

(c) *The action of the operator* $\mathcal{T}_i$ *on the fixed point basis is given by the following formula*

$$\mathcal{T}_i(\iota_w) = -\frac{1+y}{1-e^{-w\alpha_i}}\iota_w + \frac{1+ye^{-w\alpha_i}}{1-e^{-w\alpha_i}}\iota_{ws_i}.$$

(d) *The action of the adjoint operator* $\mathcal{T}_i^\vee$ *is given by*

$$\mathcal{T}_i^\vee(\iota_w) = -\frac{1+y}{1-e^{-w\alpha_i}}\iota_w + \frac{1+ye^{w\alpha_i}}{1-e^{-w\alpha_i}}\iota_{ws_i}.$$

(e) *The action of the inverse operator* $(\mathcal{T}_i^\vee)^{-1}$*is given by*

$$(\mathcal{T}_i^\vee)^{-1}(\iota_w) = -\frac{1+y^{-1}}{1-e^{w\alpha_i}}\iota_w - \frac{y^{-1}+e^{w\alpha_i}}{1-e^{-w\alpha_i}}\iota_{ws_{\alpha_i}}.$$

We also record the action of several specializations of the Demazure-Lusztig operators, see [**AMSS24**, Lemma 3.8].

LEMMA 3.4. (a) *The following specializations hold:*

$$(\mathcal{T}_i)_{y=0} = \partial_i - \mathrm{id}; \quad (\mathcal{T}_i^\vee)_{y=0} = \partial_i - \mathrm{id};$$

Further, for any $w \in W$*, the following hold:*

$$(\partial_i - \mathrm{id})(\mathcal{I}_w^T) = \begin{cases} \mathcal{I}_{ws_i}^T & \text{if } ws_i > w; \\ -\mathcal{I}_w^T & \text{if } ws_i < w. \end{cases} \quad \partial_i(\mathcal{O}_w^T) = \begin{cases} \mathcal{O}_{ws_i}^T & \text{if } ws_i > w; \\ \mathcal{O}_w^T & \text{if } ws_i < w. \end{cases}$$

(b) *Let* $w \in W$*. Then the specializations at* $y = -1$ *satisfy*

$$(\mathcal{T}_i)_{y=-1}(\iota_w) = \iota_{ws_i}.$$

In other words, this specialization is compatible with the right Weyl group multiplication.

3.3. Leading terms of DL operators. Next we use the grading induced by the equivariant Chern character to identify the 'initial terms' of the Demazure-Lusztig operators as certain operators on equivariant (co)homology related to the degenerate Hecke algebra. These operators appear as convolution operators in [**Gin98**] and determine the Chern-Schwartz-MacPherson classes of Schubert cells [**AM16**].

As usual, $X = G/B$ but we consider the extended torus $A := T \times \mathbb{C}^*$ where $\mathbb{C}^*$ acts trivially. (This is the restriction of the action of A on $T^*(X)$, where $\mathbb{C}^*$ acts by dilation. The CSM and motivic Chern classes considered later in this paper are naturally $\mathbb{C}^*$-equivariant; this justifies the use of the extended torus.) We now set $y = -e^{-\hbar}$, and therefore (cf. §2.3)

$$\mathrm{ch}_{\mathbb{C}^*}(y) = -e^{-\hbar} = -1 + \hbar + O(\hbar^2) \in \widehat{H}^*_{\mathbb{C}^*}(\mathrm{pt}).$$

We analyze the relation between the cohomological and K-theoretic DL operators.

PROPOSITION 3.5. *Let $w \in W$ and consider the Grothendieck class $\mathcal{O}_w^A \in \mathrm{K}_A(X)$ for the Schubert variety $X(w)$. Then*

$$\mathrm{ch}_A(\mathcal{T}_i(\mathcal{O}_w^A)) = \mathcal{T}_i^{H,\hbar}[X(w)]_A + \text{l.o.t.}$$

and

$$\mathrm{ch}_A(\mathcal{T}_i^\vee(\mathcal{O}_w^A)) = \mathcal{T}_i^{H,\vee,\hbar}[X(w)]_A + \text{l.o.t.},$$

where l.o.t. are terms in $\prod_{i<\ell(w)} H_{2i}^A(X)$.

PROOF. Since $X = G/B$ is non-singular, the Chern character is a ring homomorphism, thus for any invariant subvariety $Z \subseteq X$ and any equivariant line bundle $\mathcal{L}$,

$$\mathrm{ch}_A([\mathcal{O}_Z]_A \cdot \mathcal{L}) = [Z]_A + c_1^A(\mathcal{L}) \cdot [Z]_A + \text{l.o.t.}$$

We take $Z = X(w)$, and we have two cases: either $w < ws_i$ or $w > ws_i$. If $w < ws_i$ then $\partial_i(\mathcal{O}_w^A) = \mathcal{O}_{ws_i}^A$. Using this, we obtain

$$\begin{aligned}
\mathrm{ch}_A(\mathcal{T}_i(\mathcal{O}_w^A)) &= \mathrm{ch}_A(\mathcal{O}_{ws_i}^A + y\mathcal{L}_{\alpha_i} \cdot \mathcal{O}_{ws_i}^A - \mathcal{O}_w^A)\\
&= \mathrm{ch}_A(\mathcal{O}_{ws_i}^A) - e^{-\hbar} e^{c_1^A(\mathcal{L}_{\alpha_i})} \mathrm{ch}_A(\mathcal{O}_{ws_i}^A) - [X(w)]_A + \text{l.o.t.}\\
&= \mathrm{ch}_A(\mathcal{O}_{ws_i}^A) - (1-\hbar)(1 + c_1^A(\mathcal{L}_{\alpha_i})) \mathrm{ch}_A(\mathcal{O}_{ws_i}^A) - [X(w)]_A + \text{l.o.t.}\\
&= \hbar[X(ws_i)]_A - c_1^A(\mathcal{L}_{\alpha_i})[X(ws_i)]_A - [X(w)]_A + \text{l.o.t.}\\
&= \hbar\partial_i^H[X(w)]_A - (\mathrm{id} + c_1^A(\mathcal{L}_{\alpha_i})\partial_i^H)[X(w)]_A + \text{l.o.t.}
\end{aligned}$$

where l.o.t. $\in \prod_{i<\ell(w)} H_i^A(X)$. By [**AM16**, (3)] (which uses a different sign convention) the last expression equals

$$(\hbar\partial_i^H - s_i)[X(w)]_A + \text{l.o.t.} = \mathcal{T}_i^{H,\hbar}[X(w)]_A + \text{l.o.t.}$$

and we are done in this case. If $w > ws_i$ then $\partial_i(\mathcal{O}_w^A) = \mathcal{O}_w^A$, $\partial_i[X(w)]_A = 0$ and $s_i[X(w)]_A = [X(w)]_A$ by [**AM16**, (4)]. Then a similar, but simpler calculation proves the first part of the proposition.

For the second statement we start by observing that for $a \in \mathrm{K}_T(X)$, by the Grothendieck-Hirzebruch-Riemann-Roch (GHRR)

$$\mathrm{ch}_A\, \partial_i(a) = \mathrm{ch}_A\, p_i^*(p_i)_*(a) = p_i^*\, \mathrm{ch}_A((p_i)_*(a)) = \partial_i^H(\mathrm{ch}_A(a)\, \mathrm{Td}^A(T_{p_i})). \tag{3.8}$$

Then the second statement can be proved as follows. By the GHRR theorem,

$$\begin{aligned}\mathrm{ch}_A(\mathcal{T}_i^\vee(\mathcal{O}_w^A)) =& \mathrm{ch}_A\, \partial_i(\mathcal{O}_w^A + y\mathcal{L}_{\alpha_i}\cdot \mathcal{O}_w^A) - \mathrm{ch}_A(\mathcal{O}_w^A)\\ =&\partial_i^H\Big(\mathrm{ch}_A(\mathcal{O}_w^A + y\mathcal{L}_{\alpha_i}\cdot \mathcal{O}_w^A)\,\mathrm{Td}^A(T_{p_i})\Big) - \mathrm{ch}_A(\mathcal{O}_w^A)\\ =&\partial_i^H\Big(\mathrm{ch}_A(\mathcal{O}_w^A)\,\mathrm{Td}^A(T_{p_i})(1-e^{-\hbar+c_1^A(\mathcal{L}_{\alpha_i})})\Big) - \mathrm{ch}_A(\mathcal{O}_w^A)\end{aligned}$$

Observe that $\mathrm{Td}^A(T_{p_i})(1-e^{-\hbar+c_1^A(\mathcal{L}_{\alpha_i})}) = \hbar + c_1^A(\mathcal{L}_{-\alpha_i}) +$ (terms of degree ≥ 2) in *cohomology*. Then the last expression equals

$$\begin{aligned}&\partial_i^H\Big((\hbar + c_1^A(\mathcal{L}_{-\alpha_i})[X(w)]_A\Big) - [X(w)]_A + l.o.t\\ =&\hbar\partial_i^H[X(w)]_A + (\mathrm{id} + c_1^A(\mathcal{L}_{\alpha_i})\partial_i^H)[X(w)]_A + \mathrm{l.o.t.}\\ =&(\hbar\partial_i^H + s_i)[X(w)]_A + \mathrm{l.o.t.}\\ =&\mathcal{T}_i^{H,\vee,\hbar}[X(w)]_A + \mathrm{l.o.t.}\end{aligned}$$

Here the third equation follows from the definition of s_i from (3.4), and the second from the general fact that for every weight λ, $c_1^A(\mathcal{L}_\lambda)\partial_i^H = \partial_i^H c_1^A(\mathcal{L}_{s_i\lambda}) - \langle \lambda, \alpha_i^\vee\rangle$. This can be proved by e.g., adapting parts (a) and (b) of Lemma 3.3 to the cohomological context. □

4. Equivariant motivic Chern classes

The aim of this section is to introduce and recall the basic properties of the motivic Chern classes - one of the main objects in this note. In the second part of the section we focus on the motivic Chern classes of Schubert cells in flag manifolds. Aside from recalling results proved in e.g., [**AMSS24**], we prove a new divisibility property of the coefficients in the transition matrix from motivic classes to the Schubert basis (Proposition 4.15). In §4.3 we discuss the relation to point counting over finite fields; this was mentioned briefly in [**AMSS24**] and [**MS22**].

4.1. Preliminaries about motivic Chern classes. We recall the definition of the motivic Chern classes, following [**BSY10**]. For now let X be a quasi-projective, complex algebraic variety, with an action of T. First we recall the definition of the (relative) motivic Grothendieck group $\mathrm{K}_0^T(\mathrm{var}/X)$ of varieties over X, mostly following Looijenga's notes [**Loo02**]; see also Bittner [**Bit04**]. For simplicity, we only consider the T-equivariant quasi-projective context (replacing the 'goodness' assumption in [**Bit04**]), which is enough for all applications in this paper. The group $\mathrm{K}_0^T(\mathrm{var}/X)$ is the quotient of the free abelian group generated by symbols $[f : Z \to X]$ where Z is a quasi-projective T-variety and $f : Z \to X$ is a T-equivariant morphism modulo the additivity relations

$$[f : Z \to X] = [f : U \to X] + [f : Z \smallsetminus U \to X]$$

for $U \subseteq Z$ an open invariant subvariety. For every equivariant morphism $g : X \to Y$ of quasi-projective T-varieties there are well-defined push-forwards $g_! : \mathrm{K}_0^T(\mathrm{var}/X) \to \mathrm{K}_0^T(\mathrm{var}/Y)$ (given by composition) and pull-backs $g^* : \mathrm{K}_0^T(\mathrm{var}/Y) \to \mathrm{K}_0^T(\mathrm{var}/X)$ (given by fiber product); see [**Bit04**, §6]. There are also external products

$$\times : \mathrm{K}_0^T(\mathrm{var}/X) \times \mathrm{K}_0^T(\mathrm{var}/X') \to \mathrm{K}_0^T(\mathrm{var}/X\times X'); \quad [f]\times[f'] \mapsto [f \times f'],$$

which are $\mathrm{K}_0^T(\mathrm{var}/\mathrm{pt})$-bilinear and commute with push-forward and pull-back. If $X = \mathrm{pt}$, then $\mathrm{K}_0^T(\mathrm{var}/\mathrm{pt})$ is a ring with this external product, and the groups $\mathrm{K}_0^T(\mathrm{var}/X)$ also acquire by the external product a module structure over $\mathrm{K}_0^T(\mathrm{var}/\mathrm{pt})$ such that push-forward $g_!$ and pull-back g^* are $\mathrm{K}_0^T(\mathrm{var}/\mathrm{pt})$-linear.

REMARK 4.1. For any variety X, similar functors can be defined on the ring of constructible functions $\mathcal{F}(X)$, and the Grothendieck group $\mathrm{K}_0(\mathrm{var}/X)$ may be regarded as a motivic version of $\mathcal{F}(X)$. In fact, there is a map $e : \mathrm{K}_0(\mathrm{var}/X) \to \mathcal{F}(X)$ sending $[f : Y \to X] \mapsto f_!(\mathbb{1}_Y)$, where $f_!(\mathbb{1}_Y)$ is defined using compactly supported Euler characteristic of the fibers. The map e is a group homomorphism, and if $X = \mathrm{pt}$ then e is a ring homomorphism. The constructions extend equivariantly, with $\mathcal{F}^T(X) \subseteq \mathcal{F}(X)$ the subgroup of T-invariant constructible functions.

The following theorem was proved in the non-equivariant case by Brasselet, Schürmann and Yokura [**BSY10**, Theorem 2.1]. Minor changes in the argument are needed in the equivariant case – see also [**FRW21**, **AMSS24**]. In future work we will reprove the theorem below and relate equivariant motivic Chern classes to certain classes in the equivariant K-theory of the cotangent bundle as defined by Tanisaki [**Tan87**] with the help of equivariant mixed Hodge modules.

THEOREM 4.2. [**AMSS24**, Theorem 4.2] *Let X be a quasi-projective, non-singular, complex algebraic variety with an action of the torus T. There exists a unique natural transformation $\mathrm{MC}_y : \mathrm{K}_0^T(\mathrm{var}/X) \to \mathrm{K}_T(X)[y]$ satisfying the following properties:*

(1) *It is functorial with respect to push-forwards via T-equivariant proper morphisms of non-singular, quasi-projective varieties.*
(2) *It satisfies the normalization condition*

$$\mathrm{MC}_y[\mathrm{id}_X : X \to X] = \lambda_y(T^*X) = \sum y^i[\wedge^i T^*X] \in \mathrm{K}_T(X)[y].$$

The transformation MC_y satisfies the following properties:

(3) *It is determined by its image on classes $[f : Z \to X] = f_![\mathrm{id}_Z]$ where Z is a non-singular, irreducible, quasi-projective algebraic variety and f is a T-equivariant proper morphism.*
(4) *It satisfies a Verdier-Riemann-Roch (VRR) formula: for every smooth, T-equivariant morphism $\pi : X \to Y$ of quasi-projective and non-singular algebraic varieties, and every $[f : Z \to Y] \in \mathrm{K}_0^T(\mathrm{var}/Y)$, the following holds:*

$$\mathrm{MC}_y[\pi^* f : Z \times_Y X \to X] = \lambda_y(T^*_\pi) \cap \pi^*\mathrm{MC}_y[f : Z \to Y].$$

If one forgets the T-action, then the equivariant motivic Chern class above recovers the non-equivariant motivic Chern class from [**BSY10**] (either by its construction, or by the properties (1)–(3) from Theorem 4.2 and the corresponding results from [**BSY10**]).

REMARK 4.3. Theorem 4.2 and its proof work more generally for a possibly singular, quasi-projective T-equivariant base variety X, using the Grothendieck group of T-equivariant coherent $\mathcal{O}_X$-sheaves in the target (cf. [**FRW21**, Remark 2.2]), i.e.,

$$\mathrm{MC}_y : \mathrm{K}_0^T(\mathrm{var}/X) \to \mathrm{K}_0(\mathfrak{coh}^T(\mathcal{O}_X))[y].$$

Moreover, MC_y commutes with exterior products:

$$\mathrm{MC}_y[f \times f' : Z \times Z' \to X \times X'] = \mathrm{MC}_y[f : Z \to X] \boxtimes \mathrm{MC}_y[f' : Z' \to X'].$$

This follows as in the non-equivariant context [**BSY10**, Corollary 2.1] from part (3) of Theorem 4.2 and the multiplicativity of the equivariant λ_y-class for smooth and quasi-projective T-varieties X, X':

$$\lambda_y(T^*(X \times X')) = \lambda_y(T^*X) \boxtimes \lambda_y(T^*X') \in \mathrm{K}_T(X \times X')[y].$$

REMARK 4.4. The equivariant χ_y-genus of a T-variety Z is by definition

$$\chi_y(Z) := \mathrm{MC}_y([Z \to \mathrm{pt}]) \in \mathrm{K}_T(\mathrm{pt})[y].$$

By *rigidity* of the χ_y-genus (see [**FRW21**, §2.5] and [**Web16**, Theorem 7.2]), it contains no information about the action of T; it is equal to the non-equivariant χ_y-genus under the embedding $\mathbb{Z}[y] \to \mathrm{K}_T(\mathrm{pt})[y]$.

In what follows, the variety X will usually be understood from the context. If $Y \subseteq X$ is a T-invariant subvariety, not necessarily closed, we set

$$\mathrm{MC}_y(Y) := \mathrm{MC}_y[Y \hookrightarrow X].$$

If $i : Y \subseteq X$ is closed nonsingular subvariety and $Y' \subseteq Y$, then by functoriality

$$\mathrm{MC}_y[Y' \hookrightarrow X] = i_*\mathrm{MC}_y[Y' \hookrightarrow Y]$$

(K-theoretic push-forward). For instance if $Y' = Y$ then

$$\mathrm{MC}_y[i : Y \to X] = i_*(\lambda_y(T^*Y) \otimes [\mathcal{O}_Y])$$

as an element in $\mathrm{K}_T(X)$. We will often abuse notation and suppress the push-forward notation. Similarly for the other transformations discussed in later sections.

4.2. Motivic Chern classes of Schubert cells. Assume now that $X = G/B$. The following result from [**AMSS24**, Corollary 5.2, cf. Remark 5.4], allows us to calculate recursively the motivic Chern classes of Schubert cells.

THEOREM 4.5. *Let $w \in W$. Then the motivic Chern class $\mathrm{MC}_y(X(w)^\circ)$ is given by*

$$\mathrm{MC}_y(X(w)^\circ) = \mathcal{T}_{w^{-1}}(\mathcal{O}^T_{\mathrm{id}}).$$

Here, $\mathcal{T}_{w^{-1}} = \mathcal{T}_{i_k} \cdots \mathcal{T}_{i_1}$ if $w = s_{i_1} \cdots s_{i_k}$ is a reduced decomposition. This operator is well-defined by Proposition 3.2.

Following [**AMSS24**, Remark 6.4], we introduce an operator which will yield the (Poincaré) duals of motivic Chern classes. For each simple root $\alpha_i \in \Delta$, define

$$\mathfrak{L}_i = \partial_i + y(\partial_i \mathcal{L}_{\alpha_i} + \mathrm{id}) = -y(\mathcal{T}_i^\vee)^{-1} = \mathcal{T}_i^\vee + (1+y)\,\mathrm{id}\,. \tag{4.1}$$

Since $\mathcal{T}_i^\vee$ satisfy the usual braid and commutativity relations, so do these operators. Hence, $\mathfrak{L}_w$ is well-defined for all $w \in W$, where $\mathfrak{L}_w = \mathfrak{L}_{i_1} \cdots \mathfrak{L}_{i_k}$ if $w = s_{i_1} \cdots s_{i_k}$ is a reduced decomposition. Define the following elements in the equivariant K theory ring:

$$\widetilde{\mathrm{MC}}_y(Y(w)^\circ)) := \mathfrak{L}_{w^{-1}w_0}(\mathcal{O}^{w_0,T}); \quad \widetilde{\mathrm{MC}}_y(X(w)^\circ) = \mathfrak{L}_{w^{-1}}(\mathcal{O}^T_{\mathrm{id}}).$$

By definition $\widetilde{\mathrm{MC}}_y(Y(w)^\circ))$ and $\widetilde{\mathrm{MC}}_y(X(w)^\circ)$ are elements in $\mathrm{K}_T(X)[y]$.[1]

[1] If w_0^L denotes the left Weyl group action by w_0, as in [**MNS22**], then $w_0^L.\widetilde{\mathrm{MC}}_y(X(w)^\circ) = \widetilde{\mathrm{MC}}_y(Y(w_0w)^\circ)$. This generalizes the more familiar formula from Schubert calculus: $w_0^L.[X(w)]_T = [Y(w_0w)]_T$.

THEOREM 4.6. [**AMSS24**, Theorem 6.2] *The classes* $\widetilde{\mathrm{MC}}_y(Y(w)^\circ))$ *are orthogonal to the motivic Chern classes: for any* $u, v \in W$,

$$\langle \mathrm{MC}_y(X(u)^\circ), \widetilde{\mathrm{MC}}_y(Y(v)^\circ)\rangle = \delta_{u,v} \prod_{\alpha>0}(1+ye^{-\alpha}),$$

where the pairing on the left hand side is the intersection pairing defined in §2.2.

Note that $\prod_{\alpha>0}(1+ye^{-\alpha}) = \lambda_y(T^*_{w_0}X)$.

REMARK 4.7. Another family of classes dual to motivic Chern classes is given by a certain Serre dual variant of Segre motivic classes. Combining [**AMSS24**, Theorem 8.11] and [**MNS22**, Theorem 7.1]) (see also [**FRW21**]), one obtains the remarkable equality:

$$\frac{\widetilde{\mathrm{MC}}_y(Y(v)^\circ)}{\prod_{\alpha>0}(1+ye^{-\alpha})} = (-y)^{\dim G/B-\ell(v)} \frac{\mathcal{D}(\mathrm{MC}_y(Y(v)^\circ))}{\lambda_y(T^*X)} \in K_T(X)[y]_S[y^{-1}].$$

Here $\mathcal{D}$ denotes the (Grothendieck-Serre) duality, extended to the parameter y via $y^n \mapsto y^{-n}$, and $\mathrm{K}_T(X)[y]_S$ is appropriately localized so that $\lambda_y(T^*X)$ is invertible (see [**AMSS24**, Remark 8.9]). We note that one may define these 'Serre-Segre' motivic classes for any partial flag manifold G/P, and they are always dual to motivic Chern classes of Schubert cells; see [**MNS22**, Theorem 7.2]. Geometrically, the duality above is expected to arise from a transversality formula, generalizing to K-theory the results from [**Sch17**]. In cohomology, this is explained in [**AMSS17**, §7].

Consider the expansions of the equivariant motivic Chern classes,

$$\mathrm{MC}_y(X(w)^\circ) = \sum_{u\leq w} c_{u,w}(y;e^t)\mathcal{O}_u^T. \tag{4.2}$$

The equivariant K-Chevalley formula, used to multiply a Schubert class by the line bundle $\mathcal{L}_{\alpha_i}$ [**LP07, PR99**], and Theorem 4.5, gives a recursive procedure to calculate the motivic Chern classes of Schubert cells. The recursion also implies that the coefficients are polynomials $c_{u,w}(y;e^t) \in \mathbb{Z}[e^{\pm\alpha_1},\dots,e^{\pm\alpha_r}][y] \subseteq \mathrm{K}_T(\mathrm{pt})[y]$ in the characters associated to the simple roots. Note that the inclusion may be strict, reflecting the fact that the root lattice is in general strictly included in the weight lattice. Also, recall that e^t stands for the tuple $(e^{\pm t_1},\dots,e^{\pm t_r})$.

We provide next a few calculations for the motivic Chern classes of the flag manifolds $\mathbb{P}^1$ and $\mathrm{Fl}(3)$.

EXAMPLE 4.8 ($\mathrm{Fl}(2) = \mathbb{P}^1$). The equivariant motivic Chern classes for $\mathbb{P}^1$ are:

$$\mathrm{MC}_y(X(\mathrm{id})) = \mathcal{O}_{\mathrm{id}}^T; \quad \mathrm{MC}_y(X(s)^\circ) = (1+e^{-\alpha_1}y)\mathcal{O}_s^T - (1+(1+e^{-\alpha_1})y)\mathcal{O}_{\mathrm{id}}^T.$$

EXAMPLE 4.9 (The Schubert cell $X(s_1s_2)^\circ$). The equivariant motivic Chern classes for larger flag manifolds are much more complicated. For instance, the equivariant motivic Chern class of the Schubert cell $X(s_1s_2)^\circ \subseteq \mathrm{Fl}(3)$ is

$$\begin{aligned}\mathrm{MC}_y(X(s_1s_2)^\circ) =&(1+e^{-\alpha_1}y)(1+e^{-(\alpha_1+\alpha_2)}y)\mathcal{O}_{s_1s_2}^T\\ &-(1+e^{-\alpha_1}y)(1+(1+e^{-(\alpha_1+\alpha_2)})y)\mathcal{O}_{s_1}^T\\ &-(1+(1+e^{-\alpha_1})(1+e^{-\alpha_2})y+e^{-\alpha_2}(1+e^{-\alpha_1}+e^{-2\alpha_1})y^2)\mathcal{O}_{s_2}^T\\ &+(1+(2+e^{-\alpha_1}+e^{-\alpha_2}+e^{-(\alpha_1+\alpha_2)})y)\mathcal{O}_{\mathrm{id}}^T\\ &+(1+e^{-\alpha_1}+e^{-\alpha_2}+e^{-(\alpha_1+\alpha_2)}+e^{-(2\alpha_1+\alpha_2)})y^2\mathcal{O}_{\mathrm{id}}^T.\end{aligned}$$

The expressions above encode a remarkable amount of information. For instance, a simple verification in Example 4.9 shows that the expression for the *sum* of the coefficients simplifies dramatically and equals y^2, reflecting the geometric fact that we deal with a cell of dimension 2. We will prove this and more in Proposition 4.13 and Theorem 5.1 below. For now, we also provide some examples of non-equivariant motivic Chern classes. These are obtained from the equivariant ones by making the substitution $e^\lambda \mapsto 1$ for each weight λ.

EXAMPLE 4.10. The following are the non-equivariant motivic Chern classes of Schubert cells in Fl(3). (Recall that we use the notation MC_y for both the notion in ordinary and in equivariant K-theory.)

$$\begin{aligned}
\mathrm{MC}_y(X(\mathrm{id})) =& \mathcal{O}_{\mathrm{id}};\\
\mathrm{MC}_y(X(s_1)^\circ) =& (1+y)\mathcal{O}_{s_1} - (1+2y)\mathcal{O}_{\mathrm{id}};\\
\mathrm{MC}_y(X(s_2)^\circ) =& (1+y)\mathcal{O}_{s_2} - (1+2y)\mathcal{O}_{\mathrm{id}};\\
\mathrm{MC}_y(X(s_1s_2)^\circ) =& (1+y)^2\mathcal{O}_{s_1s_2} - (1+y)(1+2y)\mathcal{O}_{s_1} - (1+y)(1+3y)\mathcal{O}_{s_2}\\
&+ (5y^2+5y+1)\mathcal{O}_{\mathrm{id}};\\
\mathrm{MC}_y(X(s_2s_1)^\circ) =& (1+y)^2\mathcal{O}_{s_2s_1} - (1+y)(1+2y)\mathcal{O}_{s_2} - (1+y)(1+3y)\mathcal{O}_{s_1}\\
&+ (5y^2+5y+1)\mathcal{O}_{\mathrm{id}};\\
\mathrm{MC}_y(X(w_0)^\circ) =& (1+y)^3\mathcal{O}_{w_0} - (1+y)^2(1+2y)(\mathcal{O}_{s_1s_2}+\mathcal{O}_{s_2s_1})+\\
&(1+y)(5y^2+4y+1)(\mathcal{O}_{s_1}+\mathcal{O}_{s_2}) - (8y^3+11y^2+5y+1)\mathcal{O}_{\mathrm{id}}.
\end{aligned}$$

The classes $\widetilde{\mathrm{MC}}_y(Y(w)^\circ)$ for the Schubert cells in Fl(3) are:

$$\begin{aligned}
\widetilde{\mathrm{MC}}_y(Y(w_0)) =& \mathcal{O}^{w_0};\\
\widetilde{\mathrm{MC}}_y(Y(s_1s_2)^\circ) =& (1+y)\mathcal{O}^{s_1s_2} + y\mathcal{O}^{w_0};\\
\widetilde{\mathrm{MC}}_y(Y(s_2s_1)^\circ) =& (1+y)\mathcal{O}^{s_2s_1} + y\mathcal{O}^{w_0};\\
\widetilde{\mathrm{MC}}_y(Y(s_1)^\circ) =& (1+y)^2\mathcal{O}^{s_1} + y(1+y)\mathcal{O}^{s_1s_2} + 2y(1+y)\mathcal{O}^{s_2s_1} + y^2\mathcal{O}^{w_0};\\
\widetilde{\mathrm{MC}}_y(Y(s_2)^\circ) =& (1+y)^2\mathcal{O}^{s_2} + 2y(1+y)\mathcal{O}^{s_1s_2} + y(1+y)\mathcal{O}^{s_2s_1} + y^2\mathcal{O}^{w_0};\\
\widetilde{\mathrm{MC}}_y(Y(\mathrm{id})^\circ) =& (1+y)^3\mathcal{O}^{\mathrm{id}} + y(1+y)^2(\mathcal{O}^{s_1}+\mathcal{O}^{s_2})+\\
&2y^2(1+y)(\mathcal{O}^{s_1s_2}+\mathcal{O}^{s_2s_1}) + y^3\mathcal{O}^{w_0}.
\end{aligned}$$

An algebra check together with fact that $\langle \mathcal{O}_u, \mathcal{O}^v\rangle = 1$ if and only if $u \geq v$, shows that

$$\langle \mathrm{MC}_y(X(u)^\circ), \widetilde{\mathrm{MC}}_y(Y(v)^\circ)\rangle = (1+y)^3\delta_{u,v}$$

as stated in Theorem 4.6.

On the basis of these (and other) examples, we can conjecture that the motivic Chern classes and their duals satisfy a certain positivity property. This is discussed in section 8. At this time we note that the positivity of the dual classes in Example 4.10 does *not* extend beyond small cases. For instance, the coefficient of $\mathcal{O}^{s_3s_1s_2}$ in the expansion of $\widetilde{\mathrm{MC}}_y(Y(\mathrm{id})^\circ) \in K(\mathrm{Fl}(4))$ equals $y^2(4y-1)(1+y)^3$.

The 'top' Schubert coefficient is calculated in the following result.

LEMMA 4.11. *The coefficient $c_{w,w}(y; e^t)$ is given by*

$$c_{w,w}(y, e^t) = \prod_{\alpha>0, w\alpha<0} (1 + ye^{w\alpha}) = \lambda_y(T^*_{e_w}X(w)).$$

PROOF. The localization $\mathrm{MC}_y(X(w)^\circ)|_w$ equals $c_{w,w}(y,e^t)(\mathcal{O}_w^T)|_w$. By Lemma 3.3(c) and Theorem 4.5, we get

$$\mathrm{MC}_y(X(w)^\circ)|_w = \prod_{\alpha>0, w\alpha<0} \frac{1+ye^{w\alpha}}{1-e^{w\alpha}} \iota_w|_w.$$

However, $(\mathcal{O}_w^T)|_w = \frac{\iota_w|_w}{\lambda_{-1}(T_w^*X(w))} = \frac{\iota_w|_w}{\prod_{\alpha>0,w\alpha<0}(1-e^{w\alpha})}$. The claim follows. $\square$

We end this section with an analogue of Theorem 4.5 for the Segre motivic classes

$$\mathrm{SMC}_y(X(w)^\circ, X) := \frac{\mathrm{MC}_y(X(w)^\circ)}{\lambda_y(T^*X)}.$$

These classes live in a localization $\mathrm{K}_T(X)_S$ where the element $\prod_{\alpha>0}(1+ye^\alpha)(1+ye^{-\alpha}) \in \mathrm{K}_T(\mathrm{pt})[y]$ is invertible; see [**AMSS24**, Remark 8.9]. We recall [**MS22**, Theorem 4.2], which will be used below when discussing the Hirzebruch version of the Segre classes.

THEOREM 4.12. *For any $w \in W$ one has in $K_T(X)[y]_S$:*

$$\frac{\mathrm{MC}_y(X(w)^\circ)}{\lambda_y(T^*X)} = \frac{\mathcal{T}^\vee_{w^{-1}}(\mathcal{O}^T_{\mathrm{id}})}{\prod_{\alpha>0}(1+ye^\alpha)} = \mathcal{T}^\vee_{w^{-1}}\left(\frac{\mathcal{O}^T_{\mathrm{id}}}{\prod_{\alpha>0}(1+ye^\alpha)}\right)$$

and

$$\frac{\mathrm{MC}_y(Y(w)^\circ)}{\lambda_y(T^*X)} = \frac{\mathcal{T}^\vee_{(w_0w)^{-1}}(\mathcal{O}^{w_0,T})}{\prod_{\alpha>0}(1+ye^{-\alpha})} = \mathcal{T}^\vee_{(w_0w)^{-1}}\left(\frac{\mathcal{O}^{w_0,T}}{\prod_{\alpha>0}(1+ye^{-\alpha})}\right). \tag{4.3}$$

Note that $\prod_{\alpha>0}(1+ye^\alpha) = \lambda_y(T^*_{e_{\mathrm{id}}}X)$ and $\prod_{\alpha>0}(1+ye^{-\alpha}) = \lambda_y(T^*_{e_{w_0}}X)$.

4.3. Integrals of motivic Chern classes and point counting. By functoriality of motivic Chern classes, the integral of the motivic Chern class of a Schubert cell equals

$$\begin{aligned}\int_{G/B} \mathrm{MC}_y(X(w)^\circ) &:= \mathrm{MC}_y[X(w)^\circ \to \mathrm{pt}] = \mathrm{MC}_y[\mathbb{A}^{\ell(w)} \to \mathrm{pt}] \\ &= \mathrm{MC}_y[\mathbb{A}^1 \to \mathrm{pt}]^{\ell(w)}.\end{aligned}$$

The last equality uses the fact the the map $\mathrm{MC}_y : \mathrm{K}_0^T(\mathrm{var}/\mathrm{pt}) \to \mathrm{K}_0^T(\mathrm{pt})$ is a *ring* homomorphism, with the product given by exterior product of varieties; see [**BSY10**, Corollary 2.1] extended equivariantly in [**AMSS24**, Theorem 4.2]. One can calculate $\mathrm{MC}_y[\mathbb{A}^1 \to \mathrm{pt}]$ directly from Example 4.8:

$$\mathrm{MC}_y[\mathbb{A}^1 \to \mathrm{pt}] = \int_{\mathbb{P}^1} \mathrm{MC}_y(\mathbb{A}^1) = -y.$$

Combining all these, we deduce:

PROPOSITION 4.13. *Recall the Schubert expansion* (4.2). *Then the following hold:*

(a) $\int_{G/B} \mathrm{MC}_y(X(w)^\circ) = \sum c_{w,u}(y,e^t) = (-y)^{\ell(w)}$.
(b) *The χ_y-genus of G/B equals*

$$\chi_y(G/B) = \int_{G/B} \lambda_y(T^*(G/B)) = \sum_{w\in W} (-y)^{\ell(w)}.$$

PROOF. Part (a) follows from the considerations above and because $\int_{G/B} \mathcal{O}_w^T = 1$, since $H^i(X(w), \mathcal{O}_{G/B}) = 0$ for $i > 0$, as Schubert varieties are rational with rational singularities. Part (b) follows from (a), using the fact that $\lambda_y(T^*(G/B)) \otimes \mathcal{O}_{G/B} = \mathrm{MC}_y(G/B) = \sum_{w \in W} \mathrm{MC}_y(X(w)^\circ)$. □

If one specializes $y = -q$, this proposition shows that the χ_y-genus of a Schubert variety $X(w)$ is equal to the number of points of $X(w)$ over $\mathbb{F}_q$, the field with q elements. This type of arguments are discussed more generally for any G/P, or for smooth, projective T-varieties with finitely many fixed points in [**MS22**, Theorem A.1]; see Remark 4.14 below for a further generalization.

Utilizing again the specialization $y = -q$ and taking $G/B = \mathrm{Fl}(n)$, one recovers in a natural way q-analogues of classical formulae. In this case, $W = S_n$ (the symmetric group) and

$$\chi_{-q}(\mathrm{Fl}(n)) = \sum_{w \in S_n} q^{\ell(w)}.$$

It is known that this sum equals the q-analogue of the factorial,

$$\sum_{w \in S_n} q^{\ell(w)} = [n]_q! = (1+q)(1+q+q^2) \cdot \ldots \cdot (1+q+\ldots+q^{n-1}).$$

In fact, it is fun to work out a geometric interpretation of this formula. The natural projection $p_n : \mathrm{Fl}(n) \to \mathrm{Gr}(n-1,n)$ sending a flag $(F_1 \subseteq \ldots \subseteq F_n = \mathbb{C}^n)$ to $F_{n-1} \subseteq \mathbb{C}^n$ is a G-equivariant Zariski locally trivial fibration, with fiber isomorphic to $\mathrm{Fl}(n-1)$. By additivity and multiplicativity of motivic Chern classes over a point, it follows that

$$\begin{aligned}\chi_{-q}(\mathrm{Fl}(n)) &= \chi_{-q}(\mathrm{Fl}(n-1)) \cdot \chi_{-q}(\mathrm{Gr}(n-1,n)) \\ &= \chi_{-q}(\mathrm{Fl}(n-1)) \cdot (1+q+q^2+q^3+\ldots+q^{n-1}).\end{aligned}$$

The last equality follows from the fact that the (dual) projective space $\mathrm{Gr}(n-1,n)$ is the disjoint union of Schubert cells, each of which with contribution $q^{\dim(\mathrm{cell})}$ to the χ_{-q} genus. Then the equality follows by induction on n. A more detailed analysis of this geometric argument is carried out in §4.5.

REMARK 4.14. This relation between $\mathbb{F}_q$ point count and the χ_y-genus of a complex algebraic variety X holds more generally for X of strongly polynomial-count in the sense of Katz [**HRV08**, 6 Appendix]. By Theorem 6.1.2 in *loc. cit.* one obtains the relation $P_X(q) = E(x,y)$ for $q = xy$, with $P_X(q)$ the polynomial point count over $\mathbb{F}_q$ and E the corresponding E-polynomial of X defined in terms of the mixed Hodge numbers of the compactly supported cohomology of X. By forgetting the weight filtration one gets in these cases $E(y,1) = \chi_{-y}(X)$ by [**BSY10**, (5.3) and (5.5), p. 43-44].

4.4. Divisibility properties. Consider now the expansions of the *non-equivariant* motivic Chern classes,

$$\mathrm{MC}_y(X(w)^\circ) = \sum_{u \le w} c_{u,w}(y) \mathcal{O}_u \quad \in \mathrm{K}(X)[y]. \tag{4.4}$$

The coefficients $c_{u,w}(y)$ from (4.4) are polynomials in $\mathbb{Z}[y]$. Example 4.10 suggests a divisibility property of these coefficients.

PROPOSITION 4.15. *The coefficient $c_{u,w}(y)$ is divisible by $(1+y)^{\ell(u)}$.*

PROOF. We prove the statement by induction on $\ell(u)$. If $u = w$, Lemma 4.11 gives $c_{w,w}(y) = (1+y)^{\ell(w)}$. Now assume $u < w$ and that the statement is true for any $v \le w$ with $\ell(v) > \ell(u)$. Arguing by contradiction, suppose that $(1+y)^\ell \mid c_{u,w}(y)$ and $(1+y)^{\ell+1} \nmid c_{u,w}(y)$ for some $\ell < \ell(u)$. We use the hypotheses of Proposition 3.5, where we only keep the action of $\mathbb{C}^*$. In particular $y = -e^{-\hbar}$, therefore the initial term of $1+y$ is $\hbar$. Consider the expansion:

(4.5)

$$\mathrm{ch}_{\mathbb{C}^*}(\mathrm{MC}_{-e^{-\hbar}}(X(w)^\circ)) = \sum_{z \le w, \ell(z) < \ell(u)} c_{z,w}(-e^{-\hbar})\, \mathrm{ch}_{\mathbb{C}^*}(\mathcal{O}_z) + \sum_{z \le w, \ell(z) > \ell(u)} c_{z,w}(-e^{-\hbar})\, \mathrm{ch}_{\mathbb{C}^*}(\mathcal{O}_z) + \sum_{z \le w, \ell(z) = \ell(u)} c_{z,w}(-e^{-\hbar})\, \mathrm{ch}_{\mathbb{C}^*}(\mathcal{O}_z).$$

Theorem 4.5 and Proposition 3.5 imply that the part with the highest homological degree in $\mathrm{ch}_{\mathbb{C}^*}(\mathrm{MC}_{-e^{-\hbar}}(X(w)^\circ))$ lies in $H_0^{\mathbb{C}^*}(X)$. (In Theorem 7.1 below we will show this is the homogenized Chern-Schwartz-MacPherson class $c^\hbar_{\mathrm{SM}}(X(w)^\circ)$, in particular it is nonzero.) Since $\ell < \ell(u)$, it follows that the coefficient of $\hbar^\ell[X(u)] \in H^{\mathbb{C}^*}_{2\ell(u)-2\ell}(X)$ in the left-hand side of (4.5) is equal to 0. We analyze this coefficient on the right-hand side. The first summand has no contribution because $\ell(z) < \ell(u)$. By induction, every term in the second summand is divisible by $\hbar^{\ell(u)+1}$, thus again it does not contribute to the coefficient of $\hbar^\ell[X(u)]$. In the last summand, only the term with $z = u$ can contribute. Its contribution equals the coefficient of $\hbar^\ell$ in $c_{u,w}(-e^{-\hbar})$. This coefficient is non-zero, by the hypothesis on ℓ, and it gives a non-zero coefficient of $\hbar^\ell[X(u)]$, contradicting the previous conclusion. □

We end with the following corollary.

COROLLARY 4.16. *Consider the non-equivariant motivic Chern class*

$$\mathrm{MC}_y(X(w)^\circ) = \sum c_{u,w}(y)\mathcal{O}_u.$$

Then $c_{\mathrm{id},w}(-1) = 1$.

PROOF. By the divisibility property, $c_{u,w}(-1) = 0$ for $\ell(u) > 0$. Then

$$1 = \int_{G/B} \mathrm{MC}_{-1}(X(w)^\circ) = c_{\mathrm{id},w}(-1)$$

by Proposition 4.13. □

We invite the reader to verify this corollary for the motivic Chern classes in Fl(3) from Example 4.10 above.

4.5. The parabolic case. Consider the (generalized) partial flag manifold G/P, and let $\pi : G/B \to G/P$ be the natural projection. This is a G-equivariant locally trivial fibration in the Zariski topology, with fiber $F := \pi^{-1}(1.P) = P/B$. This fiber is the flag manifold $L/(B \cap L)$, where L is the Levi subgroup of P. The Schubert varieties in F are indexed by the elements in W_P. Furthermore, the image $\pi(X(w)^\circ)$ equals $X(wW_P)^\circ$, and the restriction of π to $X(w)^\circ$ is a trivial fibration, showing that $X(w)^\circ \simeq X(wW_P)^\circ \times (w.F \cap X(w)^\circ)$. It follows that in the Grothendieck group $\mathrm{K}_0(\mathrm{var}/(G/P))$,

$$[X(w)^\circ \to G/P] = [w.F \cap X(w)^\circ \to wP] \boxtimes [X(wW_P)^\circ \to G/P]. \tag{4.6}$$

The intersection $w.F \cap X(w)^\circ$ is the Schubert cell in $w.F$ indexed by $w_2 \in W_P$, where $w = w_1 w_2$ is the parabolic factorization of w with respect to P; cf. [**BCMP22**, Theorem 2.8]. This argument allows us calculate the push-forwards of motivic Chern classes.

PROPOSITION 4.17. *The following hold:*

(a) $\pi_* \mathrm{MC}_y(X(w)^\circ) = (-y)^{\ell(w)-\ell(wW_P)} \mathrm{MC}_y(X(wW_P)^\circ)$ *in* $\mathrm{K}_T(G/P)$.

(b) *More generally, let* $P \subseteq Q$ *be two standard parabolic subgroups, and* $\pi' : G/P \to G/Q$ *the natural projection. Then*

$$\pi'_* \mathrm{MC}_y(X(wW_P)^\circ) = (-y)^{\ell(wW_P)-\ell(wW_Q)} \mathrm{MC}_y(X(wW_Q)^\circ).$$

PROOF. Equation (4.6) and functoriality of motivic Chern classes imply that

$$\pi_* \mathrm{MC}_y(X(w)^\circ) = \mathrm{MC}_y[w.F \cap X(w)^\circ \to wP] \cdot \mathrm{MC}_y(X(wW_P)^\circ),$$

Then the claim follows from Proposition 4.13(a). Part (b) may be obtained by applying (a) to the composition of projections $G/B \to G/Q \to G/P$. □

We end this section by pointing out that the argument from Proposition 4.13 extends *verbatim* to the parabolic case. One obtains:

PROPOSITION 4.18. *Let* $w \in W^P$ *and consider the Schubert expansion* $\mathrm{MC}_y(X(wW_P)^\circ) = \sum_{u \in W^P, u \le w} c_{u,w}(y, e^t) \mathrm{MC}_y(X(uW_P)^\circ)$. *Then the following hold:*

(a) $\int_{G/P} \mathrm{MC}(X(wW_P)^\circ) = \sum c_{w,u}(y, e^t) = (-y)^{\ell(w)}$.

(b) *The* χ_y*-genus of* G/P *equals* $\chi_y(G/P) = \sum_{w \in W^P} (-y)^{\ell(w)}$.

To illustrate, let $G/P = \mathrm{Gr}(k, n)$ be the Grassmann manifold of subspaces of dimension k in $\mathbb{C}^n$. The set W^P corresponds to partitions $\lambda = (\lambda_1, \dots, \lambda_k)$ included in the $k \times (n-k)$ rectangle such that $\dim X(\lambda) = |\lambda| = \lambda_1 + \dots + \lambda_k$. For $q = -y$, the χ_{-q} genus is

$$\chi_{-q}(\mathrm{Gr}(k, n)) = \sum_\lambda q^{|\lambda|} = \binom{n}{k}_q,$$

the q-analogue of the binomial coefficient.

5. The parameter y in motivic Chern classes

In this section we discuss some key combinatorial properties of the Schubert expansion of the motivic Chern classes, including specializations of the parameter y and their geometric interpretation.

THEOREM 5.1. *Let* P *be a parabolic subgroup of* G, $X = G/P$, *and* $w \in W^P$. *Then the following hold:*

(a) *The specialization* $y = -1$ *gives* $\mathrm{MC}_{-1}(X(wW_P)^\circ) = \iota_{wW_P}$, *the equivariant class of the unique torus fixed point in* $X(wW_P)^\circ$.

(b) *The specialization* $y = 0$ *gives* $\mathrm{MC}_0(X(wW_P)^\circ) = \mathcal{I}^T_{wW_P}$, *the class of the ideal sheaf* $\mathcal{O}_{X(wW_P)}(-\partial X(wW_P))$.

(c) *The degree of* $\mathrm{MC}_y(X(wW_P)^\circ)$ *with respect to* y *is equal to* $\ell(w)$, *and the coefficient of* $y^{\ell(w)}$ *in* $\mathrm{MC}_y(X(wW_P)^\circ)$ *is the class of* $\omega_{X(wW_P)}$, *the dualizing sheaf on* $X(wW_P)$.

PROOF. We first prove all the statements for $P = B$. Parts (a) and (b) follow from Theorem 4.5 and the specializations at $y = -1$ and $y = 0$ of the operator $\mathcal{T}_i$, from Lemma 3.4. Let $\theta : Z \to X(w)$ denote the Bott-Samelson resolution of the Schubert variety $X(w)$ used, e.g., in [**AM16**]. By functoriality,

$$\mathrm{MC}_y[X(w)^\circ \to G/B] = \theta_* \mathrm{MC}_y[X(w)^\circ \to Z].$$

The restriction of θ to $\theta^{-1}(X(w)^\circ)$ is an isomorphism, and it is known that the complement of $\theta^{-1}(X(w)^\circ)$ in Z is a simple normal crossing divisor. Since Z is smooth, by inclusion-exclusion it follows that the term with highest power of y in $\mathrm{MC}_y(Z \smallsetminus \theta^{-1}(X(w)^\circ))$ is the same as the one in $\mathrm{MC}_y(Z) = \lambda_y(T^*(Z))$, namely $y^{\ell(w)} \wedge^{\dim Z} T^*(Z) = y^{\ell(w)} \omega_Z$, a multiple of the canonical bundle of Z. This finishes the proof of (c), since $\theta_*(\omega_Z) = \omega_{X(w)}$ as $X(w)$ has rational singularities; see e.g., [**BK05**].

We now turn to the general G/P situation. Since $w \in W^P$, the projection $\pi : G/B \to G/P$ restricts to a birational map $\pi : X(w) \to X(wW_P)$ which is an isomorphism over the Schubert cell $X(wW_P)^\circ$. By Proposition 4.17, $\pi_* \mathrm{MC}_y(X(w)^\circ) = \mathrm{MC}_y(X(wW_P)^\circ)$. Then each claim follows from the corresponding statement for G/B, taking into consideration that $\pi_*(\iota_w) = \iota_{wW_P}$, $\pi_* \mathcal{I}_w^T = \mathcal{I}_{wW_P}^T$ (cf. [**Bri02**, Proof of Lemma 4]), and $\pi_* \omega_{X(w)} = \omega_{X(wW_P)}$ (since $X(wW_P)$ has rational singularities; see [**BK05**]). □

REMARK 5.2. In the non-equivariant case, the result in (b) can also be proved using the fact that, for the Schubert *variety*, we have $\mathrm{MC}_0(X(w)) = \mathcal{O}_w^T$ since $X(w)$ has rational, hence DuBois, singularities [**BSY10**, Example 3.2]. The equivariant generalization of this argument would use either an equivariant version of the DuBois complex, which is not available at this time in the literature, or a corresponding equality in $K_0(\mathfrak{coh}^T(\mathcal{O}_X))$.

The duality in Theorem 4.6 allows us to calculate the specializations of $\widetilde{\mathrm{MC}}_y(Y(w)^\circ)$.

COROLLARY 5.3. *Let $w \in W$. Then the following hold:*

(a) *The specialization $y = -1$ gives*

$$\widetilde{\mathrm{MC}}_{-1}(Y(w)^\circ) = \frac{\lambda_{-1}(T^*_{w_0}(G/B))}{\lambda_{-1}(T^*_w(G/B))} \iota_w = \frac{\prod_{\alpha>0}(1 - e^{-\alpha})}{\prod_{\alpha>0}(1 - e^{w\alpha})} \iota_w.$$

(b) *The specialization $y = 0$ gives $\widetilde{\mathrm{MC}}_0(Y(w)^\circ) = \mathcal{O}^{w,T}$.*

PROOF. Theorem 4.6 and Theorem 5.1 imply that $\kappa := \widetilde{\mathrm{MC}}_{-1}(Y(w)^\circ)$ is a class in $\mathrm{K}_T(G/B)$ with localizations

$$\kappa|_u = \delta_{u,w} \lambda_{-1}(T^*_{w_0}(G/B)).$$

Then (a) follows from the injectivity of the localization map, see [**Nie74**, Theorem 3.2], and the fact that $\langle \iota_w, \iota_u \rangle = \delta_{u,w} \lambda_{-1}(T^*_w(G/B))$. Part (b) is a consequence of the duality between the ideal sheaves and structure sheaves recalled in (3.1), combined with Theorem 5.1(b). □

6. Equivariant Hirzebruch classes

6.1. The equivariant Hirzebruch transformation. In the non equivariant case, it is useful to define and study several transformations associated with motivic Chern classes. These transformations provide a unified point of view from which, for instance, the classical Chern-Schwartz-MacPherson (CSM) classes may be obtained as a subproduct of a 'normalized' Todd transformation together with Grothendieck-Riemann-Roch. This process is explained in [**BSY10**, §0 and §3]. In this paper we use the equivariant Riemann-Roch theorem proved by Edidin and Graham [**EG00**, Theorem 3.1] to explain the equivariant counterparts of these 'unnormalized' and 'normalized' transformations. We note that Weber [**Web16, Web17**] first studied the equivariant unnormalized Hirzebruch class we will consider below.

Since many of the results explained here are not available in this generality in the literature, we find it useful to recall the precise hypotheses we utilize. We hope this section may be used as a reference in the future. We will consider more in detail the case of (equivariant) CSM classes in §7. There we will also provide an alternative way to obtain CSM classes from motivic Chern classes.

By X we denote a complex algebraic variety. For a commutative ring R, the completions $\widehat{A}_*^T(X,R), \widehat{H}_*^T(X,R)$ denote the product of the equivariant Chow groups [**EG98**], respectively the equivariant Borel-Moore homology groups (where the degree is doubled), with coefficients in R:

$$\widehat{A}_*^T(X,R) := \prod_{i\leq \dim X} A_i^T(X)\otimes R \quad \text{or} \quad \widehat{H}_*^T(X,R) := \prod_{i\leq \dim X} H_{2i}^T(X)\otimes R \quad .$$

The Chow and Borel-Moore homology are related by a cycle map $cl : A_i^T(X) \to H_{2i}^T(X)$, and one may work directly in the Chow context, or in the image under this map; cf. [**EG98**, §2.8]. In what follows we have chosen to work in the Borel-Moore context. The coefficients ring R will be mostly $\mathbb{Z}, \mathbb{Q}, \mathbb{Q}[y]$ or $\mathbb{Q}[y,(1+y)^{-1}]$. In case no coefficients are mentioned, we are using $R=\mathbb{Q}$ (as before).

For a T-variety X let

$$\mathrm{td}_* : K_0(\mathfrak{coh}^T(\mathcal{O}_X)) \to \widehat{H}_*^T(X)$$

be the equivariant Todd class transformation to the completion $\widehat{H}_*^T(X)$, constructed in [**EG00**, §3.2]. Then td_* is covariant for proper T-equivariant morphisms. Also note that

$$\mathrm{td}_*(\mathcal{O}_X) = \mathrm{Td}(TX)\cap [X]_T \tag{6.1}$$

for X smooth by [**EG00**, Theorem 3.1(d)(i)] and [**EG05**, Remark 6.10 and Lemma A.1], since T is abelian so that the adjoint action of T on its Lie algebra is trivial (see also [**MS15**, p. 2218-2219] and compare with [**AMSS17**, Proof of Theorem 3.3] for the counterpart of Ohmoto's equivariant Chern class transformation). We also have the equivariant Chern character (§2.3)

$$\mathrm{ch} : \mathrm{K}_T(X) \to \widehat{H}_T^*(X).$$

This is a contravariant ring homomorphism for T-equivariant morphisms. Then the Todd transformation satisfies the module property

$$\mathrm{td}_*([E]_T\otimes[\mathcal{F}]_T) = \mathrm{ch}([E]_T)\cap \mathrm{td}_*([\mathcal{F}]_T) \tag{6.2}$$

for $[E]_T \in \mathrm{K}_T(X)$ and $[\mathcal{F}]_T \in \mathrm{K}_0(\mathfrak{coh}^T(\mathcal{O}_X))$ ([**EG00**, Theorem 3.1]). Recall that for a T-equivariant vector bundle E, the cohomological Todd class $\mathrm{Td}(E) :=$

$\mathrm{Td}([E]_T)$ is multiplicative in short exact sequences, and for an equivariant line bundle $\mathcal{L}$ it is defined by

$$\mathrm{Td}(\mathcal{L}) := \frac{c_1^T(\mathcal{L})}{1 - e^{-c_1^T(\mathcal{L})}} = 1 + \frac{1}{2}c_1^T(\mathcal{L}) + \dots$$

Define the **unnormalized** (respectively **normalized**) **cohomological Hirzebruch class** $\widetilde{\mathrm{Td}}_y$ (resp., Td_y) by

$$\widetilde{\mathrm{Td}}_y(\mathcal{L}) := \mathrm{ch}(\lambda_y(\mathcal{L}^\vee))\,\mathrm{Td}(\mathcal{L}) = \frac{c_1^T(\mathcal{L})(1 + ye^{-c_1^T(\mathcal{L})})}{1 - e^{-c_1^T(\mathcal{L})}}$$

and

$$\mathrm{Td}_y(\mathcal{L}) := \frac{\widetilde{\mathrm{Td}}_y((1+y)[\mathcal{L}]_T)}{1+y} = \frac{c_1^T(\mathcal{L})(1 + ye^{-c_1^T(\mathcal{L})(1+y)})}{1 - e^{-c_1^T(\mathcal{L})(1+y)}}$$

Then extend these definitions to any equivariant vector bundle using the splitting principle, by requiring that they be multiplicative on short exact sequences. Note the specializations:

$$\widetilde{\mathrm{Td}}_{y=0}(\mathcal{L}) = \mathrm{Td}_{y=0}(\mathcal{L}) = \mathrm{Td}(\mathcal{L}), \tag{6.3}$$

and

$$\widetilde{\mathrm{Td}}_{y=-1}(\mathcal{L}) = c_1^T(\mathcal{L}) \quad , \quad \mathrm{Td}_{y=-1}(\mathcal{L}) = c^T(\mathcal{L}) = 1 + c_1^T(\mathcal{L}). \tag{6.4}$$

The power series $\mathrm{Td}_y(\mathcal{L}) = 1 + \dots$ has constant coefficient 1, whereas $\widetilde{\mathrm{Td}}_y(\mathcal{L}) = 1 + y + \dots$ has constant coefficient $1 + y$, explaining the name 'unnormalized'.

Combining Theorem 4.2 with the equivariant Riemann-Roch theorem proved by Edidin and Graham [**EG00**, Theorem 3.1] one obtains the following results about the unnormalized equivariant Hirzebruch class transformation.

THEOREM 6.1. *Let X be a quasi-projective, non-singular, complex algebraic variety with an action of the torus T. The unnormalized (equivariant) Hirzebruch transformation*

$$\widetilde{\mathrm{Td}}_{y,*} := \mathrm{td}_* \circ \mathrm{MC}_y : \mathrm{K}_0^T(\mathrm{var}/X) \to \widehat{H}_*^T(X)[y] \subseteq \widehat{H}_*^T(X; \mathbb{Q}[y, (1+y)^{-1}])$$

is the unique natural transformation satisfying the following properties:

(a) *It is functorial with respect to T-equivariant proper morphisms of non-singular, quasi-projective varieties.*
(b) *It satisfies the normalization condition*

$$\widetilde{\mathrm{Td}}_{y,*}([\mathrm{id}_X]) = \widetilde{\mathrm{Td}}_y(TX) \cap [X]_T.$$

(c) *It is determined by its image on classes $[f : Z \to X] = f_![\mathrm{id}_Z]$ where Z is a non-singular, irreducible, quasi-projective algebraic variety and f is a T-equivariant proper morphism.*
(d) *It satisfies a Verdier-Riemann-Roch (VRR) formula: for any smooth, T-equivariant morphism $\theta : X \to Y$ of quasi-projective and non-singular algebraic varieties, and any $[f : Z \to Y] \in \mathrm{K}_0^T(\mathrm{var}/Y)$,*

$$\widetilde{\mathrm{Td}}_y(T_\theta^*) \cap \theta^* \,\widetilde{\mathrm{Td}}_{y,*}[f : Z \to Y] = \widetilde{\mathrm{Td}}_{y,*}[\theta^* f : Z \times_Y X \to X].$$

If one forgets the T-action, then the unnormalized equivariant Hirzebruch transformation above recovers the corresponding non-equivariant transformation $\widetilde{T}_{y,*}$ from [**BSY10**] (either by its construction, or by the properties (a)–(c) from Theorem 6.1 and the corresponding results from [**BSY10**].)

REMARK 6.2. Theorem 6.1 and its proof work more generally for a possibly singular, quasi-projective T-equivariant base variety X. Moreover, $\widetilde{\mathrm{Td}}_{y,*}$ commutes with exterior products:

$$\widetilde{\mathrm{Td}}_{y,*}[f \times f' : Z \times Z' \to X \times X'] = \widetilde{\mathrm{Td}}_{y,*}[f : Z \to X] \boxtimes \widetilde{\mathrm{Td}}_{y,*}[f' : Z' \to X'].$$

This follows as in the non-equivariant context [**BSY10**, Corollary 3.1] from part (3) of Theorem 6.1 and the multiplicativity of the corresponding equivariant cohomological class for smooth and quasi-projective T-varieties X, X':

$$\widetilde{\mathrm{Td}}_y(T^*(X \times X')) = \widetilde{\mathrm{Td}}_y(T^*X) \boxtimes \widetilde{\mathrm{Td}}_y(T^*X') \in \widehat{H}^*_T(X \times X')[y].$$

One can also define a *normalized* equivariant Hirzebruch class transformation. With this aim, we introduce the following functorial **Adams operations** (in (co)homology):

$$\psi^*_{1+y} : \widehat{H}^*_T(X, \mathbb{Q}[y]) \to \widehat{H}^*_T(X, \mathbb{Q}[y])$$

and

$$\psi_*^{1+y} : \widehat{H}^T_*(X, \mathbb{Q}[y]) \to \widehat{H}^T_*(X, \mathbb{Q}[y, (1+y)^{-1}])$$

given by multiplication with $(1+y)^i$ on $H^{2i}_T(X, \mathbb{Q}[y])$ respectively $(1+y)^{-j}$ on $H^T_{2j}(X, \mathbb{Q}[y])$, and satisfying the module and ring properties

(6.5)
$$\psi_*^{1+y}(- \cap -) = \psi^*_{1+y}(-) \cap \psi_*^{1+y}(-) \quad \text{and} \quad \psi^*_{1+y}(- \cup -) = \psi^*_{1+y}(-) \cup \psi^*_{1+y}(-).$$

(In the Chow context, the (co)homological grading will not be doubled.)

REMARK 6.3. These module and ring properties also hold for the functorial (co)homological duality transformations

$$\psi^*_{-1} : \widehat{H}^*_T(X, \mathbb{Q}) \to \widehat{H}^*_T(X, \mathbb{Q}) \quad \text{and} \quad \psi_*^{-1} : \widehat{H}^T_*(X, \mathbb{Q}) \to \widehat{H}^T_*(X, \mathbb{Q})$$

given by multiplication with $(-1)^i$ on $H^{2i}_T(X, \mathbb{Q})$ resp. $(-1)^j$ on $H^T_{2j}(X, \mathbb{Q})$. If X is smooth, these are consistent with the corresponding duality involutions in K-theory:

$$\mathrm{ch} \circ (-)^\vee = \psi^*_{-1} \circ \mathrm{ch}(-) : \mathrm{K}_T(X) \to \widehat{H}^*_T(X, \mathbb{Q}),$$

since $c_1^T(\mathcal{L}^\vee) = -c_1^T(\mathcal{L})$ for a T-equivariant line bundle $\mathcal{L}$. Similarly, for the Grothendieck-Serre duality $\mathcal{D}$,

$$\mathrm{td}_* \circ \mathcal{D}(-) = \psi_*^{-1} \circ \mathrm{td}_*(-) : \mathrm{K}_T(X) \to \widehat{H}^T_*(X, \mathbb{Q}),$$

since

$$\begin{aligned} \mathrm{td}_*(\omega_X) &= \mathrm{ch}(\omega_X)\,\mathrm{Td}(TX) \cap [X]_T = \mathrm{Td}(T^*X) \cap [X]_T \\ &= (-1)^{\dim X} \psi_*^{-1}(\mathrm{Td}(TX) \cap [X]_T). \end{aligned}$$

By definition, the (co)homological Adams operations satisfy

$$\psi^*_{1+y}(c_1^T(\mathcal{L})) = (1+y)c_1^T(\mathcal{L}); \quad \psi^*_{1+y}(\widetilde{\mathrm{Td}}_y(\mathcal{L})) = (1+y)\,\mathrm{Td}_y(\mathcal{L})$$

and $\psi_*^{1+y}([X]_T) = (1+y)^{-d}[X]_T$ for X of pure dimension d. It follows that

(6.6) $$\psi_*^{1+y}(\widetilde{\mathrm{Td}}_y(TX) \cap [X]_T) = \psi^*_{1+y}(\widetilde{\mathrm{Td}}_y(TX)) \cap \psi_*^{1+y}([X]_T) = \mathrm{Td}_y(TX) \cap [X]_T$$

for X smooth and pure dimensional. In particular,

(6.7) $$\psi_*^{1+y}(\widetilde{\mathrm{Td}}_y(TX) \cap [X]_T) \in \widehat{H}^T_*(X, \mathbb{Q}[y]) \subseteq \widehat{H}^T_*(X, \mathbb{Q}[y, (1+y)^{-1}]).$$

This motivates the following:

DEFINITION 6.4. The **normalized equivariant Hirzebruch transformation** $\mathrm{Td}_{y,*}$ is defined by:

(6.8)

$$\mathrm{Td}_{y,*} := \psi_*^{1+y} \circ \widetilde{\mathrm{Td}}_{y,*} : \mathrm{K}_0^T(\mathrm{var}/X) \to \widehat{H}_*^T(X, \mathbb{Q}[y]) \subseteq \widehat{H}_*^T(X, \mathbb{Q}[y, (1+y)^{-1}]).$$

This transformation satisfies the same properties listed in Theorem 6.1 and Remark 6.2, with unnormalized classes replaced throughout by normalized classes. Furthermore, the normalized transformation has values in $\widehat{H}_*^T(X, \mathbb{Q}[y])$, by Theorem 6.1(c) and Equation (6.6) above, cf. (6.7).

REMARK 6.5. The equivariant χ_y-genus of a T-variety Z may be calculated by

$$\chi_y(Z) = \widetilde{\mathrm{Td}}_y([Z \to \mathrm{pt}]) \in \widehat{H}_T^*(\mathrm{pt})[y].$$

By *rigidity* of the χ_y-genus (see [**Web16**, Theorem 7.2]), these quantities contain no information about the action of T, i.e., both are equal to the non-equivariant χ_y-genus under the embedding $\mathbb{Z}[y] \to \mathbb{Q}[y] \to \widehat{H}_T^*(\mathrm{pt})[y]$. (Cf. Remark 4.4, where the same conclusion is reached in K-theory.) In particular one also gets

$$\chi_y(Z) = \mathrm{Td}_y([Z \to \mathrm{pt}]) \in \widehat{H}_T^*(\mathrm{pt})[y].$$

The definition of the Hirzebruch classes implies that

$$\mathrm{Td}_{y=-1}(TX) \cap [X]_T = c^T(TX) \cap [X]_T$$

and

$$\mathrm{Td}_{y=0}(TX) \cap [X]_T = \mathrm{Td}(TX) \cap [X]_T$$

for X smooth and pure dimensional. As in the non-equivariant context of [**BSY10**] (for the non-equivariant normalized Hirzebruch class $T_{y,*}$) this implies the following:

COROLLARY 6.6. *The equivariant Hirzebruch transformation* $\mathrm{Td}_{y,*}$ *fits into the following commutative diagram of natural transformations:*

$$\begin{array}{ccccc} \mathcal{F}^T(X) & \xleftarrow{\;e\;} & \mathrm{K}_0^T(\mathrm{var}/X) & \xrightarrow{\mathrm{MC}_{y=0}} & \mathrm{K}_0(\mathfrak{coh}^T(\mathcal{O}_X)) \\ {\scriptstyle c_*^T \otimes \mathbb{Q}}\downarrow & & {\scriptstyle \mathrm{Td}_{y,*}}\downarrow & & \downarrow{\scriptstyle \mathrm{td}_*} \\ \widehat{H}_*^T(X) & \xleftarrow{\;y=-1\;} & \widehat{H}_*^T(X, \mathbb{Q}[y]) & \xrightarrow{\;y=0\;} & \widehat{H}_*^T(X). \end{array}$$

Here c_*^T is the *equivariant Chern class* transformation defined by Ohmoto [**Ohm06**], cf. §7.1. We note that c_*^T has values in the *integral* homology, and also the completion is not needed, i.e., $c_*^T : \mathcal{F}^T(X) \to H_*^T(X, \mathbb{Z}) \subseteq \widehat{H}_*^T(X, \mathbb{Z})$. The rational coefficients are needed due to the use of the Chern character. Finally $e([\mathrm{id}_X]) = 1\!\!1_X$ even for X singular, so that the equivariant normalized Hirzebruch class $\mathrm{Td}_{y,*}(X) := \mathrm{Td}_{y,*}([\mathrm{id}_X])$ specializes for $y = -1$ also for a singular X to the (rationalized) equivariant Chern-Schwartz-MacPherson (CSM) class $c_*^T(X) := c_*^T(1\!\!1_X)$ of X, and similarly for a locally closed T-invariant subvariety $Z \subseteq X$:

$$\mathrm{Td}_{y=-1,*}[Z \hookrightarrow X] = c_*^T(1\!\!1_Z) \otimes \mathbb{Q}. \tag{6.9}$$

Note that if $H_*^T(X, \mathbb{Z})$ is torsion free, no information about $c_*^T(1\!\!1_Z)$ is lost by switching to rational coefficients; this is the case for the flag manifolds $X = G/P$. In §7 we will explain a more direct procedure yielding the equivariant CSM classes as the leading terms of motivic Chern classes of Schubert cells.

If $Z \subseteq X$ we will refer to the classes $\widetilde{\mathrm{Td}}_{y,*}[Z \hookrightarrow X]$ and $\mathrm{Td}_{y,*}[Z \hookrightarrow X]$ in $\widehat{H}_*^T(X, \mathbb{Q}[y])$ as the **unnormalized**, respectively the **normalized Hirzebruch class** of Z in X.

6.2. Hirzebruch classes for flag manifolds. We turn now to the study of the Hirzebruch transformation in the case when $X = G/B$. There is the following commutative diagram:

$$\begin{array}{ccccc} \mathrm{K}_T(X) \times \mathrm{K}_0(\mathfrak{coh}^T(\mathcal{O}_X)) & \xrightarrow{\otimes} & \mathrm{K}_0(\mathfrak{coh}^T(\mathcal{O}_X)) & \xrightarrow{\int_X} & \mathrm{K}_T(\mathrm{pt}) \\ \downarrow{\scriptstyle \mathrm{ch}\otimes\mathrm{td}_*} & & \downarrow{\scriptstyle \mathrm{td}_*} & & \downarrow{\scriptstyle \mathrm{ch}=\mathrm{td}_*} \\ \widehat{H}_T^*(X) \times \widehat{H}_*^T(X) & \xrightarrow{\cap} & \widehat{H}_*^T(X) & \xrightarrow{\int_X} & \widehat{H}_T^*(\mathrm{pt}). \end{array}$$

Since X is smooth, from (6.1) and (6.2) we have that

$$\mathrm{td}_*(-) = \mathrm{ch}(-)\,\mathrm{Td}(TX) \cap [X]_T.$$

The functoriality of td_* gives the equivariant version of the Grothendieck-Hirzebruch-Riemann-Roch theorem for an equivariant morphism of smooth T-varieties; cf. [**EG00**, Theorem 3.1]. In particular, the GHRR theorem and (3.1) imply that

$$\langle \mathrm{td}_*(\mathcal{O}_u^T), \mathrm{ch}(\mathcal{I}^{v,T})\rangle = \delta_{u,v} = \langle \mathrm{td}_*(\mathcal{I}_u^T), \mathrm{ch}(\mathcal{O}^{v,T})\rangle. \tag{6.10}$$

As a consequence of the fact that $\{[X(w)]_T\}_{w\in W}$ is a $H_T^*(\mathrm{pt})$-basis of $H_T^*(X)$, it is not difficult to show that the Todd classes $\mathrm{td}_*(X(w)) := \mathrm{td}_*(\mathcal{O}_w^T)$ of the Schubert varieties $X(w)$, respectively the Todd classes of the corresponding ideal sheaves $\mathrm{td}_*(\mathcal{I}_w)$, give two bases of $\widehat{H}_*^T(X)$ as a $\widehat{H}_T^*(\mathrm{pt})$-module. The key point is that the corresponding coefficient matrix is triangular with respect to the Bruhat ordering, with units on the diagonal; this follows from the functoriality of td_* for a closed inclusion $X(w) \hookrightarrow X$. The corresponding dual bases are given by the Chern characters $\mathrm{ch}(\mathcal{I}^{v,T})$, respectively $\mathrm{ch}(\mathcal{O}^{v,T})$ of the opposite Schubert varieties for $v \in W$. Specializing all the equivariant parameters to 0, one recovers the natural map $\widehat{H}_*^T(X) \to H_*(X)$ forgetting the T-action and mapping the equivariant Todd transformation to the ordinary one. In this case $\mathrm{td}_*(\mathcal{O}_w) = [X(w)] + \text{l.o.t.}$ and $\mathrm{td}_*(\mathcal{I}_w) = [X(w)] + \text{l.o.t.}$ (lower order terms).

Recall from Equation (3.8) that the Demazure and BGG operators are related by

$$\mathrm{ch}(\partial_i(-)) = \partial_i^H(\mathrm{Td}(T_{p_i})\,\mathrm{ch}(-)) : \mathrm{K}_T(X) \to \widehat{H}_T^*(X). \tag{6.11}$$

Similarly, the equivariant Verdier-Riemann-Roch theorem (VRR) of [**EG00**, Theorem 3.1(d)] implies

$$\mathrm{td}_*(\partial_i(-)) = \mathrm{Td}(T_{p_i})\partial_i^H(\mathrm{td}_*(-)) : \mathrm{K}_0(\mathfrak{coh}^T(\mathcal{O}_X)) \to \widehat{H}_*^T(X). \tag{6.12}$$

Then identities (6.11) and (6.12), combined with (3.3), translate into

$$\partial_i^H(\mathrm{Td}(T_{p_i})\,\mathrm{ch}(\mathcal{O}_w^T)) = \begin{cases} \mathrm{ch}(\mathcal{O}_{ws_i}^T) & \text{if } ws_i > w; \\ \mathrm{ch}(\mathcal{O}_w^T) & \text{otherwise.} \end{cases} \tag{6.13}$$

and

$$\mathrm{Td}(T_{p_i})\partial_i^H(\mathrm{td}_*(\mathcal{O}_w^T)) = \begin{cases} \mathrm{td}_*(\mathcal{O}_{ws_i}^T) & \text{if } ws_i > w; \\ \mathrm{td}_*(\mathcal{O}_w^T) & \text{otherwise} \end{cases}$$

Similarly, Lemma 3.4(a) translates into

$$(\partial_i^H \operatorname{Td}(T_{p_i}) - \operatorname{id})(\operatorname{ch}(\mathcal{I}_w^T)) = \begin{cases} \operatorname{ch}(\mathcal{I}_{ws_i}^T) & \text{if } ws_i > w; \\ -\operatorname{ch}(\mathcal{I}_w^T) & \text{otherwise.} \end{cases} \tag{6.14}$$

and

$$(\operatorname{Td}(T_{p_i})\partial_i^H - \operatorname{id})(\operatorname{td}_*(\mathcal{I}_w^T)) = \begin{cases} \operatorname{td}_*(\mathcal{I}_{ws_i}^T) & \text{if } ws_i > w; \\ -\operatorname{td}_*(\mathcal{I}_w^T) & \text{otherwise} \end{cases} \tag{6.15}$$

The specializations from Corollary 6.6 and the last two equations motivate the definition of the **unnormalized (ordinary and dual) Hirzebruch operators** $\widetilde{\mathcal{T}}_i^{\mathrm{Hir}}, \widetilde{\mathcal{T}}_i^{\mathrm{Hir},\vee} : \widehat{H}_T^*(X,\mathbb{Q})[y] \to \widehat{H}_T^*(X,\mathbb{Q})[y]$ by

$$\widetilde{\mathcal{T}}_i^{\mathrm{Hir}} := \widetilde{\operatorname{Td}}_y(T_{p_i})\partial_i^H - \operatorname{id}; \quad \widetilde{\mathcal{T}}_i^{\mathrm{Hir},\vee} = \partial_i^H \circ (\widetilde{\operatorname{Td}}_y(T_{p_i}) \cup (-)) - \operatorname{id}.$$

Similarly we define the **normalized Hirzebruch operators** $\mathcal{T}_i^{\mathrm{Hir}}, \mathcal{T}_i^{\mathrm{Hir},\vee} : \widehat{H}_T^*(X,\mathbb{Q})[y] \to \widehat{H}_T^*(X,\mathbb{Q})[y]$ by

$$\mathcal{T}_i^{\mathrm{Hir}} := \operatorname{Td}_y(T_{p_i})\partial_i^H - \operatorname{id}; \quad \mathcal{T}_i^{\mathrm{Hir},\vee} = \partial_i^H \circ (\operatorname{Td}_y(T_{p_i}) \cup (-)) - \operatorname{id}.$$

All these operators are $\widehat{H}_T^*(\mathrm{pt})$-linear. In the statements that follow we explain in detail their relation to the K-theoretic Demazure-Lusztig operators $\mathcal{T}_i, \mathcal{T}_i^\vee$.

LEMMA 6.7. *The Hirzebruch operators satisfy the following commutativity relations:*

(a) *As operators* $\mathrm{K}_T(X)[y] \to \widehat{H}_T^*(X,\mathbb{Q})[y]$,

$$\operatorname{td}_* \mathcal{T}_i = \widetilde{\mathcal{T}}_i^{\mathrm{Hir}} \operatorname{td}_* \quad \textit{and} \quad \operatorname{ch} \mathcal{T}_i^\vee = \widetilde{\mathcal{T}}_i^{\mathrm{Hir},\vee} \operatorname{ch}. \tag{6.16}$$

(b) *As operators* $\mathrm{K}_T(X)[y] \to \widehat{H}_T^*(X,\mathbb{Q}[y,(1+y)^{-1}])$,

$$\psi_*^{1+y} \operatorname{td}_* \mathcal{T}_i = \mathcal{T}_i^{\mathrm{Hir}} \psi_*^{1+y} \operatorname{td}_* \quad \textit{and} \quad \psi_{1+y}^* \operatorname{ch} \mathcal{T}_i^\vee = \mathcal{T}_i^{\mathrm{Hir},\vee} \psi_{1+y}^* \operatorname{ch}. \tag{6.17}$$

PROOF. The first commutativity relation in part (a) follows from the following sequence of equalities:

$$\begin{aligned} \widetilde{\mathcal{T}}_i^{\mathrm{Hir}} \operatorname{td}_* &= \widetilde{\operatorname{Td}}_y(T_{p_i})\partial_i^H \operatorname{td}_* - \operatorname{td}_* \\ &= \operatorname{ch}(\lambda_y(T_{p_i}^*)) \operatorname{Td}(T_{p_i})\partial_i^H \operatorname{td}_* - \operatorname{td}_* \\ &= \operatorname{ch}(\lambda_y(T_{p_i}^*)) \operatorname{td}_* \partial_i - \operatorname{td}_* \\ &= \operatorname{td}_*(\lambda_y(T_{p_i}^*)\partial_i - \operatorname{id}) \\ &= \operatorname{td}_* \mathcal{T}_i. \end{aligned}$$

Here the third equality follows from (6.12), and the fourth from the module property (6.2) of the Todd transformation. The second commutativity relation in (a) follows from

$$\begin{aligned} \widetilde{\mathcal{T}}_i^{\mathrm{Hir},\vee} \operatorname{ch} &= \partial_i^H(\widetilde{\operatorname{Td}}_y(T_{p_i}) \operatorname{ch}) - \operatorname{ch} \\ &= \partial_i^H(\operatorname{Td}(T_{p_i}) \operatorname{ch}(\lambda_y(T_{p_i}^*) \cup (-))) - \operatorname{ch} \\ &= \operatorname{ch} \partial_i(\lambda_y(T_{p_i}^*) \cup (-)) - \operatorname{ch} \\ &= \operatorname{ch}(\partial_i \lambda_y(T_{p_i}^*) - \operatorname{id}) \\ &= \operatorname{ch} \mathcal{T}_i^\vee. \end{aligned}$$

In this case, the second equality follows since ch is a ring homomorphism, the third from (6.11), and the rest from the definitions. Part (b) follows from (a) and the identities in Lemma 6.8 below. □

LEMMA 6.8. *As operators* $\widehat{H}_T^*(X)[y] \to \widehat{H}_T^*(X, \mathbb{Q}[y,(1+y)^{-1}])$,

$$\psi_*^{1+y}\widetilde{\mathcal{T}}_i^{\mathrm{Hir}} = \mathcal{T}_i^{\mathrm{Hir}}\psi_*^{1+y} \quad \textit{and} \quad \psi_{1+y}^*\widetilde{\mathcal{T}}_i^{\mathrm{Hir},\vee} = \mathcal{T}_i^{\mathrm{Hir},\vee}\psi_{1+y}^*.$$

PROOF. The first commutativity relation follows similarly to the proof of Lemma 6.7(a) above from the following sequence of equalities:

$$\begin{aligned}
\psi_*^{1+y}\widetilde{\mathcal{T}}_i^{\mathrm{Hir}} &= \psi_*^{1+y}(\widetilde{\mathrm{Td}}_y(T_{p_i})\partial_i^H - \mathrm{id}) \\
&= \psi_{1+y}^*(\widetilde{\mathrm{Td}}_y(T_{p_i}^*))\psi_*^{1+y}\partial_i^H - \psi_*^{1+y} \\
&= \frac{\psi_{1+y}^*(\widetilde{\mathrm{Td}}_y(T_{p_i}^*))}{1+y}\partial_i^H\psi_*^{1+y} - \psi_*^{1+y} \\
&= \mathrm{Td}_y(T_{p_i}^*))\partial_i^H\psi_*^{1+y} - \psi_*^{1+y} \\
&= \mathcal{T}_i^{\mathrm{Hir}}\psi_*^{1+y}.
\end{aligned}$$

Here the second equality follows from the module property (6.5) of the Adams transformation, and the third uses the property

$$\psi_*^{1+y}\circ\partial_i^H = (1+y)^{-1}\partial_i^H\circ\psi_*^{1+y},$$

since ∂_i^H shifts the complex homological degree by one. The second commutativity relation follows similarly from:

$$\begin{aligned}
\psi_{1+y}^*\widetilde{\mathcal{T}}_i^{\mathrm{Hir},\vee} &= \psi_{1+y}^*(\partial_i^H\,\widetilde{\mathrm{Td}}_y(T_{p_i}) - \mathrm{id}) \\
&= \partial_i^H\left(\frac{\psi_{1+y}^*}{1+y}(\widetilde{\mathrm{Td}}_y(T_{p_i})\cup(-))\right) - \psi_{1+y}^* \\
&= \partial_i^H\left(\frac{\psi_{1+y}^*\,\widetilde{\mathrm{Td}}_y(T_{p_i})}{1+y}\psi_{1+y}^*(-)\right) - \psi_{1+y}^* \\
&= \partial_i^H(\mathrm{Td}_y(T_{p_i}^*)\psi_{1+y}^*(-)) - \psi_{1+y}^* \\
&= \mathcal{T}_i^{\mathrm{Hir},\vee}\psi_{1+y}^*.
\end{aligned}$$

In this case, the second equality uses the property

$$\psi_{1+y}^*\circ\partial_i^H = \partial_i^H\circ((1+y)^{-1}\psi_{1+y}^*),$$

since ∂_i^H shifts the complex cohomological degree by minus one. The third equality follows since ψ_{1+y}^* is a ring homomorphism. □

LEMMA 6.9.

(a) *The ordinary (normalized/unnormalized) Hirzebruch operators are adjoint to the dual operators, i.e. for any* $a, b \in \widehat{H}_T^*(X,\mathbb{Q})[y]$,

$$\langle\mathcal{T}_i^{\mathrm{Hir}}(a), b\rangle = \langle a, \mathcal{T}_i^{\mathrm{Hir},\vee}(b)\rangle; \quad \langle\widetilde{\mathcal{T}}_i^{\mathrm{Hir}}(a), b\rangle = \langle a, \widetilde{\mathcal{T}}_i^{\mathrm{Hir},\vee}(b)\rangle,$$

where the pairing is extended by $\mathbb{Q}[y]$*-linearity.*[2]

---

[2]Later in the context of Segre classes, we will tacitly extend this pairing by linearity from coefficients in $\mathbb{Q}[y]$ to $\mathbb{Q}[y,(1+y)^{-1}]$.

(b) *Each family of the (ordinary / dual / (un)normalized) Hirzebruch operators satisfies the same relations as the K-theoretic Demazure-Lusztig operators; cf. Proposition 3.2.*

PROOF. Part (a) can be proved by using the self-adjointness of ∂_i^H and the operators of multiplication by $\mathrm{Td}_y(T_{p_i})$, as follows. By definition,

$$\begin{aligned}\langle \mathcal{T}_i^{\mathrm{Hir}}(a), b\rangle =&\langle (\mathrm{Td}_y(T_{p_i})\partial_i^H - \mathrm{id})(a), b\rangle \\ =&\langle a, (\partial_i^H \mathrm{Td}_y(T_{p_i}) - \mathrm{id})b\rangle \\ =&\langle a, \mathcal{T}_i^{\mathrm{Hir},\vee}(b)\rangle.\end{aligned}$$

A similar proof works for the unnormalized operators.

We turn to the relations in (b). First, we deduce from (a) that it suffices to prove the statements for the ordinary operators. Then we use Lemma 6.7(a) again to show that for $a' \in \mathrm{K}_T(G/B)$,

$$\widetilde{\mathcal{T}}_{i_1}^{\mathrm{Hir}}\widetilde{\mathcal{T}}_{i_2}^{\mathrm{Hir}}\ldots\widetilde{\mathcal{T}}_{i_k}^{\mathrm{Hir}}(\mathrm{td}_*(a')) = \mathrm{td}_*\,\mathcal{T}_{i_1}\mathcal{T}_{i_2}\ldots\mathcal{T}_{i_k}(a').$$

Since td_* is surjective (after appropriately extending the coefficients via $\mathrm{ch} : \mathrm{K}_T(\mathrm{pt}) \to \widehat{H}_T^*(\mathrm{pt})$), this implies the claim for the unnormalized Hirzebruch operators. The same proof works for the normalized operators, using classes of the form $\psi_*^{1+y}\,\mathrm{td}_*(a')$, Lemma 6.7(b) and Lemma 6.8. We also note that in order to prove the statement for the dual operators, instead of adjointness, one may alternatively work with classes of type $\mathrm{ch}(a')$ and $\psi_{1+y}^*\,\mathrm{ch}(a')$. □

REMARK 6.10. The results of Lemma 6.9(b) also follow from the following argument. Regard $\widehat{H}_T^*(\mathrm{pt})[y]$ as a $\mathrm{K}_T(\mathrm{pt})[y]$-algebra via the (injective) equivariant Chern character map. Then the transformations td_* and ch induce injective homomorphisms of $\mathrm{K}_T(\mathrm{pt})[y]$-algebras

$$\overline{\mathrm{td}_*}, \overline{\mathrm{ch}} : \mathrm{End}_{\mathrm{K}_T(\mathrm{pt})[y]}(\mathrm{K}_T(X)[y]) \to \mathrm{End}_{\widehat{H}_T^*(\mathrm{pt})[y]}(\widehat{H}_T^*(X)[y]).$$

The injectivity part follows by using suitable bases, such as images of Schubert classes $\mathcal{O}_w$ under the Todd or Chern character maps. Then Lemma 6.7 may be interpreted as giving the identities

$$\overline{\mathrm{td}_*}(\mathcal{T}_i) = \widetilde{\mathcal{T}}_i^{\mathrm{Hir}} \quad \text{and} \quad \overline{\mathrm{ch}}(\mathcal{T}_i^\vee) = \widetilde{\mathcal{T}}_i^{\mathrm{Hir},\vee}.$$

Since $\overline{\mathrm{td}_*}, \overline{\mathrm{ch}}$ are algebra homomorphisms, they will preserve relations satisfied by $\mathcal{T}_i, \mathcal{T}_i^\vee$, proving the claim for the unnormalized operators.

One may argue similarly in the case of normalized operators. Start by changing the coefficient ring using the algebra isomorphism

$$\widehat{H}_T^*(X)[y] \otimes_{\widehat{H}_T^*(\mathrm{pt})[y]} \widehat{H}_T^*(\mathrm{pt}, \mathbb{Q}[y,(1+y)^{-1}]) \xrightarrow{\ \sim\ } \widehat{H}_T^*(X, \mathbb{Q}[y,(1+y)^{-1}]).$$

Then the Adams operations

$$\psi_{1+y}^* : \widehat{H}_T^*(\mathrm{pt})[y] \to \widehat{H}_T^*(\mathrm{pt})[y] \text{ and } \psi_*^{1+y} : \widehat{H}_*^T(\mathrm{pt})[y] \to \widehat{H}_*^T(\mathrm{pt}, \mathbb{Q}[y,(1+y)^{-1}])$$

induce algebra isomorphisms

$$\begin{aligned}\overline{\psi_{1+y}^*} : \mathrm{End}_{\widehat{H}_T^*(\mathrm{pt},\mathbb{Q}[y,(1+y)^{-1}])}(\widehat{H}_T^*(X, \mathbb{Q}[y,(1+y)^{-1}])) \to \\ \mathrm{End}_{\widehat{H}_T^*(\mathrm{pt},\mathbb{Q}[y,(1+y)^{-1}])}(\widehat{H}_T^*(X, \mathbb{Q}[y,(1+y)^{-1}]))\end{aligned}$$

and

$$\overline{\psi_*^{1+y}} : \mathrm{End}_{\widehat{H}_T^*(\mathrm{pt},\mathbb{Q}[y,(1+y)^{-1}])}(\widehat{H}_*^T(X,\mathbb{Q}[y,(1+y)^{-1}])) \to \mathrm{End}_{\widehat{H}_T^*(\mathrm{pt},\mathbb{Q}[y,(1+y)^{-1}])}(\widehat{H}_*^T(X,\mathbb{Q}[y,(1+y)^{-1}]))$$

which by Lemma 6.8 satisfy

$$\overline{\psi_*^{1+y}}(\widetilde{\mathcal{T}}_i^{\mathrm{Hir}}) = \mathcal{T}_i^{\mathrm{Hir}} \quad \text{and} \quad \overline{\psi_{1+y}^*}(\widetilde{\mathcal{T}}_i^{\mathrm{Hir},\vee}) = \mathcal{T}_i^{\mathrm{Hir},\vee}.$$

Therefore the relations satisfied by the unnormalized operators are the same as those for the normalized ones.

As a consequence of Lemma 6.9 (b), we can set $\widetilde{\mathcal{T}}_w^{\mathrm{Hir}} = \widetilde{\mathcal{T}}_{i_1}^{\mathrm{Hir}} \cdots \widetilde{\mathcal{T}}_{i_k}^{\mathrm{Hir}}$, $\mathcal{T}_w^{\mathrm{Hir}} = \mathcal{T}_{i_1}^{\mathrm{Hir}} \cdots \mathcal{T}_{i_k}^{\mathrm{Hir}}$ for $w = s_{i_1} \cdots s_{i_k}$ a reduced decomposition in W. We now come to the main results about equivariant Hirzebruch classes of Schubert cells, starting with their recursion.

THEOREM 6.11. *Let $w \in W$ and s_i a simple reflection such that $ws_i > w$.*

(a) *The unnormalized Hirzebruch operators satisfy:*

$$\widetilde{\mathcal{T}}_i^{\mathrm{Hir}}(\widetilde{\mathrm{Td}}_{y,*}(X(w)^\circ)) = \widetilde{\mathrm{Td}}_{y,*}(X(ws_i)^\circ);$$

(b) *The normalized Hirzebruch operators satisfy:*

$$\mathcal{T}_i^{\mathrm{Hir}}(\mathrm{Td}_{y,*}^T(X(w)^\circ)) = \mathrm{Td}_{y,*}^T(X(ws_i)^\circ).$$

(c) *In particular, for $w \in W$, the Hirzebruch classes of Schubert cells are given by:*

$$\widetilde{\mathrm{Td}}_{y,*}(X(w)^\circ) = \widetilde{\mathcal{T}}_{w^{-1}}^{\mathrm{Hir}}[e_{\mathrm{id}}]_T \quad \textit{and} \quad \mathrm{Td}_{y,*}^T(X(w)^\circ) = \mathcal{T}_{w^{-1}}^{\mathrm{Hir}}[e_{\mathrm{id}}]_T.$$

PROOF. Part (a) follows from Theorem 4.5 (more specifically [**AMSS24**, Corollary 5.2]) together with the commutation relations (6.16). The same argument applies to (b), using the definition of the normalized Hirzebruch transformation from (6.8), Theorem 4.5 again, and the commutation relations (6.17). Part (c) is a consequence of (a) and (b), taking into account that, by functoriality, $\widetilde{\mathrm{Td}}_{y,*}(e_{\mathrm{id}}) = \mathrm{Td}_{y,*}(e_{\mathrm{id}}) = [e_{\mathrm{id}}]_T$. □

Next we formulate the corresponding orthogonality results for the equivariant Hirzebruch classes of Schubert cells. In analogy with the definition of the operators $\mathfrak{L}_i$ giving the dual classes of the motivic Chern classes, define the operators $\mathfrak{L}_i^{\mathrm{Hir}}$ and $\widetilde{\mathfrak{L}}_i^{\mathrm{Hir}}$ by

$$\mathfrak{L}_i^{\mathrm{Hir}} := -y(\mathcal{T}_i^{\mathrm{Hir},\vee})^{-1} = \mathcal{T}_i^{\mathrm{Hir},\vee} + (1+y)\,\mathrm{id}$$

$$\widetilde{\mathfrak{L}}_i^{\mathrm{Hir}} := -y(\widetilde{\mathcal{T}}_i^{\mathrm{Hir},\vee})^{-1} = \widetilde{\mathcal{T}}_i^{\mathrm{Hir},\vee} + (1+y)\,\mathrm{id}\,.$$

THEOREM 6.12. *For any $u, v \in W$,*

$$\langle \widetilde{\mathrm{Td}}_{y*}(X(u)^\circ), \widetilde{\mathfrak{L}}_{v^{-1}w_0}^{\mathrm{Hir}}([e_{w_0}]_T)\rangle_H = \delta_{u,v}\,\widetilde{\mathrm{Td}}_y(T_{w_0}X) \tag{6.18}$$

and

$$\langle \mathrm{Td}_{y*}(X(u)^\circ), \mathfrak{L}_{v^{-1}w_0}^{\mathrm{Hir}}([e_{w_0}]_T)\rangle_H = \delta_{u,v}\,\mathrm{Td}_y(T_{w_0}X). \tag{6.19}$$

PROOF. We apply the equivariant Todd class transformation to both sides in the expression from Theorem 4.6 to obtain

$$\begin{aligned}
\delta_{u,v}\,\mathrm{td}_*\left(\prod_{\alpha>0}(1+ye^{-\alpha})\right) &= \mathrm{td}_*\left(\langle \mathrm{MC}_y(X(u)^\circ), \widetilde{\mathrm{MC}}_y(Y(v)^\circ)\rangle_K\right)\\
&= \mathrm{td}_*\left(\int_X^K \mathrm{MC}_y(X(u)^\circ)\cdot \widetilde{\mathrm{MC}}_y(Y(v)^\circ)\right)\\
&= \int_X^{H^*} \mathrm{td}_*(\mathrm{MC}_y(X(u)^\circ))\cdot \mathrm{ch}(\widetilde{\mathrm{MC}}_y(Y(v)^\circ))\\
&= \int_X^{H^*} \widetilde{\mathrm{Td}}_{y,*}(X(u)^\circ)\cdot \mathrm{ch}(\mathfrak{L}_{v^{-1}w_0}(\mathcal{O}^{w_0,T}))\\
&= \int_X^{H^*} \widetilde{\mathrm{Td}}_{y,*}(X(u)^\circ)\cdot \widetilde{\mathfrak{L}}^{\mathrm{Hir}}_{v^{-1}w_0}(\mathrm{ch}(\mathcal{O}^{w_0,T}))\\
&= \langle \widetilde{\mathrm{Td}}_{y,*}(X(u)^\circ), \widetilde{\mathfrak{L}}^{\mathrm{Hir}}_{v^{-1}w_0}(\mathrm{ch}(\mathcal{O}^{w_0,T}))\rangle_H.
\end{aligned}$$

Here we use K and H^* to indicate where the operation is taken, and the fifth equality follows from Lemma 6.7. Given this, the claim in (6.18) follows because

$$\mathrm{ch}(\mathcal{O}^{w_0,T}) = \frac{[e_{w_0}]_T}{\mathrm{Td}(TX)} = \frac{[e_{w_0}]_T}{\mathrm{Td}(T_{w_0}X])},$$

$$\widetilde{\mathrm{Td}}_y(T_{w_0}X) = \mathrm{ch}\left(\prod_{\alpha>0}(1+ye^{-\alpha})\right)\mathrm{Td}(T_{w_0}X),$$

and the fact that $\widetilde{\mathfrak{L}}^{\mathrm{Hir}}_i$ is $\widehat{H}^*_T(\mathrm{pt})[y]$-linear. The equality from (6.19) follows from (6.18) by application of the Adams operation ψ_*^{1+y}:

$$\begin{aligned}
\delta_{u,v}\psi^*_{1+y}\left(\widetilde{\mathrm{Td}}_y(T_{w_0}X)\right) &= \psi_*^{1+y}\left(\langle \widetilde{\mathrm{Td}}_{y*}(X(u)^\circ), \widetilde{\mathfrak{L}}^{\mathrm{Hir}}_{v^{-1}w_0}([e_{w_0}]_T)\rangle_H\right)\\
&= \langle \psi_*^{1+y}\left(\widetilde{\mathrm{Td}}_{y*}(X(u)^\circ)\right), \psi^*_{1+y}\left(\widetilde{\mathfrak{L}}^{\mathrm{Hir}}_{v^{-1}w_0}([e_{w_0}]_T)\right)\rangle_H\\
&= \langle \mathrm{Td}_{y*}(X(u)^\circ), \mathfrak{L}^{\mathrm{Hir}}_{v^{-1}w_0}\psi^*_{1+y}([e_{w_0}]_T)\rangle_H.
\end{aligned}$$

Then the claim follows because

$$\psi^*_{1+y}\left(\widetilde{\mathrm{Td}}_y(T_{e_{w_0}}X)\right) = (1+y)^{\dim X}\,\mathrm{Td}_y(T_{e_{w_0}}X)$$

and

$$\psi^*_{1+y}([e_{w_0}]_T) = (1+y)^{\dim X}[e_{w_0}]_T,$$

with $[e_{w_0}]_T$ viewed as an equivariant cohomology class of complex degree $\dim X$ (by equivariant Poincaré duality). □

We finish this section with the counterpart of Theorem 4.12, using now the (un)normalized Segre version of the Hirzebruch classes:

$$\frac{\widetilde{\mathrm{Td}}_{y*}(X(w)^\circ)}{\widetilde{\mathrm{Td}}_y(TX)} \quad \text{and} \quad \frac{\mathrm{Td}_{y*}(X(w)^\circ)}{\mathrm{Td}_y(TX)}.$$

Observe that for any smooth X, the class $\widetilde{\mathrm{Td}}_y(X)$ is invertible in the completed ring $\widehat{H}^*_T(X,\mathbb{Q}[y,(1+y)^{-1}])$, since its leading term is the invertible element $1+y$. Similarly $\mathrm{Td}_y(X)$ is invertible in $\widehat{H}^*_T(X,\mathbb{Q}[y])$, as its leading term is 1.

THEOREM 6.13. *For any $w \in W$ one has in $\widehat{H}^*_T(X, \mathbb{Q}[y,(1+y)^{-1}])$, resp., $\widehat{H}^*_T(X, \mathbb{Q}[y])$:*

$$\frac{\widetilde{\mathrm{Td}}_{y*}(X(w)^\circ)}{\widetilde{\mathrm{Td}}_y(TX)} = \widetilde{\mathcal{T}}^{\mathrm{Hir},\vee}_{w^{-1}}\left(\frac{[e_{\mathrm{id}}]_T}{\widetilde{\mathrm{Td}}_y(T_{e_{\mathrm{id}}}X)}\right)$$

and

$$\frac{\mathrm{Td}_{y*}(X(w)^\circ)}{\mathrm{Td}_y(TX)} = \mathcal{T}^{\mathrm{Hir},\vee}_{w^{-1}}\left(\frac{[e_{\mathrm{id}}]_T}{\mathrm{Td}_y(T_{e_{\mathrm{id}}}X)}\right),$$

as well as

$$\frac{\widetilde{\mathrm{Td}}_{y*}(Y(w)^\circ)}{\widetilde{\mathrm{Td}}_y(TX)} = \widetilde{\mathcal{T}}^{\mathrm{Hir},\vee}_{(w_0w)^{-1}}\left(\frac{[e_{w_0}]_T}{\widetilde{\mathrm{Td}}_y(T_{e_{w_0}}X)}\right)$$

and

$$\frac{\mathrm{Td}_{y*}(Y(w)^\circ)}{\mathrm{Td}_y(TX)} = \mathcal{T}^{\mathrm{Hir},\vee}_{(w_0w)^{-1}}\left(\frac{[e_{w_0}]_T}{\mathrm{Td}_y(T_{e_{w_0}}X)}\right).$$

PROOF. We only explain the proof for the opposite Schubert cells $Y(w)^\circ$, since the result for the Schubert cells $X(w)^\circ$ are shown in exactly the same way. (Alternatively, one may apply the automorphism w_0^L obtained by left multiplication by the longest element $w_0 \in W$; see [**AMSS17**, §5.2] or [**MNS22**, §3.1].) We start with the unnormalized classes. The application of ch to (4.3) together with Lemma 6.7(a) imply:

$$\begin{aligned}\frac{\mathrm{ch}(\mathrm{MC}_y(Y(w)^\circ))}{\mathrm{ch}(\lambda_y T^*X)} &= \widetilde{\mathcal{T}}^{\mathrm{Hir},\vee}_{(w_0w)^{-1}}\left(\frac{\mathrm{ch}(\mathcal{O}^{w_0,T})}{\mathrm{ch}\left(\prod_{\alpha>0}(1+ye^{-\alpha})\right)}\right)\\ &= \widetilde{\mathcal{T}}^{\mathrm{Hir},\vee}_{(w_0w)^{-1}}\left(\frac{[e_{w_0}]_T}{\widetilde{\mathrm{Td}}_y(T_{e_{w_0}}X)}\right),\end{aligned}$$

with the last equality as in the proof of Theorem 6.12. Then the result follows from

$$\widetilde{\mathrm{Td}}_{y*}(Y(w)^\circ) = \mathrm{td}_*(\mathrm{MC}_y(Y(w)^\circ)) = \mathrm{ch}(\mathrm{MC}_y(Y(w)^\circ))\,\mathrm{Td}(TX)$$

and

$$\widetilde{\mathrm{Td}}_y(TX) = \mathrm{ch}(\lambda_y T^*X)\,\mathrm{Td}(TX).$$

To deduce the result for the normalized classes, we further apply the Adams transformation ψ^*_{1+y}, with $\mathrm{td}_*(\mathrm{MC}_y(Y(w)^\circ))$ and $[e_{e_0}]_T$ viewed as an equivariant cohomology class as before (by equivariant Poincaré duality, with $[e_{w_0}]_T$ of complex degree $\dim X$):

$$\begin{aligned}\psi^*_{1+y}\left(\frac{\widetilde{\mathrm{Td}}_{y,*}(Y(w)^\circ)}{\widetilde{\mathrm{Td}}_y(TX)}\right) &= \left(\psi^*_{1+y}\widetilde{\mathcal{T}}^{\mathrm{Hir},\vee}_{(w_0w)^{-1}}\right)\left(\frac{[e_{w_0}]_T}{\widetilde{\mathrm{Td}}_y(T_{e_{w_0}}X)}\right)\\ &= \left(\widetilde{\mathcal{T}}^{\mathrm{Hir},\vee}_{(w_0w)^{-1}}\psi^*_{1+y}\right)\left(\frac{[e_{w_0}]_T}{\widetilde{\mathrm{Td}}_y(T_{e_{w_0}}X)}\right)\\ &= \widetilde{\mathcal{T}}^{\mathrm{Hir},\vee}_{(w_0w)^{-1}}\left(\frac{(1+y)^{\dim X}[e_{w_0}]_T}{(1+y)^{\dim X}\,\mathrm{Td}_y(T_{e_{w_0}}X)}\right).\end{aligned}$$

Then the result follows from

$$\psi^*_{1+y}\left(\widetilde{\mathrm{Td}}_y(TX)\right) = (1+y)^{\dim X}\,\mathrm{Td}_y(TX)$$

and

$$\psi^*_{1+y}\left(\widetilde{\mathrm{Td}}_{y*}(Y(w)^\circ)\right) = (1+y)^{\dim X}\,\mathrm{Td}_{y*}(Y(w)^\circ),$$

since

$$\begin{aligned}\psi^*_{1+y}(-)\cap[X]_T &= \psi^*_{1+y}(-)\cap\psi^{1+y}_*((1+y)^{\dim X}[X]_T)\\ &= (1+y)^{\dim X}\psi^{1+y}_*(-\cap[X]_T).\end{aligned}$$

by the module property (6.5). □

6.3. Specializations of (dual) Hirzebruch operators and Hirzebruch classes of Schubert cells. We take this opportunity to record the specializations at $y=-1$ and $y=0$ for the (un)normalized Hirzebruch operators $\widetilde{\mathcal{T}}_i^{\mathrm{Hir}}$ and $\mathcal{T}_i^{\mathrm{Hir}}$ and their (shifted) dual operators $\widetilde{\mathcal{T}}_i^{\mathrm{Hir},\vee}, \mathcal{T}_i^{\mathrm{Hir},\vee}$ and $\widetilde{\mathfrak{L}}_i^{\mathrm{Hir}}, \mathfrak{L}_i^{\mathrm{Hir}}$. These follow from the definitions of these objects, using the corresponding specializations of the Hirzebruch classes from (6.3) and (6.4), and are stated in the next proposition.

Proposition 6.14. *The following hold:*

(a) *The specializations at $y=0$ of the (un)normalized Hirzebruch operators are given by*

$$(\widetilde{\mathcal{T}}_i^{\mathrm{Hir}})_{y=0} = (\mathcal{T}_i^{\mathrm{Hir}})_{y=0} = \mathrm{Td}(T_{p_i})\partial_i^H - \mathrm{id},$$

and for their duals by

$$(\widetilde{\mathcal{T}}_i^{\mathrm{Hir},\vee})_{y=0} = (\mathcal{T}_i^{\mathrm{Hir},\vee})_{y=0} = \partial_i^H\,\mathrm{Td}(T_{p_i}) - \mathrm{id}$$

so that

$$(\widetilde{\mathfrak{L}}_i^{\mathrm{Hir}})_{y=0} = (\mathfrak{L}_i^{\mathrm{Hir}})_{y=0} = \partial_i^H\,\mathrm{Td}(T_{p_i}).$$

(b) *The specializations at $y=-1$ of the (un)normalized Hirzebruch operators is given by:*

$$(\widetilde{\mathcal{T}}_i^{\mathrm{Hir}})_{y=-1} = -s_i;\quad (\mathcal{T}_i^{\mathrm{Hir}})_{y=-1} = \mathcal{T}_i^H,$$

and for their (shifted) duals by

$$(\widetilde{\mathcal{T}}_i^{\mathrm{Hir},\vee})_{y=-1} = (\widetilde{\mathfrak{L}}_i^{\mathrm{Hir}})_{y=-1} = -s_i^\vee = s_i;\quad (\mathcal{T}_i^{\mathrm{Hir},\vee})_{y=-1} = (\mathfrak{L}_i^{\mathrm{Hir}})_{y=-1} = \mathcal{T}_i^{H,\vee},$$

where the operators s_i and $\mathcal{T}_i^H, \mathcal{T}_i^{H,\vee}$ are defined in (3.4), *respectively* (3.5).

Using these specializations of the (shifted dual) Hirzebruch operators, we can specialize in a similar way the corresponding results from the Theorems 6.11, 6.12 and 6.13 to $y=0$ and $y=-1$. First we consider the case $y=0$. We obtain:

$$\mathrm{Td}_{y=0,*}(X(w)^\circ) = \widetilde{\mathrm{Td}}_{y=0,*}(X(w)^\circ) = \mathrm{td}_*(\mathrm{MC}_0(X(w)^\circ)) = \mathrm{td}_*(\mathcal{I}_w^T) \tag{6.20}$$

by Theorem 5.1(b). Then Theorem 6.11 specializes for $y=0$ to the recursion

$$(\mathrm{Td}(T_{p_i})\partial_i^H - \mathrm{id})(\mathrm{td}_*(\mathcal{I}_w^T)) = \mathrm{td}_*(\mathcal{I}_{ws_i}^T) \text{ for } ws_i > w$$

from (6.15) for $w\in W$ and s_i a simple reflection. Similarly Theorem 6.12 specializes for $y=0$ to

$$\langle \mathrm{Td}_*(\mathcal{I}_u^T), (\mathfrak{L}^{\mathrm{Hir}}_{v^{-1}w_0})_{y=0}([e_{w_0}]_T)\rangle_H = \delta_{u,v}\,\mathrm{Td}(T_{e_{w_0}}X) \tag{6.21}$$

for $u,v\in W$. Since $\mathrm{ch}(\mathcal{O}^{w_0,T}) = \frac{[e_{w_0}]_T}{\mathrm{Td}(T_{e_{w_0}}X)}$, this translates into

$$\langle \mathrm{Td}_*(\mathcal{I}_u^T), (\mathfrak{L}^{\mathrm{Hir}}_{v^{-1}w_0})_{y=0}(\mathrm{ch}(\mathcal{O}^{w_0,T}))\rangle_H = \delta_{u,v},$$

from which we deduce by (6.10) (for $v\in W$) that:

$$\mathrm{ch}(\mathcal{O}^{v,T}) = (\mathfrak{L}^{\mathrm{Hir}}_{v^{-1}w_0})_{y=0}(\mathrm{ch}(\mathcal{O}^{w_0,T})), \tag{6.22}$$

with $(\mathfrak{L}_i^{\text{Hir}})_{y=0} = \partial_i^H \operatorname{Td}(T_{p_i})$. This recovers Equation (6.13).

Finally, since $\operatorname{td}_*(-) = \operatorname{ch}(-)\operatorname{Td}(TX)$ for $y = 0$, Theorem 6.13 specializes to

$$\operatorname{ch}(\mathcal{I}_w^T)) = \left(\mathcal{T}_{w^{-1}}^{\text{Hir},\vee}\right)_{y=0} (\operatorname{ch}(\mathcal{I}_{\text{id}}^T))$$

and

$$\operatorname{ch}(\mathcal{I}^{w,T})) = \left(\mathcal{T}_{(w_0 w)^{-1}}^{\text{Hir},\vee}\right)_{y=0} (\operatorname{ch}(\mathcal{I}^{w_0,T}))$$

for $w \in W$, with $(\mathcal{T}_i^{\text{Hir},\vee})_{y=0} = \partial_i^H \operatorname{Td}(T_{p_i}) - \text{id}$, consistent with Equation (6.14).

Next we record the specializations of Theorems 6.11, 6.12 and 6.13 for $y = -1$. For simplicity we only consider the more interesting case of normalized classes and operators. Note that

$$\operatorname{Td}_{y=-1,*}(X(w)^\circ) = c_*^T(1\!\!1_{X(w)^\circ}) =: c_{\text{SM}}^T(X(w)^\circ) \in H_*^T(G/B, \mathbb{Z})$$

by (6.9), since $H_*^T(G/B,\mathbb{Z})$ is torsion free. Recall that $c_{\text{SM}}^T(X(w)^\circ)$ is *Chern-Schwartz-MacPherson* (CSM) class of the Schubert cell $X(w)^\circ$; this class is discussed in more detail in the next section. Theorem 6.11(b) for the normalized Hirzebruch classes *implies* for $y = -1$ the following recursion from [**AM16**, Theorem 6.4] (formulated in Equation (7.1) in the next section in terms of homogenized classes):

$$\mathcal{T}_i^H(c_{\text{SM}}^T(X(w)^\circ)) = (c^T(T_{p_i})\partial_i^H - \text{id})(c_{\text{SM}}^T(X(w)^\circ)) = c_{\text{SM}}^T(X(ws_i)^\circ) \tag{6.23}$$

for $w \in W$ and s_i a simple reflection, with $ws_i > w$.

Similarly, Theorem 6.12 for the normalized Hirzebruch classes *implies* for $y = -1$ the corresponding **Hecke orthogonality** of [**AMSS17**, Theorem 7.2] (with their equivariant parameter $\hbar \in H^2_{\mathbb{C}^*}(\text{pt},\mathbb{Z})$ specialized here to $\hbar = 1$):

$$\langle c_{\text{SM}}^T(X(u)^\circ), (\mathfrak{L}_{v^{-1}w_0}^{\text{Hir}})_{y=-1}([e_{w_0}]_T)\rangle = \delta_{u,v} c^T(T_{e_{w_0}}X) = \delta_{u,v} \prod_{\alpha>0}(1+\alpha) \tag{6.24}$$

for $u, v \in W$. Here

$$(\mathfrak{L}_{v^{-1}w_0}^{\text{Hir}})_{y=-1}([e_{w_0}]_T) = \mathcal{T}_{v^{-1}w_0}^{H,\vee}([e_{w_0}]_T) = c_{\text{SM}}^{T,\vee}(Y(v)^\circ)$$

is the *dual Chern-Schwartz-MacPherson* class from [**AMSS17**, Equation (14)]. Note that, in terms of the duality operators from Remark 6.3,

$$c_{\text{SM}}^{T,\vee}(Y(v)^\circ) = (-1)^{\dim X - \ell(v)} \psi_*^{-1}(c_{\text{SM}}^T(Y(v)^\circ),$$

since the homogenized operators satisfy

$$\mathcal{T}_i^{H,\vee,\hbar} = \hbar\partial_i^H + s_i = -(-\hbar\partial_i^H - s_i) = -\mathcal{T}_i^{H,\hbar}|_{\hbar\mapsto -h}. \tag{6.25}$$

Finally Theorem 6.13 for the normalized Hirzebruch classes implies for $y = -1$:

$$\frac{c_{\text{SM}}^T(X(w)^\circ)}{c^T(TX)} = \mathcal{T}_{w^{-1}}^{H,\vee}\left(\frac{[e_{\text{id}}]_T}{\prod_{\alpha>0}(1-\alpha)}\right) = \frac{c_{\text{SM}}^{T,\vee}(X(w)^\circ)}{\prod_{\alpha>0}(1-\alpha)} \tag{6.26}$$

and

$$\frac{c_{\text{SM}}^T(Y(w)^\circ)}{c^T(TX)} = \mathcal{T}_{(w_0 w)^{-1}}^{H,\vee}\left(\frac{[e_{w_0}]_T}{\prod_{\alpha>0}(1+\alpha)}\right) = \frac{c_{\text{SM}}^{T,\vee}(Y(w)^\circ)}{\prod_{\alpha>0}(1+\alpha)}.$$

This recovers [**AMSS17**, Theorem 7.5], which is one of the main results of that paper (again with the equivariant parameter $\hbar \in H^2_{\mathbb{C}^*}(\text{pt},\mathbb{Z})$ specialized to $\hbar = 1$).

6.4. Parabolic Hirzebruch classes. We now consider the (generalized) partial flag manifold G/P, and we let $\pi : G/B \to G/P$ be the natural projection. The Schubert varieties $X(wW_P)^\circ$ in G/P are indexed by the elements in $w \in W^P$, with the image $\pi(X(w)^\circ) = X(wW_P)^\circ$ for $w \in W$. Applying td_* and $\psi_*^{1+y}\,\mathrm{td}_*$ to the equalities from Proposition 4.17 implies by functoriality the following counterpart for the Hirzebruch classes.

PROPOSITION 6.15. *The following hold for $w \in W$:*

(a) *For P parabolic:*

$$\pi_* \widetilde{\mathrm{Td}}_{y*}(X(w)^\circ) = (-y)^{\ell(w)-\ell(wW_P)} \widetilde{\mathrm{Td}}_{y*}(X(wW_P)^\circ) \in \widehat{H}_*^T(G/P)[y]$$

and

$$\pi_* \mathrm{Td}_{y*}(X(w)^\circ) = (-y)^{\ell(w)-\ell(wW_P)} \mathrm{Td}_{y*}(X(wW_P)^\circ) \in \widehat{H}_*^T(G/P)[y].$$

(b) *More generally, let $P \subseteq Q$ be two standard parabolic subgroups, and $\pi' : G/P \to G/Q$ the natural projection. Then*

$$\pi'_* \widetilde{\mathrm{Td}}_{y*}(X(wW_P)^\circ) = (-y)^{\ell(wW_P)-\ell(wW_Q)} \pi_* \widetilde{\mathrm{Td}}_{y*}(X(wW_Q)^\circ) \in \widehat{H}_*^T(G/Q)[y]$$

and

$$\pi'_* \mathrm{Td}_{y*}(X(wW_P)^\circ) = (-y)^{\ell(wW_P)-\ell(wW_Q)} \pi_* \mathrm{Td}_{y*}(X(wW_Q)^\circ) \in \widehat{H}_*^T(G/Q)[y].$$

Specializing to $y = 0$ (with the convention $0^0 = 1$), we get by Theorem 5.1(b):

$$\pi_* \mathrm{Td}_*(\mathcal{I}_w^T) = \begin{cases} \mathrm{Td}_*(\mathcal{I}_{wW_P}^T) & \text{if } \ell(w) = \ell(wW_P); \\ 0 & \text{otherwise.} \end{cases}$$

Similarly, specializing the normalized classes to $y = -1$, we get for $w \in W$:

(6.27)

$$\pi_* c_{\mathrm{SM}}^T(X(w)^\circ) = c_{\mathrm{SM}}^T(X(wW_P)^\circ) \text{ and } \pi'_* c_{\mathrm{SM}}^T(X(wW_P)^\circ) = c_{\mathrm{SM}}^T(X(wW_Q)^\circ).$$

These equalities hold in $H_*^T(G/P, \mathbb{Z})$ and $H_*^T(G/Q, \mathbb{Z})$, since these are torsion free.

7. The Chern-Schwartz-MacPherson classes as leading terms

We have seen in the Corollary 6.6 that the Chern-Schwartz-MacPherson (CSM) classes may be recovered from the Hirzebruch transformation by specializing at $y = -1$. In this section we take a different route, and recover the CSM classes directly, by identifying them as the leading terms of the motivic Chern classes; cf. Theorem 7.1. We will illustrate this process for complete flag varieties, since these are the main object of study in this paper.

7.1. Chern-Schwartz-MacPherson classes. We first recall the context leading to the definition of CSM classes. According to a conjecture attributed to Deligne and Grothendieck, there is a unique natural transformation $c_* : \mathcal{F}(-) \to H_*(-, \mathbb{Z})$ from the functor of constructible functions on a complex algebraic variety X to homology (i.e., even degree Borel-Moore homology, or Chow groups), where all morphisms are proper, such that if X is smooth then $c_*(1\!\!1_X) = c(TX) \cap [X]$. This conjecture was proved by MacPherson [**Mac74**]; the class $c_*(1\!\!1_X)$ for possibly singular varieties X was shown to coincide via Alexander duality with a class defined earlier by M.-H. Schwartz [**Sch65a, Sch65b, BS81**]. For any constructible subset $W \subseteq X$, the class $c_{\mathrm{SM}}(W) := c_*(1\!\!1_W) \in H_*(X, \mathbb{Z})$ is called the *Chern-Schwartz-MacPherson* (CSM) class of W in X. If X is a T-variety, an equivariant version of the group of constructible functions $\mathcal{F}^T(X)$ and a Chern class transformation $c_*^T : \mathcal{F}^T(X) \to H_*(X;\mathbb{Z})$ were defined by Ohmoto [**Ohm06**]; see [**AMSS17**, §3.2] for a summary of Ohmoto's definition and a discussion of alternative (equivalent) definitions. This is the notion we consider in this paper.

7.2. The homogenized CSM class via the motivic Chern class. We now consider the case of flag varieties, so $X = G/B$. If

$$c_{\mathrm{SM}}^T(X(w)^\circ) = \sum_i c_{\mathrm{SM}}^T(X(w)^\circ)_i \in H_*^T(G/B,\mathbb{Z}),$$

where $c_{\mathrm{SM}}^T(X(w)^\circ)_i \in H_{2i}^T(G/B,\mathbb{Z})$, the homogenized CSM class is defined to be

$$c_{\mathrm{SM}}^{T,\hbar}(X(w)^\circ) := \sum_i \hbar^i c_{\mathrm{SM}}^T(X(w)^\circ)_i \in H_0^{T\times\mathbb{C}^*}(G/B,\mathbb{Z}).$$

Here $\mathbb{C}^*$ acts trivially on G/B and $\hbar \in H^2_{\mathbb{C}^*}(\mathrm{pt},\mathbb{Z})$ is a generator. Consider the Schubert expansion of the homogenized CSM class:

$$c_{\mathrm{SM}}^{T,\hbar}(X(w)^\circ) = \sum_{u\le w} c'_{u,w}(\hbar,t)[X(u)]_T \in H_0^{T\times\mathbb{C}^*}(X),$$

where $c'_{u,w}(\hbar,t) \in H^*_{T\times\mathbb{C}^*}(\mathrm{pt},\mathbb{Z})$ is a homogeneous polynomial of degree $\ell(u)$. As usual $t=(t_1,\dots,t_s)$ stands for a sequence of variables corresponding to a basis of the character group of T; see §2.1. We also recall that α_i denote the simple roots, regarded as elements of $H_T^2(\mathrm{pt})$. With $\mathcal{T}_i^{H,\hbar}$ as in (3.6), we have

$$\mathcal{T}_i^{H,\hbar}(c_{\mathrm{SM}}^{T,\hbar}(X(w)^\circ)) = c_{\mathrm{SM}}^{T,\hbar}(X(ws_i)^\circ); \tag{7.1}$$

this is the homogenized version of (6.23) above. We will now verify that, combined with Proposition 3.5, (7.1) implies that the CSM class of the Schubert cell is the 'initial term' of the motivic Chern class $\mathrm{MC}_y(X(w)^\circ)$, where $y=-e^{-\hbar}$.

THEOREM 7.1. *Let $w\in W$ and consider the Schubert expansions*

$$\mathrm{MC}_y(X(w)^\circ) = \sum_{u\le w} c_{u,w}(y,e^t)\mathcal{O}_u^T \in \mathrm{K}_T(G/B)[y]$$

and

$$c_{\mathrm{SM}}^{T,\hbar}(X(w)^\circ) = \sum_{u\le w} c'_{u,w}(\hbar,t)[X(u)]_T \in H_0^{T\times\mathbb{C}^*}(G/B,\mathbb{Z}),$$

where $c'_{u,w}(\hbar,t) \in H^{2\ell(u)}_{T\times\mathbb{C}^*}(\mathrm{pt})$. With $y=-e^{-\hbar}$ (and hence $\mathrm{ch}_{\mathbb{C}^*}(y) = -e^{-\hbar}$ cf. §2.3), the following hold:

(a) *The image $\mathrm{ch}_{T\times\mathbb{C}^*}(c_{u,w}(y,e^t))$ of the coefficient $c_{u,w}(y,e^t)$ under the Chern character belongs to $\prod_{i\ge\ell(u)} H^{2i}_{T\times\mathbb{C}^*}(\mathrm{pt})$.*
(b) *The coefficient $c'_{u,w}(\hbar,t)$ equals the term of degree $\ell(u)$ in $c_{u,w}(y,e^t)$, i.e.,*

$$c'_{u,w}(\hbar,t) = (\mathrm{ch}_{T\times\mathbb{C}^*}(c_{u,w}(y,e^t)))_{\ell(u)}.$$

Equivalently,

$$c_{\mathrm{SM}}^{T,\hbar}(X(w)^\circ) = \textit{degree } 0 \textit{ component of } \mathrm{ch}(\mathrm{MC}_y(X(w)^\circ)).$$

PROOF. Both parts follow by induction on $\ell(w)$, using the recursion calculating $\mathrm{MC}_y(X(w)^\circ)$ from Theorem 4.5 combined with Proposition 3.5; in part (b) we utilize the recursion for $c_{\mathrm{SM}}^{T,\hbar}(X(w)^\circ)$ from (7.1). □

EXAMPLE 7.2. Consider the equivariant motivic Chern class in $K_T(\mathbb{P}^1)[y]$:

$$\mathrm{MC}_y(X(s_1)^\circ) = (1+e^{-\alpha_1}y)\mathcal{O}^T_{s_1} - (1+(1+e^{-\alpha_1})y)\mathcal{O}^T_{\mathrm{id}}.$$

The specialization $y = -e^{-\hbar}$ in the coefficient $c_{s_1,s_1}(y, e^t)$ gives:

$$c_{s_1,s_1}(-e^{-\hbar}, e^t) = 1 - e^{-\alpha_1-\hbar} = \hbar + \alpha_1 + \text{h.o.t.}$$

(higher order terms). The term of degree 1 is $c'_{s_1,s_1} = \hbar + \alpha_1$. Similarly, the specialization of $c_{\mathrm{id},s_1}(y, e^t)$ gives

$$c_{\mathrm{id},s_1}(y, e^t) = -(1-(1+e^{-\alpha_1})e^{-\hbar}) = -1 + e^{-\hbar} + e^{-\hbar-\alpha_1} = -1 + 2 + \text{h.o.t.}$$

By Theorem 7.1,

$$c^{T,\hbar}_{\mathrm{SM}}(X(s_1)^\circ) = (\hbar+\alpha_1)[X(s_1)]_T + [X(\mathrm{id})]_T,$$

consistent with a positivity property from Conjecture 1 below.

Consider now the non-equivariant case, i.e., in the expansions from Theorem 7.1 we set $\alpha = 0$, so $e^\alpha \mapsto 1$. In this case we denote the coefficients in the two expansion by $c_{u,w}(y)$ respectively $c'_{u,w}(\hbar)$. Note that $c_{u,w}(y) \in \mathbb{Z}[y]$ and $c'_{u,w}(\hbar) \in \mathbb{Z}[\hbar]$. Furthermore, by homogeneity,

$$c'_{u,w}(\hbar) = \overline{c}_{u,w}\hbar^{\ell(u)},$$

where $\overline{c}_{u,w} \in \mathbb{Z}$ is an integer. Next we give a more direct relation between these coefficients, using that the polynomial $c_{u,w}(y)$ is divisible by $(1+y)^{\ell(u)}$ by Proposition 4.15.

COROLLARY 7.3. *The coefficient $\overline{c}_{u,w}$ equals the specialization at $y = -1$ of $\frac{c_{u,w}(y)}{(1+y)^{\ell(u)}}$:*

$$\overline{c}_{u,w} = \left(\frac{c_{u,w}(y)}{(1+y)^{\ell(u)}}\right)_{y\mapsto -1}.$$

PROOF. Let $Q_{u,w}(y) := \sum a_i y^i$ in $\mathbb{Z}[y]$ be the quotient $\frac{c_{u,w}(y)}{(1+y)^{\ell(u)}}$. By Theorem 7.1, the coefficient $\overline{c}_{u,w}$ equals the term of degree 0 in the specialization $Q_{u,w}(-e^{-\hbar})$. Since $-e^{-\hbar} = -1+$ higher order terms, $\overline{c}_{u,w} = Q_{u,w}(-1)$ as claimed. □

EXAMPLE 7.4. Consider the non-equivariant version of Example 7.2:

$$\mathrm{MC}_y(X(s_1)^\circ) = (1+y)\mathcal{O}_{s_1} - (1+2y)\mathcal{O}_{\mathrm{id}}.$$

According to Corollary 7.3, we need to divide each coefficient $c_{u,s_1}(y)$ by $(1+y)^{\ell(u)}$ and then specialize at $y = -1$. We obtain:

$$c^{\hbar}_{\mathrm{SM}}(X(s_1)^\circ) = \hbar[X(s_1)] + [X(\mathrm{id})].$$

The non-homogenized class is obtained by setting $\hbar$ to 1.

EXAMPLE 7.5. Consider the motivic Chern class in $K(\mathrm{Fl}(3))[y]$:

$$\begin{aligned}\mathrm{MC}_y(X(s_1s_2)^\circ) =& (1+y)^2\mathcal{O}_{s_1s_2} - (1+y)(1+2y)\mathcal{O}_{s_1}\\ &- (1+y)(1+3y)\mathcal{O}_{s_2} + (5y^2+5y+1)\mathcal{O}_{\mathrm{id}}.\end{aligned}$$

As before, we need to divide each coefficient $c_{u,s_1s_2}(y)$ by $(1+y)^{\ell(u)}$ and then specialize at $y = -1$. We obtain:

$$c_{\mathrm{SM}}(X(s_1s_2)^\circ) = [X(s_1s_2)] + [X(s_1)] + 2[X(s_2)] + [X(\mathrm{id})].$$

8. Positivity, unimodality, and log-concavity conjectures

In this section we record several conjectures involving Schubert expansions of the motivic Chern and CSM classes and about the structure constants of the CSM classes. Some of these conjectures have been made by other authors; our goal is to collect all these statements in a single place.

We start with the CSM classes, since this is the case when we have the most (partial) results.

8.1. Positivity of Schubert expansions of CSM classes. Consider the (non-equivariant) CSM class of a Schubert cells in a generalized flag manifolds G/P:

$$c_{\mathrm{SM}}(X(wW_P)^\circ) = \sum_{vW_P \leq wW_P} c_{v,w}[X(vW_P)],$$

with $c_{v,w} \in \mathbb{Z}$. For $G/P = \mathrm{Gr}(k;n)$, it was conjectured in [**AM09**] that the coefficients $c_{v;w}$ are nonnegative; this was proved in some special cases in *loc.cit.* and in [**Mih15, Jon10, Str11**], and in full generality (for Grassmannians) by J. Huh [**Huh16**]. The recursive algorithm from [**AM16**] yielded calculations of CSM classes of Schubert cells in any G/P, and provided supporting evidence that the CSM classes of Schubert cells in all flag manifolds are effective. This conjecture was proved in [**AMSS17**, Corollary 1.4].

Equivariantly, the numerical evidence supports the following conjecture.

CONJECTURE 1 (Equivariant Positivity). *Let $X(wW_P)^\circ \subseteq G/P$ be any Schubert cell and consider the Schubert expansion of the equivariant CSM class:*

$$c^T_{\mathrm{SM}}(X(wW_P)^\circ) = \sum_{v \leq w} c_{v,w}(\alpha)[X(vW_P)]_T \quad \in H^T_*(G/P).$$

Then $c_{v;w}(\alpha)$ is a polynomial in positive roots α with non-negative coefficients.

In the non-equivariant case, while we have proved in [**AMSS17**] that $c_{v,w} \geq 0$, the evidence suggests a stronger result.

CONJECTURE 2 (Strong positivity). *Let $X(wW_P)^\circ \subseteq G/P$ be any Schubert cell and consider the Schubert expansion:*

$$c_{\mathrm{SM}}(X(wW_P)^\circ) = \sum_{v \leq w} c_{v,w}[X(vW_P)] \quad \in H_*(G/P;\mathbb{Z}).$$

Then $c_{v,w} > 0$ for all $v \leq w$.

Huh's result for Grassmannians in [**Huh16**] shows that each homogeneous component $c_{\mathrm{SM}}(X(wW_P)^\circ)_k$ of the CSM class is represented by a non-empty irreducible variety. This is slightly weaker than the requirement in Conjecture 2. On the other hand, if this variety may be chosen to be T-stable, then Huh's result and the positivity results of Graham [**Gra01**] would imply Conjecture 1 for Grassmannians.

For any parabolic subgroup P, let $\pi : G/B \to G/P$ be the natural projection. Since $\pi_*(c_{\mathrm{SM}}(X(w)^\circ)) = c_{\mathrm{SM}}(X(wW_P)^\circ)$ (see e.g., (6.27)) it follows that if Conjecture 1 or Conjecture 2 holds for cells in G/B, then it also holds in G/P.

8.2. Positivity of CSM/SM structure constants. The CSM classes $c_{\mathrm{SM}}(X(w)^\circ)$ may be viewed as deformations of the fundamental classes $[X(w)]$; we consider analogues of the Littlewood-Richardson coefficients in the context of these classes. Since $c_{\mathrm{SM}}(X(w)^\circ)$ is naturally a homology class, we focus on structure constants for their Poincaré duals, the Segre-MacPherson (SM) classes. **In this section we only consider the non-equivariant context.** In general, if Z is a subvariety of a nonsingular variety X, we set

$$s_{\mathrm{SM}}(Z,X) = \frac{c_{\mathrm{SM}}(Z)}{c(T(X))}$$

in the homology or Chow group of Z; see e.g., [**AMSS22**]. We will implicitly push-forward this class to $H_*(X)$.

Since $c(T(G/B)) \cdot c(T^*(G/B)) = 1$, as proved in [**AMSS17**, Lemma 8.2], in G/B we have

$$s_{\mathrm{SM}}(Y(w)^\circ, G/B) = \frac{c_{\mathrm{SM}}(Y(w)^\circ)}{c(T(G/B))} = c(T^*(G/B)) \cap c_{\mathrm{SM}}(Y(w)^\circ).$$

We also proved that

$$s_{\mathrm{SM}}(X(ws_i)^\circ, G/B) = \mathcal{T}_i^{H,\vee} s_{\mathrm{SM}}(X(w)^\circ, G/B)$$

(cf. (6.26) above for the equivariant version of this equality). Poincaré duality states that for any parabolic $P \supset B$,

$$\langle s_{\mathrm{SM}}(Y(vW_P)^\circ, G/P), c_{\mathrm{SM}}(X(wW_P)^\circ)\rangle = \delta_{v,w}, \tag{8.1}$$

cf. [**AMSS17**, Theorem 7.1]. This can be proved using a transversality formula due to Schürmann [**Sch17**], extended equivariantly in [**AMSS17**, Corollary 10.3].

As a consequence of (6.26) (cf. [**AMSS17**, Eq. (36)]), the Schubert expansions of $c_{\mathrm{SM}}(X(w)^\circ)$ and $s_{\mathrm{SM}}(X(w)^\circ, G/B)$ in G/B are related by changing signs. More precisely, if

$$s_{\mathrm{SM}}(X(w)^\circ, G/B) = \sum f_{v;w}[X(v)] \quad \in H_*(G/B),$$

then with notation as above $f_{v;w} = (-1)^{\ell(w)-\ell(v)} c_{v;w}$. This follows because the homogenized operator $\mathcal{T}_i^{H,\hbar} = \hbar\partial_i - s_i$, giving CSM classes, and its adjoint $\mathcal{T}_i^{H,\vee,\hbar} = \hbar\partial_i + s_i$, giving SM classes, differ by a sign; see also (6.25) for the more general equivariant statement.

Consider now the structure constants

$$s_{\mathrm{SM}}(Y(u)^\circ, G/B) \cdot s_{\mathrm{SM}}(Y(v)^\circ, G/B) = \sum e_{u,v}^w s_{\mathrm{SM}}(Y(w)^\circ, G/B). \tag{8.2}$$

Schürmann's transversality theorem [**Sch17**] shows that

$$e_{u,v}^w = \chi(g_1 Y(u)^\circ \cap g_2 Y(v)^\circ \cap g_3 X(w)^\circ),$$

the topological Euler characteristic of the intersection of three Schubert cells translated in general position via $g_1, g_2, g_3 \in G$. This interpretation of the structure constants holds for any G/P, although the relation between the Schubert expansions of CSM and SM classes does not extend beyond G/B. (Still, the SM classes are known to be Schubert alternating; see [**AMSS22**].)

Due to its statement involving only 'classical' objects, perhaps the most remarkable positivity conjecture in this paper is the next.

CONJECTURE 3 (Alternation of Euler characteristic). *The Euler characteristic of the intersection of three Schubert cells in general position in G/P is alternating, i.e., for any $u, v, w \in W^P$,*

$$(-1)^{\ell(u)+\ell(v)+\ell(w)}\chi(g_1Y(uW_P)^\circ \cap g_2Y(vW_P)^\circ \cap g_3X(wW_P)^\circ) \geq 0.$$

Utilizing deep connections between SM classes to the theory of integrable systems, this was proved by Knutson and Zinn-Justin for d-step flag manifolds with $d \leq 3$, and it was conjectured to hold for $d = 4$; cf. [**KZJ21**, p. 43]. Independently, and based on multiplications of SM classes from [**AMSS17**], the authors of this paper stated this conjecture in several conference and seminar talks, for partial flag manifolds G/P in arbitrary Lie type[3].

S. Kumar [**Kum22**] conjectured that the CSM class of the Richardson cells are Schubert positive, that is, if

$$c_{\mathrm{SM}}(Y(u)^\circ \cap X(v)^\circ) = \sum_w f_{u,v}^w [Y(w)],$$

then $f_{u,v}^w \geq 0$. (We also learned about this conjecture independently from Rui Xiong, and it is now stated in [**FGX22**, Conjecture 9.2].) It is shown in [**Kum22**] that this implies Conjecture 3. Note that the *Segre* class of the Richardson cell

$$s_{\mathrm{SM}}(R_v^{u,\circ}, G/B) = s_{\mathrm{SM}}(Y(u)^\circ \cap X(v)^\circ, G/B)$$

is Schubert alternating by [**AMSS22**, Theorem 1.1], since the inclusion of $Y(u)^\circ \cap X(v)^\circ$ is an affine morphism.

For G/B, the absolute value of the structure constants $e_{u,v}^w$ from (8.2) give the structure constants to multiply CSM classes of Schubert cells. This generalizes the positivity in ordinary Schubert Calculus: if $\ell(uW_P) + \ell(vW_P) = \ell(wW_P)$, then the intersection in question is 0 dimensional and reduced, and the Euler characteristic counts the number of points in the intersection. A different algorithm to calculate the SM structure constants is given in [**Su21**].

We end this section by proving a property of the sum of the coefficients $e_{u,v}^w$.

PROPOSITION 8.1. *Consider the multiplication*

$$s_{\mathrm{SM}}(Y(uW_P)^\circ, G/P) \cdot s_{\mathrm{SM}}(Y(vW_P)^\circ, G/P) = \sum_w e_{u,v}^w s_{\mathrm{SM}}(Y(wW_P)^\circ, G/P).$$

Then $\sum_w e_{u,v}^w = \delta_{w_0uW_P, vW_P}$.

PROOF. By the transversality formula from [**Sch17**], and omitting the ambient G/P for short,

$$\begin{aligned} s_{\mathrm{SM}}(Y(uW_P)^\circ) \cdot s_{\mathrm{SM}}(Y(vW_P)^\circ) &= s_{\mathrm{SM}}(X(w_0uW_P)^\circ) \cdot s_{\mathrm{SM}}(Y(vW_P)^\circ) \\ &= s_{\mathrm{SM}}(X(w_0uW_P)^\circ \cap Y(vW_P)^\circ). \end{aligned}$$

[3] After this paper was submitted, a proof of the conjecture was given in [**SSW23**].

By Poincaré duality (8.1) and the equality $c(T(G/P)) = \sum_w c_{\mathrm{SM}}(X(wW_P)^\circ)$, the sum of the coefficients $e^w_{u,v}$ equals

$$\begin{aligned}\int_{G/P} & s_{\mathrm{SM}}(Y(uW_P)^\circ) \cdot s_{\mathrm{SM}}(Y(vW_P)^\circ) \cdot c(T(G/P)) \\ &= \int_{G/P} s_{\mathrm{SM}}(X(w_0uW_P)^\circ \cap Y(vW_P)^\circ) \cdot c(T(G/P)) \\ &= \int_{G/P} c_{\mathrm{SM}}(X(w_0uW_P)^\circ \cap Y(vW_P)^\circ) = \delta_{w_0uW_P,vW_P}.\end{aligned}$$

Here the last equality follows because the Richardson cell $X(w_0uW_P)^\circ \cap Y(vW_P)^\circ$ is torus-stable, therefore its Euler characteristic equals the Euler characteristic of the fixed locus; see [**BB73**, Corollary 2], applied for a general $\mathbb{C}^* \subseteq T$. In this case the fixed locus is empty or one point, giving $\delta_{w_0uW_P,vW_P}$. □

EXAMPLE 8.2. Take G is Lie type G_2 and write $s_{\mathrm{SM}}(-)$ for $s_{\mathrm{SM}}(-, G/B)$. Then

$$\begin{aligned} s_{\mathrm{SM}}(Y(\mathrm{id})^\circ) \cdot s_{\mathrm{SM}}(Y(\mathrm{id})^\circ) &= s_{\mathrm{SM}}(Y(\mathrm{id})^\circ) - s_{\mathrm{SM}}(Y(s_1)^\circ) - s_{\mathrm{SM}}(Y(s_2)^\circ) \\ &+ 2s_{\mathrm{SM}}(Y(s_2s_1)^\circ) + 4s_{\mathrm{SM}}(Y(s_1s_2)^\circ) \\ &- 9s_{\mathrm{SM}}(Y(s_1s_2s_1)^\circ) - 11s_{\mathrm{SM}}(Y(s_2s_1s_2)^\circ) \\ &+ 22s_{\mathrm{SM}}(Y(s_2s_1s_2s_1)^\circ) + 34s_{\mathrm{SM}}(Y(s_1s_2s_1s_2)^\circ) \\ &- 57s_{\mathrm{SM}}(Y(s_1s_2s_1s_2s_1)^\circ) - 51s_{\mathrm{SM}}(Y(s_2s_1s_2s_1s_2)^\circ) \\ &+ 67s_{\mathrm{SM}}(Y(s_2s_1s_2s_1s_2s_1)^\circ).\end{aligned}$$

Observe that these structure constants are alternating, and add up to 0, confirming Conjecture 3 and Proposition 8.1 in this case.

8.3. Unimodality and log-concavity for CSM polynomials. Following [**Sta89**], a sequence $a_0, \dots, a_n$ is **unimodal** if there exists i_0 such that

$$a_0 \le a_1 \le \dots \le a_{i_0} \ge a_{i_0+1} \ge \dots \ge a_n.$$

The sequence is **log-concave** if for any $1 \le i \le n-1$,

$$a_i^2 \ge a_{i-1}a_{i+1}.$$

A log-concave sequence of nonnegative integers with no internal zeros is unimodal. A polynomial $P(x) = \sum a_i x^i$ is unimodal, resp., log-concave, if its sequence of coefficients satisfies the corresponding property.

Consider any class $\kappa = \sum c_w[X(wW_P)]$ in $H_*(G/P)$. We define the H-**polynomial** associated to κ by

$$H(\kappa) := \sum c_w x^{\ell(w)} = \sum c_i x^i,$$

determining the coefficients c_i. For $w \in W^P$ we denote by

$$H_w(x) := H(c_{\mathrm{SM}}(X(wW_P)))$$

the H-polynomial of the CSM class of the corresponding Schubert *variety*.

CONJECTURE 4 (Unimodality and log-concavity). *Let $X(w) \subseteq G/P$ be any Schubert variety. Then the following hold:*

(a) *The polynomial H_w is unimodal with no internal zeros.*
(b) *If G is of Lie type A (i.e., $G/P = \mathrm{Fl}(i_1, \dots, i_k; n)$ is a partial flag manifold) then H_w is log-concave.*

A similar conjecture for Mather classes can be found in [**MS20**].

EXAMPLE 8.3. Consider the Grassmannian $\mathrm{Gr}(3,6)$ and the Schubert variety $X_{(2,1)}$ of dimension 3. Then

$$c_{\mathrm{SM}}(X_{(2,1)}) = [X_{(2,1)}] + 3[X_2] + 3[X_{1,1}] + 8[X_1] + 5[X_0].$$

Its H-polynomial is

$$x^3 + 6x^2 + 8x + 5,$$

which is log-concave.

For the 5-dimensional quadric Q^5, the H-polynomial of $c(TQ^5)$ is

$$x^5 + 5x^4 + 11x^3 + 26x^2 + 18x + 6$$

which is unimodal, but not log-concave.

8.4. Conjectures about the motivic Chern classes. Given the (proved and conjectural) positivity properties of the CSM classes of Schubert cells, it is natural to conjecture analogous properties for the motivic Chern classes of Schubert cells. The following conjecture was stated by Fehér, Rimányi, and Weber [**FRW17**] in type A, and in [**AMSS24**, Conjecture 1] for arbitrary Lie type.

CONJECTURE 5 (Positivity of MC classes). *Consider the Schubert expansion:*

$$\mathrm{MC}_y(X(w)^\circ) = \sum_{u \le w} c_{u,w}(y, e^t)\mathcal{O}_u^T \in K_T(G/B)[y].$$

Then for any $u \le w \in W$, we have

$$(-1)^{\ell(w)-\ell(u)} c_{w,u}(y, e^t) \in \mathbb{Z}_{\ge 0}[y][e^{-\alpha_1}, \dots, e^{-\alpha_r}] \subseteq \mathrm{K}_T(\mathrm{pt})[y],$$

i.e., the coefficients $(-1)^{\ell(w)-\ell(u)} c_{u,w}(y, e^t)$ are polynomials in the variables y and characters $e^{-\alpha_1}, \dots, e^{-\alpha_r}$ in simple roots with non-negative coefficients.

The conjecture implies that the coefficients of the non-equivariant motivic Chern classes of Schubert cells are sign-alternating:

$$(-1)^{\ell(w)-\ell(u)} c_{u,w}(y) \in \mathbb{Z}_{\ge 0}[y].$$

Conjecture 5 holds after specializing $y = 0$, as a consequence of Theorem 5.1(b) and the fact that the ideal sheaves are alternating in Schubert classes; see (3.2) above, and e.g., [**Bri05**, Proposition 4.3.2]. More evidence for Conjecture 5 is available, since some particular coefficients $c_{u,w}(y, e^t)$ are known to be positive. For instance, the coefficient

$$c_{w,w}(y, e^t) = \prod_{\alpha > 0, w(\alpha) < 0} (1 + y e^{w\alpha})$$

calculated in Lemma 4.11 is positive. The specialization at $y = -1$ gives the equivariant class ι_w of the structure sheaf of the fixed point of w as in Theorem 5.1 above (see Proposition 5.1). This is consistent with the conjecture; the class ι_w is known to be Schubert alternating, by e.g., the positivity in equivariant K-theory proved by Anderson, Griffeth and Miller [**AGM11**]. Finally, we verified Conjecture 5 for flag manifolds of type A_n for $n \le 5$ (i.e., up to $\mathrm{Fl}(5)$), and for the Lie types B_2, C_2, D_3 and G_2, by means of a computer calculation.

Fehér, Rimányi, and Weber [**FRW17**] also observed a conjectural log-concavity property for motivic Chern classes in Lie type A. We confirmed their observations in many examples, and also checked additional examples in other Lie types.

CONJECTURE 6 (Log concavity for MC classes). *Let $X(w)^\circ \subseteq G/B$ and consider the Schubert expansion of the non-equivariant motivic Chern class:*

$$\mathrm{MC}(X(w)^\circ) = \sum_{v \leq w} c_{v;w}(y)\mathcal{O}_v.$$

Then $c_{v;w}(y)$ is log-concave.

EXAMPLE 8.4. Consider G of Lie type G_2. The coefficient of $\mathcal{O}_{\mathrm{id}}$ in the expansion of $\mathrm{MC}_y(X(w_0)^\circ)$ is

$$64y^6 + 141y^5 + 125y^4 + 69y^3 + 29y^2 + 8y + 1.$$

This is a log-concave polynomial. Its specialization at $y = -1$ gives 1, reflecting the fact that it calculates the Euler characteristic of the big cell in G_2/B.

REMARK 8.5. To simplify the notation, we only stated Conjecture 6 in the non-equivariant case. As pointed out in [**FRW17**], the conjecture has a natural extension to the equivariant setting.

8.5. An interpretation in the Hecke algebra. We can use Hecke algebra to give a combinatorial interpretation of the coefficients $c_{u,w}(y)$ in Conjecture 5. For the cohomology case, see [**Lee18**, Theorem 6.2].

Recall the K-theoretic BGG operator ∂_i satisfies $\partial_i^2 = \partial_i$ and the braid relations. The operators $\mathcal{T}_i$ (and $\mathcal{T}_i^\vee$) satisfy the finite Hecke algebra relation, see Proposition 3.2. Besides, we also have:

LEMMA 8.6. *For any simple root α and torus weight λ,*

$$\partial_{s_\alpha}\mathcal{L}_\lambda - \mathcal{L}_{s_\alpha\lambda}\partial_{s_\alpha} = \frac{\mathcal{L}_\lambda - \mathcal{L}_{s_\alpha\lambda}}{1-\mathcal{L}_\alpha} \in \mathrm{End}_{K_T(\mathrm{pt})[y]}\mathrm{K}_T(G/B)[y],$$

$$\mathcal{T}_{s_\alpha}\mathcal{L}_\lambda - \mathcal{L}_{s_\alpha\lambda}\mathcal{T}_{s_\alpha} = (1+y)\frac{\mathcal{L}_{s_\alpha\lambda} - \mathcal{L}_\lambda}{1-\mathcal{L}_{-\alpha}} \in \mathrm{End}_{K_T(\mathrm{pt})[y]}\mathrm{K}_T(G/B)[y]$$

and

$$\mathcal{T}_{s_\alpha}^\vee\mathcal{L}_\lambda - \mathcal{L}_{s_\alpha\lambda}\mathcal{T}_{s_\alpha}^\vee = (1+y)\frac{\mathcal{L}_{s_\alpha\lambda} - \mathcal{L}_\lambda}{1-\mathcal{L}_{-\alpha}} \in \mathrm{End}_{\mathrm{K}_T(\mathrm{pt})[y]}\mathrm{K}_T(G/B)[y].$$

Here $\frac{\mathcal{L}_\lambda - \mathcal{L}_{s_\alpha\lambda}}{1-\mathcal{L}_\alpha}$ is defined as follows. Suppose $\frac{e^\lambda - e^{s_\alpha\lambda}}{1-e^\alpha} = \sum_\mu e^\mu$, then $\frac{\mathcal{L}_\lambda - \mathcal{L}_{s_\alpha\lambda}}{1-\mathcal{L}_\alpha} := \sum_\mu \mathcal{L}_\mu$.

PROOF. We can check the equalities on the fixed point basis. Then all of them follow from Lemma 3.3 and the equality $\mathcal{L}_\lambda \otimes \iota_w = e^{w\lambda}\iota_w$. □

Let us recall the definition of the K-theoretic Kostant-Kumar Hecke algebra [**KK90**] and the affine Hecke algebra. Let P denote the weight lattice of G.

DEFINITION 8.7.

(1) The Kostant-Kumar Hecke algebra $\mathcal{H}$ is a free $\mathbb{Z}$ module with basis $\{D_w e^\lambda | w \in W, \lambda \in P\}$, such that
   - For any $\lambda, \mu \in P$, $e^\lambda e^\mu = e^{\lambda+\mu}$.
   - For any simple root α, $D_{s_\alpha}^2 = D_{s_\alpha}$.
   - For any $w, y \in W$ such that $\ell(wy) = \ell(w) + \ell(y)$, $D_w D_y = D_{wy}$
   - For any simple root α and $\lambda \in P$,

$$e^\lambda D_i - D_i e^{s_i\lambda} = \frac{e^\lambda - e^{s_i\lambda}}{1-e^{\alpha_i}}.$$

(2) The affine Hecke algebra $\mathbb{H}$ is a free $\mathbb{Z}[q,q^{-1}]$ module with basis $\{T_w e^\lambda | w \in W, \lambda \in P\}$, such that
- For any $\lambda, \mu \in P$, $e^\lambda e^\mu = e^{\lambda+\mu}$.
- For any simple root α, $(T_{s_\alpha}+1)(T_{s_\alpha}-q)=0$.
- For any $w, y \in W$ such that $\ell(wy)=\ell(w)+\ell(y)$, $T_w T_y = T_{wy}$
- For any simple root α and $\lambda \in P$,

$$T_\alpha e^{s_\alpha \lambda} - e^\lambda T_\alpha = (1-q)\frac{e^\lambda - e^{s_\alpha\lambda}}{1-e^{-\alpha}}.$$

It follows from Lemma 8.6 that the Kostant-Kumar Hecke algebra $\mathcal{H}$ acts on $\mathrm{K}_T(G/B)$ by sending D_i to ∂_i and e^λ to $\mathcal{L}_\lambda$, see [**KK90**]; the affine Hecke algebra $\mathbb{H}$ acts on $\mathrm{K}_T(G/B)[y,y^{-1}]$ by sending q to $-y$, T_i to $\mathcal{T}_i$ (or $\mathcal{T}_i^\vee$), and e^λ to $\mathcal{L}_\lambda$, see [**Lus85**]. In the rest of the section, we always identify q with $-y$.

For any simple root α_i, let

$$T_i := (1+ye^{\alpha_i})D_i - 1 = D_i(1+ye^{-\alpha_i}) - (1+y+ye^{-\alpha_i}) \in \mathcal{H}[y].$$

Then these T_i and e^λ satisfies the relations in the affine Hecke algebra $\mathbb{H}$. Therefore, T_w is well-defined for all $w \in W$.

For any w, we can expand $T_{w^{-1}}$ as a linear combination of terms $D_{u^{-1}}$,

$$T_{w^{-1}} := \sum_{u \le w} D_{u^{-1}} a_{u,w}(y;e^t), \tag{8.3}$$

for some $a_{u,w}(y,e^t) \in \mathbb{C}[T][y]$. It is easy to compute

$$a_{w,w}(y;e^t) = \prod_{\alpha>0, w\alpha<0} (1+ye^{w\alpha}),$$

which equals the coefficient $c_{w,w}(y;e^t)$ by Lemma 4.11. In fact, we have the following more general relation.

PROPOSITION 8.8. *For any $u \le w \in W$, we have*

$$a_{u,w}(y;e^t) = c_{u,w}(y;e^t).$$

PROOF. Under the action of the Kostant-Kumar Hecke algebra $\mathcal{H}$ on $\mathrm{K}_T(G/B)$, T_i is sent to the DL operator $\mathcal{T}_i$ from (3.7). By Theorem 4.5, Equation (3.3) and the equality $\mathcal{L}_\lambda \otimes \mathcal{O}^T_{\mathrm{id}} = e^\lambda \mathcal{O}^T_{\mathrm{id}}$, applying (8.3) to $\iota_{\mathrm{id}} \in \mathrm{K}_T(G/B)$ we get

$$\mathrm{MC}_y(X(w)^\circ) = \sum_{u \le w} a_{u,w}(y,e^t)\mathcal{O}^T_u.$$

Therefore, $a_{u,w}(y;e^t) = c_{u,w}(y;e^t)$. □

Proposition 8.8 provides a purely combinatorial way to compute the coefficients $c_{u,w}(y;e^t)$. In particular, we can check Conjecture 5 in the case when $\ell(w) \le 2$ as follows:

(1) $\ell(w)=1$. From $T_i = D_i(1+ye^{-\alpha_i})-(1+y+ye^{-\alpha_i})$, we get $c_{\mathrm{id},s_i} = -(1+y+ye^{-\alpha_i})$ and $c_{s_i,s_i} = 1+ye^{-\alpha_i}$. (This is consistent with Example 4.8.)

(2) $\ell(w)=2$. Pick two simple roots α_i,α_j. We calculate

$$\begin{aligned}
T_iT_j &= \big(D_i(1+ye^{-\alpha_i})-(1+y+ye^{-\alpha_i})\big)\big(D_j(1+ye^{-\alpha_j})-(1+y+ye^{-\alpha_j})\big)\\
&= D_iD_j(1+ye^{-\alpha_j})(1+ye^{-s_j\alpha_i})-D_j(1+ye^{-\alpha_j})(1+y+ye^{-s_j\alpha_i})\\
&-D_i\left((1+ye^{-\alpha_i})(1+y+ye^{-\alpha_j})-y(1+ye^{-\alpha_j})\frac{e^{-\alpha_i}-e^{-s_j\alpha_i}}{1-e^{\alpha_j}}\right)\\
&+(1+y+ye^{-\alpha_j})(1+y+ye^{-\alpha_i})-y(1+ye^{-\alpha_j})\frac{e^{-\alpha_i}-e^{-s_j\alpha_i}}{1-e^{\alpha_j}},
\end{aligned}$$

where

$$\frac{e^{-\alpha_i}-e^{-s_j\alpha_i}}{1-e^{\alpha_j}}=-e^{-s_j\alpha_i}-e^{-s_j\alpha_i+\alpha_j}-\cdots-e^{-\alpha_i-\alpha_j}.$$

This verifies Conjecture 5 when $\ell(w)=2$.

To illustrate this, consider $G=SL(3,\mathbb{C})$, $i=2$, and $j=1$. Then

$$\begin{aligned}
T_2T_1 &= D_2D_1(1+ye^{-\alpha_1})(1+ye^{-s_1\alpha_2})-D_1(1+ye^{-\alpha_1})(1+y+ye^{-s_1\alpha_2})\\
&-D_2\left((1+ye^{-\alpha_2})(1+y+ye^{-\alpha_1})-y(1+ye^{-\alpha_1})\frac{e^{-\alpha_2}-e^{-s_1\alpha_2}}{1-e^{\alpha_1}}\right)\\
&+(1+y+ye^{-\alpha_1})(1+y+ye^{-\alpha_2})-y(1+ye^{-\alpha_1})\frac{e^{-\alpha_2}-e^{-s_1\alpha_2}}{1-e^{\alpha_1}}\\
&= D_2D_1(1+ye^{-\alpha_1})(1+ye^{-\alpha_2-\alpha_1})-D_1(1+ye^{-\alpha_1})(1+y+ye^{-\alpha_2-\alpha_1})\\
&-D_2\left((1+ye^{-\alpha_2})(1+y+ye^{-\alpha_1})+y(1+ye^{-\alpha_1})e^{-\alpha_1-\alpha_2}\right)\\
&+(1+y+ye^{-\alpha_1})(1+y+ye^{-\alpha_2})+y(1+ye^{-\alpha_1})e^{-\alpha_1-\alpha_2}.
\end{aligned}$$

Under the above 'dictionary' between the Hecke algebra elements and operators, this recovers Example 4.9.

9. Star duality

Recall that by 'star duality' we mean the involution $\star:\mathrm{K}_T(X)\to\mathrm{K}_T(X)$ which sends (the class of) a vector bundle $[E]$ to $[E^\vee]=[\mathrm{Hom}_{\mathcal{O}_X}(E,\mathcal{O}_X)]$. This is not an involution of $\mathrm{K}_T(\mathrm{pt})$-algebras, but it satisfies $\star(\mathbb{C}_\lambda)=\mathbb{C}_{-\lambda}$, where $\mathbb{C}_\lambda$ denotes the trivial line bundle with weight λ. We extend $\star$ to $\mathrm{K}_T(X)[y,y^{-1}]$ by linearity, requiring $\star(y^i)=y^i$ for $i\in\mathbb{Z}$.

The goal of this section is to study the effect of this duality on the motivic Chern classes for Schubert cells in $X=G/B$. We are motivated by a result of Brion [**Bri05**, Proposition 4.3.4] who proved that in $\mathrm{K}(G/B)$,

$$\mathcal{L}_{-\rho}\otimes\mathcal{I}_w=(-1)^{\operatorname{codim}X(w)}\star(\mathcal{O}_w).$$

where (recall) ρ denotes the half sum of the positive roots. (See also Proposition 9.2 below.) We will upgrade this to the (equivariant) motivic Chern classes $\mathrm{MC}_y(X(w)^\circ)$ and the (equivariant, normalized) classes $\widetilde{\mathrm{MC}}_y(X(w)^\circ)$. The reason behind this choice is that (non-equivariantly) the specialization $y=0$ gives the ideal sheaves $\mathcal{I}_w$, respectively the structure sheaves $\mathcal{O}_w$; cf. Theorem 5.1 and Theorem 4.6. (By Theorem 4.6, the opposite classes $\widetilde{\mathrm{MC}}_y(Y(w)^\circ)$ are orthogonal to the motivic classes.)

Recall the DL operators $\mathcal{T}_i$ and $\mathfrak{L}_i$ from (3.7) respectively (4.1). These determine recursively the motivic Chern classes $\mathrm{MC}_y(X(w)^\circ)=\mathcal{T}_{w^{-1}}(\mathcal{O}_{\mathrm{id}}^T)$ (Theorem 4.5) and the (normalized) classes $\widetilde{\mathrm{MC}}_y(X(w)^\circ)=\mathfrak{L}_{w^{-1}}(\mathcal{O}_{\mathrm{id}}^T)$.

We state next the main result in this section.

THEOREM 9.1. *Let $w \in W$. Then the following hold:*

(a) $\mathbb{C}_{-\rho} \otimes \mathcal{L}_{-\rho} \otimes \mathrm{MC}_y(X(w)^\circ) = (-1)^{\operatorname{codim} X(w)} \star (\widetilde{\mathrm{MC}}_y(X(w)^\circ))$.

(b) *Consider the Schubert expansions*

$$\mathrm{MC}_y(X(w)^\circ) = \sum_{u \leq w} c_{u,w}(y; e^t)\mathcal{O}_u^T; \quad \widetilde{\mathrm{MC}}_y(X(w)^\circ) = \sum_{u \leq w} d_{u,w}(y; e^t)\mathcal{I}_u^T.$$

Then $c_{u,w}(y; e^t) = (-1)^{\ell(u)-\ell(w)} \star (d_{u,w}(y; e^t))$*, or, equivalently,*

$$\langle \mathrm{MC}_y(X(w)^\circ), \mathcal{I}^u \rangle = (-1)^{\ell(w)-\ell(u)} \star \langle \widetilde{\mathrm{MC}}_y(X(w)^\circ), \mathcal{O}^{u,T} \rangle.$$

(c) *Consider the Schubert expansions*

$$\mathrm{MC}_y(X(w)^\circ) = \sum_{u \leq w} a_{u,w}(y; e^t)\mathcal{I}_u^T; \quad \widetilde{\mathrm{MC}}_y(X(w)^\circ) = \sum_{u \leq w} b_{u,w}(y; e^t)\mathcal{O}_u^T.$$

Then $a_{u,w}(y; e^t) = (-1)^{\ell(u)-\ell(w)} b_{u,w}(y; e^t)$*, or, equivalently,*

$$\langle \mathrm{MC}_y(X(w)^\circ), \mathcal{O}^{u,T} \rangle = (-1)^{\ell(w)-\ell(u)} \langle \widetilde{\mathrm{MC}}_y(X(w)^\circ), \mathcal{I}^{u,T} \rangle.$$

Before we give the proof of this theorem, we recall that the $y = 0$ specialization recovers a known relation between the ideal sheaves and structure sheaves; see Proposition 9.2 below and compare to [**Bri05**, Proposition 4.3.4]. Brion proves the result in the non-equivariant case, and for completeness we sketch a proof for the equivariant generalization. Aside from the intrinsic interest, we also note that we use this result in the proof of Theorem 9.1.

PROPOSITION 9.2 (Brion). *Let $w \in W$. Then the following holds in* $\mathrm{K}_T(X)$*:*

$$\mathbb{C}_{-\rho} \otimes \mathcal{L}_{-\rho} \otimes \mathcal{I}_w^T = (-1)^{\operatorname{codim} X(w)} \star (\mathcal{O}_w^T).$$

PROOF. Following Brion's proof, as equivariant sheaves,

$$\star(\mathcal{O}_w^T) = (-1)^{\operatorname{codim} X(w)} \omega_{X(w)} \cdot \omega_X^{-1};$$

this uses the fact that Schubert varieties are irreducible and Cohen-Macaulay - see [**Bri05**, §3.3]. The difference in the equivariant case is that the canonical sheaf of $X(w)$ needs to be twisted by a trivial bundle:

$$\omega_{X(w)} = \mathcal{O}_{X(w)}(-\partial X(w)) \otimes \mathcal{L}_\rho \otimes \mathbb{C}_{-\rho}.$$

This follows from [**BK05**, Proposition 2.2.2] to which one applies the projection formula. (Note that our convention defining $\mathcal{L}_\lambda$ is opposite to the one from [**BK05**].) If $X(w) = G/B$, then $\omega_X = \mathcal{L}_{2\rho}$ (see e.g., [**BK05**, (8) in §2.1]). The result follows from this. □

Define the $\mathbb{Z}$-linear endomorphism

$$\Psi : \mathrm{K}_T(X) \to \mathrm{K}_T(X); \quad [E]_T \mapsto \mathbb{C}_\rho \otimes \mathcal{L}_\rho \otimes \star[E]_T.$$

The previous proposition shows that

$$\Psi(\mathcal{I}_w^T) = (-1)^{\operatorname{codim} X(w)} \mathcal{O}_w^T. \tag{9.1}$$

We need the following lemma.

LEMMA 9.3. *Let $w \in W$. Then*

$$\Psi(\iota_w) = \frac{(-1)^{\dim G/B}}{e^{w(\rho)-\rho}} \iota_w.$$

PROOF. We start by observing that

$$(\star(\iota_w))|_u = \star((\iota_w)|_u) = \delta_{u,w}\lambda_{-1}T_w(G/B).$$

Then

$$\begin{aligned}\Psi(\iota_w)|_u &= (\mathbb{C}_\rho \otimes \mathcal{L}_\rho \otimes \star(\iota_w))|_u \\ &= \delta_{w,u} e^{\rho+w(\rho)}\lambda_{-1}(T_w(G/B)) \\ &= \delta_{w,u} e^{\rho+w(\rho)} \prod_{\alpha>0}(1-e^{-w\alpha}) \\ &= \delta_{w,u}\frac{(-1)^{\dim G/B}}{e^{w(\rho)-\rho}} \prod_{\alpha>0}(1-e^{w\alpha}) \\ &= \frac{(-1)^{\dim G/B}}{e^{w(\rho)-\rho}}\iota_w|_u.\end{aligned}$$

The claim follows from the injectivity of the localization map. □

The map Ψ intertwines with the Hecke algebra action in the following way.

THEOREM 9.4. *For any* $a \in \mathrm{K}_T(X)$,

$$\Psi(\mathcal{T}_i(a)) = -\mathfrak{L}_i(\Psi(a)).$$

In particular, if $w \in W$, *then*

$$\Psi((\partial_i - \mathrm{id})(\mathcal{I}_w^T)) = (-1)^{\operatorname{codim} X(w)+1}\partial_i(\mathcal{O}_w^T).$$

PROOF. The last statement follows from the first after specializing at $y=0$ and using (9.1) and Lemma 3.4. Therefore it suffices to prove the first statement. By localization, it suffices to prove this for the fixed point basis elements $a := \iota_w$. Let $n = \dim G/B$. We use the formulae from Lemma 3.3 and Lemma 9.3 to calculate:

$$\Psi(\mathcal{T}_i(\iota_w)) = (-1)^n\frac{-(1+y)}{e^{w(\rho)-\rho}(1-e^{w(\alpha_i)})}\iota_w + (-1)^n\frac{1+ye^{w(\alpha_i)}}{e^{ws_i(\rho)-\rho}(1-e^{w(\alpha_i)})}\iota_{ws_i}. \tag{9.2}$$

By definiton of $\mathfrak{L}_i$,

$$-\mathfrak{L}_i\Psi(\iota_w) = -(\mathcal{T}_i^\vee + (1+y)\,\mathrm{id})\Psi(\iota(w)) = \frac{(-1)^{n+1}}{e^{w(\rho)-\rho}}(\mathcal{T}_i^\vee(\iota_w) + (1+y)\iota_w).$$

Using now the action of $\mathcal{T}_i^\vee$ from Lemma 3.3 we calculate the last term as

$$\frac{(-1)^n}{e^{w(\rho)-\rho}}\left((1+y)\left(\frac{1}{1-e^{-w(\alpha_i)}}-1\right)\iota_w - \frac{1+ye^{w(\alpha_i)}}{1-e^{-w(\alpha_i)}}\iota_{ws_i}\right). \tag{9.3}$$

A simple algebra calculation shows that the coefficients of ι_w in both (9.2) and (9.3) are equal. The equality of the coefficients of ι_{ws_i} is proved similarly, using in addition that $s_i(\rho) = \rho - \alpha_i$. □

REMARK 9.5. Theorem 9.4 has a particularly natural interpretation in terms of the Kostant-Kumar Hecke algebra $\mathcal{H}$. We keep the notation from §8.5. There is a Hecke algebra automorphism $A : \mathcal{H} \to \mathcal{H}$ sending $D_i \mapsto 1 - D_i$ and $e^\lambda \mapsto e^{-\lambda}$. Let $L_i := D_i(1+ye^{\alpha_i}) + y$. Then it follows from the definition that

$$A(T_i) = A((1+ye^{\alpha_i})D_i - 1) = -L_i.$$

Therefore, Theorem 9.4 shows that $\Psi : \mathrm{K}_T(X) \to \mathrm{K}_T(X)$ commutes with the Hecke automorphism A.

PROOF OF THEOREM 9.1. Observe first that $\mathcal{O}_{\mathrm{id}}^T = \mathcal{I}_{\mathrm{id}}^T = \iota_{\mathrm{id}}$, and that $\Psi(\iota_{\mathrm{id}}) = (-1)^{\dim G/B}\iota_{\mathrm{id}}$. Then, by Theorem 9.4,

$$\Psi(\mathcal{T}_{w^{-1}}(\mathcal{O}_{\mathrm{id}}^T)) = (-1)^{\ell(w^{-1})}\mathfrak{L}_{w^{-1}}\Psi(\iota_{\mathrm{id}}) = (-1)^{\operatorname{codim} X(w)}\mathfrak{L}_{w^{-1}}(\iota_{\mathrm{id}})$$
$$= (-1)^{\operatorname{codim} X(w)}\widetilde{\mathrm{MC}}_y(X(w)^\circ).$$

Since $\mathcal{T}_{w^{-1}}(\mathcal{O}_{\mathrm{id}}^T) = \mathrm{MC}_y(X(w)^\circ)$ by Theorem 4.5, this proves (a).

The equality

$$c_{u,w}(y;e^t) = (-1)^{\ell(u)-\ell(w)} \star (d_{u,w}(y;e^t)),$$

which, by (3.1), is equivalent to

$$\langle \mathrm{MC}_y(X(w)^\circ), \mathcal{I}^{u,T}\rangle = (-1)^{\ell(w)-\ell(u)} * \langle \widetilde{\mathrm{MC}}_y(X(w)^\circ), \mathcal{O}^{u,T}\rangle,$$

follows by applying Ψ to both sides of

$$\mathrm{MC}_y(X(w)^\circ) = \sum_{u\le w} c_{u,w}(y;e^t)\mathcal{O}_u^T$$

and using Proposition 9.2. Finally, for part (c) we use that $\mathcal{O}^{u,T} = \sum_{v\ge u}\mathcal{I}^{v,T}$ (proved by Möbius inversion). Then:

$$\begin{aligned}\langle \mathrm{MC}_y(X(w)^\circ), \mathcal{O}^{u,T}\rangle =& \langle \mathrm{MC}_y(X(w)^\circ), \sum_{v\ge u}\mathcal{I}^{v,T}\rangle \\ =& \sum_{v\ge u}(-1)^{\ell(w)-\ell(v)} \star \langle \widetilde{\mathrm{MC}}_y(X(w)^\circ), \mathcal{O}^{v,T}\rangle \\ =& (-1)^{\ell(w)-\ell(u)}\langle \widetilde{\mathrm{MC}}_y(X(w)^\circ), \mathcal{I}^{u,T}\rangle.\end{aligned}$$

This finishes the proof. □

Acknowledgments

We would like to thank Mahir Can and Jörg Feldvoss for organizing the AMS Special Session on 'Combinatorial and Geometric Representation Theory', and for their interest in this work. This material is partly based upon work supported by the National Science Foundation under Grant No. DMS-1929284 while LM was in residence at the Institute for Computational and Experimental Research in Mathematics in Providence, RI, during the Combinatorial Algebraic Geometry program. LM was also supported in part by NSF grant DMS-2152294 and a Simons Collaboration Grant. PA is supported in part by Simons Collaboration Grant #625561. J. Schürmann is funded by the Deutsche Forschungsgemeinschaft (DFG, German Research Foundation) Project-ID 427320536 – SFB 1442, as well as under Germany's Excellence Strategy EXC 2044 390685587, Mathematics Münster: Dynamics – Geometry – Structure.

Throughout this project we used the Maple package `Equivariant Schubert Calculator`, written by Anders Buch.[4]

[4]The package is available at `https://sites.math.rutgers.edu/~asbuch/equivcalc/`

References

[AF23] D. Anderson and W. Fulton, *Equivariant cohomology in algebraic geometry*, Cambridge Studies in Advanced Mathematics, vol. 210, Cambridge University Press, Cambridge, 2023. MR4655919

[AGM11] D. Anderson, S. Griffeth, and E. Miller, *Positivity and Kleiman transversality in equivariant K-theory of homogeneous spaces*, J. Eur. Math. Soc. (JEMS) **13** (2011), no. 1, 57–84, DOI 10.4171/JEMS/244. MR2735076

[AM09] P. Aluffi and L. C. Mihalcea, *Chern classes of Schubert cells and varieties*, J. Algebraic Geom. **18** (2009), no. 1, 63–100, DOI 10.1090/S1056-3911-08-00482-7. MR2448279

[AM16] P. Aluffi and L. C. Mihalcea, *Chern-Schwartz-MacPherson classes for Schubert cells in flag manifolds*, Compos. Math. **152** (2016), no. 12, 2603–2625, DOI 10.1112/S0010437X16007685. MR3594289

[AMSS17] P. Aluffi, L. C. Mihalcea, J. Schürmann, and C. Su, *Shadows of characteristic cycles, Verma modules, and positivity of Chern-Schwartz-MacPherson classes of Schubert cells*, Duke Math. J. **172** (2023), no. 17, 3257–3320, DOI 10.1215/00127094-2022-0101. MR4688156

[AMSS22] P. Aluffi, L. C. Mihalcea, J. Schürmann, and C. Su, *Positivity of Segre-MacPherson classes*, Facets of algebraic geometry. Vol. I, London Math. Soc. Lecture Note Ser., vol. 472, Cambridge Univ. Press, Cambridge, 2022, pp. 1–28. MR4381895

[AMSS24] P. Aluffi, L. C. Mihalcea, J. Schürmann, and C. Su, *Motivic Chern classes of Schubert cells, Hecke algebras, and applications to Casselman's problem* (English, with English and French summaries), Ann. Sci. Éc. Norm. Supér. (4) **57** (2024), no. 1, 87–141. MR4732679

[And12] D. Anderson, *Introduction to equivariant cohomology in algebraic geometry*, Contributions to algebraic geometry, EMS Ser. Congr. Rep., Eur. Math. Soc., Zürich, 2012, pp. 71–92, DOI 10.4171/114-1/3. MR2976939

[BB73] A. Białynicki-Birula, *On fixed point schemes of actions of multiplicative and additive groups*, Topology **12** (1973), 99–103, DOI 10.1016/0040-9383(73)90024-4. MR313261

[BBL15] B. Brubaker, D. Bump, and A. Licata, *Whittaker functions and Demazure operators*, J. Number Theory **146** (2015), 41–68, DOI 10.1016/j.jnt.2014.01.001. MR3267111

[BCMP22] Anders S. Buch, Pierre-Emmanuel Chaput, Leonardo C. Mihalcea, and Nicolas Perrin, *Positivity of minuscule quantum K theory*, Preprint, `arXiv:2205.08630`.

[BGG73] I. N. Bernšteĭn, I. M. Gel′fand, and S. I. Gel′fand, *Schubert cells, and the cohomology of the spaces G/P* (Russian), Uspehi Mat. Nauk **28** (1973), no. 3(171), 3–26. MR429933

[Bit04] F. Bittner, *The universal Euler characteristic for varieties of characteristic zero*, Compos. Math. **140** (2004), no. 4, 1011–1032, DOI 10.1112/S0010437X03000617. MR2059227

[BK05] M. Brion and S. Kumar, *Frobenius splitting methods in geometry and representation theory*, Progress in Mathematics, vol. 231, Birkhäuser Boston, Inc., Boston, MA, 2005. MR2107324

[Bri02] M. Brion, *Positivity in the Grothendieck group of complex flag varieties*, J. Algebra **258** (2002), no. 1, 137–159, DOI 10.1016/S0021-8693(02)00505-7. Special issue in celebration of Claudio Procesi's 60th birthday. MR1958901

[Bri05] M. Brion, *Lectures on the geometry of flag varieties*, Topics in cohomological studies of algebraic varieties, Trends Math., Birkhäuser, Basel, 2005, pp. 33–85, DOI 10.1007/3-7643-7342-3_2. MR2143072

[BS81] J.-P. Brasselet and M.-H. Schwartz, *Sur les classes de Chern d'un ensemble analytique complexe* (French), The Euler-Poincaré characteristic (French), Astérisque, vol. 82, Soc. Math. France, Paris, 1981, pp. 93–147. MR629125

[BSY10] J.-P. Brasselet, J. Schürmann, and S. Yokura, *Hirzebruch classes and motivic Chern classes for singular spaces*, J. Topol. Anal. **2** (2010), no. 1, 1–55, DOI 10.1142/S1793525310000239. MR2646988

[CG09] N. Chriss and V. Ginzburg, *Representation theory and complex geometry*, Modern Birkhäuser Classics, Birkhäuser Boston, Ltd., Boston, MA, 2010. Reprint of the 1997 edition, DOI 10.1007/978-0-8176-4938-8. MR2838836

[Dem74] M. Demazure, *Désingularisation des variétés de Schubert généralisées* (French), Ann. Sci. École Norm. Sup. (4) **7** (1974), 53–88. MR354697

[EG98] D. Edidin and W. Graham, *Equivariant intersection theory*, Invent. Math. **131** (1998), no. 3, 595–634, DOI 10.1007/s002220050214. MR1614555

[EG00] D. Edidin and W. Graham, *Riemann-Roch for equivariant Chow groups*, Duke Math. J. **102** (2000), no. 3, 567–594, DOI 10.1215/S0012-7094-00-10239-6. MR1756110

[EG05] D. Edidin and W. Graham, *Nonabelian localization in equivariant K-theory and Riemann-Roch for quotients*, Adv. Math. **198** (2005), no. 2, 547–582, DOI 10.1016/j.aim.2005.06.010. MR2183388

[FGX22] Neil Fan, Peter Guo, and Rui Xiong, *Pieri and Murnaghan–Nakayama type rules for Chern classes of Schubert cells*. arXiv:2211.06802

[FRW17] L. M. Fehér, R. Rimányi, and A. Weber, *Characteristic classes of orbit stratifications, the axiomatic approach*, Schubert calculus and its applications in combinatorics and representation theory, Springer Proc. Math. Stat., vol. 332, Springer, Singapore, [2020] ©2020, pp. 223–249. MR4167518

[FRW21] L. M. Fehér, R. Rimányi, and A. Weber, *Motivic Chern classes and K-theoretic stable envelopes*, Proc. Lond. Math. Soc. (3) **122** (2021), no. 1, 153–189, DOI 10.1112/plms.12374. MR4210261

[Ful84] W. Fulton, *Intersection theory*, Ergebnisse der Mathematik und ihrer Grenzgebiete (3) [Results in Mathematics and Related Areas (3)], vol. 2, Springer-Verlag, Berlin, 1984, DOI 10.1007/978-3-662-02421-8. MR732620

[Gin98] V. Ginzburg, *Geometric methods in the representation theory of Hecke algebras and quantum groups*, Representation theories and algebraic geometry (Montreal, PQ, 1997), NATO Adv. Sci. Inst. Ser. C: Math. Phys. Sci., vol. 514, Kluwer Acad. Publ., Dordrecht, 1998, pp. 127–183. Notes by Vladimir Baranovsky [V. Yu. Baranovskiĭ]. MR1649626

[GK08] W. Graham and S. Kumar, *On positivity in T-equivariant K-theory of flag varieties*, Int. Math. Res. Not. IMRN, posted on 2008, Art. ID rnn 093, 43 pp., DOI 10.1093/imrn/rnn093. MR2439542

[Gra01] W. Graham, *Positivity in equivariant Schubert calculus*, Duke Math. J. **109** (2001), no. 3, 599–614, DOI 10.1215/S0012-7094-01-10935-6. MR1853356

[Har77] R. Hartshorne, *Algebraic geometry*, Graduate Texts in Mathematics, No. 52, Springer-Verlag, New York-Heidelberg, 1977. MR463157

[HRV08] T. Hausel and F. Rodriguez-Villegas, *Mixed Hodge polynomials of character varieties*, Invent. Math. **174** (2008), no. 3, 555–624, DOI 10.1007/s00222-008-0142-x. With an appendix by Nicholas M. Katz. MR2453601

[Huh16] J. Huh, *Positivity of Chern classes of Schubert cells and varieties*, J. Algebraic Geom. **25** (2016), no. 1, 177–199, DOI 10.1090/jag/646. MR3419959

[Jon10] B. F. Jones, *Singular Chern classes of Schubert varieties via small resolution*, Int. Math. Res. Not. IMRN **8** (2010), 1371–1416, DOI 10.1093/imrn/rnp174. MR2628830

[KK90] B. Kostant and S. Kumar, *T-equivariant K-theory of generalized flag varieties*, J. Differential Geom. **32** (1990), no. 2, 549–603. MR1072919

[Knu] Allen Knutson, *A Schubert Calculus recurrence from the noncomplex W-action on G/B*, Preprint, arXiv:math/0306304v1.

[Kum02] S. Kumar, *Kac-Moody groups, their flag varieties and representation theory*, Progress in Mathematics, vol. 204, Birkhäuser Boston, Inc., Boston, MA, 2002, DOI 10.1007/978-1-4612-0105-2. MR1923198

[Kum22] S. Kumar, *Conjectural positivity of Chern-Schwartz-Macpherson classes for Richardson cells*, Int. Math. Res. Not. IMRN **2** (2024), 1154–1165, DOI 10.1093/imrn/rnad036. MR4692369

[KZJ21] A. Knutson and P. Zinn-Justin, *Schubert puzzles and integrability II: multiplying motivic Segre classes*, Preprint, arXiv:2102.00563

[Lee18] S. J. Lee, *Chern class of Schubert cells in the flag manifold and related algebras*, J. Algebraic Combin. **47** (2018), no. 2, 213–231, DOI 10.1007/s10801-017-0773-3. MR3775220

[LH09] J. Lipman and M. Hashimoto, *Foundations of Grothendieck duality for diagrams of schemes*, Lecture Notes in Mathematics, vol. 1960, Springer-Verlag, Berlin, 2009, DOI 10.1007/978-3-540-85420-3. MR2531717

[LLL17] Kyu-Hwan Lee, Cristian Lenart, and Dongwen Liu, *Whittaker functions and Demazure characters*, Journal of the Institute of Mathematics of Jussieu (2017), 1–23.

[LLT96] A. Lascoux, B. Leclerc, and J.-Y. Thibon, *Twisted action of the symmetric group on the cohomology of a flag manifold*, Parameter spaces (Warsaw, 1994), Banach Center Publ., vol. 36, Polish Acad. Sci. Inst. Math., Warsaw, 1996, pp. 111–124. MR1481484

[Loo02] E. Looijenga, *Motivic measures*, Astérisque **276** (2002), 267–297. Séminaire Bourbaki, Vol. 1999/2000. MR1886763

[LP07] C. Lenart and A. Postnikov, *Affine Weyl groups in K-theory and representation theory*, Int. Math. Res. Not. IMRN **12** (2007), Art. ID rnm038, 65 pp., DOI 10.1093/imrn/rnm038. MR2344548

[Lus85] G. Lusztig, *Equivariant K-theory and representations of Hecke algebras*, Proc. Amer. Math. Soc. **94** (1985), no. 2, 337–342, DOI 10.2307/2045401. MR784189

[Mac74] R. D. MacPherson, *Chern classes for singular algebraic varieties*, Ann. of Math. (2) **100** (1974), 423–432, DOI 10.2307/1971080. MR361141

[Mih15] L. C. Mihalcea, *Binomial determinants and positivity of Chern-Schwartz-MacPherson classes*, Australas. J. Combin. **62** (2015), 155–171. MR3357054

[MNS22] L. C. Mihalcea, H. Naruse, and C. Su, *Left Demazure-Lusztig operators on equivariant (quantum) cohomology and K-theory*, Int. Math. Res. Not. IMRN **16** (2022), 12096–12147, DOI 10.1093/imrn/rnab049. MR4465997

[MS15] L. G. Maxim and J. Schürmann, *Characteristic classes of singular toric varieties*, Comm. Pure Appl. Math. **68** (2015), no. 12, 2177–2236, DOI 10.1002/cpa.21553. MR3417881

[MS20] L. C. Mihalcea and R. Singh, *Mather classes and conormal spaces of Schubert varieties in cominuscule spaces*, Algebr. Geom. **10** (2023), no. 5, 554–575, DOI 10.14231/ag-2023-019. MR4636283

[MS22] L. C. Mihalcea and C. Su, *Whittaker functions from motivic Chern classes*, Transform. Groups **27** (2022), no. 3, 1045–1067, DOI 10.1007/s00031-022-09731-x. With an appendix by Mihalcea, Su and Dave Anderson. MR4475486

[Nie74] H. A. Nielsen, *Diagonalizably linearized coherent sheaves* (English, with French summary), Bull. Soc. Math. France **102** (1974), 85–97. MR366928

[Ohm06] T. Ohmoto, *Equivariant Chern classes of singular algebraic varieties with group actions*, Math. Proc. Cambridge Philos. Soc. **140** (2006), no. 1, 115–134, DOI 10.1017/S0305004105008820. MR2197579

[PR99] H. Pittie and A. Ram, *A Pieri-Chevalley formula in the K-theory of a G/B-bundle*, Electron. Res. Announc. Amer. Math. Soc. **5** (1999), 102–107, DOI 10.1090/S1079-6762-99-00067-0. MR1701888

[RTV15] R. Rimányi, V. Tarasov, and A. Varchenko, *Trigonometric weight functions as K-theoretic stable envelope maps for the cotangent bundle of a flag variety*, J. Geom. Phys. **94** (2015), 81–119, DOI 10.1016/j.geomphys.2015.04.002. MR3350271

[Sch65a] M.-H. Schwartz, *Classes caractéristiques définies par une stratification d'une variété analytique complexe. I* (French), C. R. Acad. Sci. Paris **260** (1965), 3262–3264. MR212842

[Sch65b] M.-H. Schwartz, *Classes caractéristiques définies par une stratification d'une variété analytique complexe* (French), C. R. Acad. Sci. Paris **260** (1965), 3535–3537. MR184254

[Sch09] Jörg Schürmann, *Characteristic classes of mixed Hodge modules*, Topology of Stratified Spaces, MSRI Publications **58** (2009), 419–471.

[Sch17] J. Schürmann, *Chern classes and transversality for singular spaces*, Singularities in geometry, topology, foliations and dynamics, Trends Math., Birkhäuser/Springer, Cham, 2017, pp. 207–231, DOI 10.1007/978-3-319-39339-1_13. MR3706222

[SGA71] *Théorie des intersections et théorème de Riemann-Roch* (French), Lecture Notes in Mathematics, Vol. 225, Springer-Verlag, Berlin-New York, 1971. Séminaire de Géométrie Algébrique du Bois-Marie 1966–1967 (SGA 6); Dirigé par P. Berthelot, A. Grothendieck et L. Illusie. Avec la collaboration de D. Ferrand, J. P. Jouanolou, O. Jussila, S. Kleiman, M. Raynaud et J. P. Serre. MR354655

[SSW23] Jörg Schürmann, Connor Simpson, and Botong Wang, *A new generic vanishing theorem on homogeneous varieties and the positivity conjecture for triple intersections of Schubert cells*, Preprint, `arXiv:2303.13833`, 2023, to appear in Compositio Math.

[Sta89] R. P. Stanley, *Log-concave and unimodal sequences in algebra, combinatorics, and geometry*, Graph theory and its applications: East and West (Jinan, 1986), Ann. New

York Acad. Sci., vol. 576, New York Acad. Sci., New York, 1989, pp. 500–535, DOI 10.1111/j.1749-6632.1989.tb16434.x. MR1110850

[Str11] J. P. Stryker III, *Chern-Schwartz-MacPherson classes of graph hypersurfaces and Schubert varieties*, ProQuest LLC, Ann Arbor, MI, 2011. Thesis (Ph.D.)–The Florida State University. MR2949825

[Su21] C. Su, *Structure constants for Chern classes of Schubert cells*, Math. Z. **298** (2021), no. 1-2, 193–213, DOI 10.1007/s00209-020-02595-5. MR4257082

[Tan87] T. Tanisaki, *Hodge modules, equivariant K-theory and Hecke algebras*, Publ. Res. Inst. Math. Sci. **23** (1987), no. 5, 841–879, DOI 10.2977/prims/1195176035. MR934674

[Web16] A. Weber, *Equivariant Hirzebruch class for singular varieties*, Selecta Math. (N.S.) **22** (2016), no. 3, 1413–1454, DOI 10.1007/s00029-015-0214-x. MR3518555

[Web17] A. Weber, *Hirzebruch class and Białynicki-Birula decomposition*, Transform. Groups **22** (2017), no. 2, 537–557, DOI 10.1007/s00031-016-9388-3. MR3649467

Mathematics Department, Florida State University, Tallahassee Florida 32306
Email address: aluffi@math.fsu.edu

Department of Mathematics, Virginia Tech University, Blacksburg, Virginia 24061
Email address: lmihalce@vt.edu

Mathematisches Institut, Universität Münster, Germany
Email address: jschuerm@uni-muenster.de

Yau Mathematical Sciences Center, Tsinghua University, 100084, Beijing, People's Republic of China
Email address: changjiansu@mail.tsinghua.edu.cn

Contemporary Mathematics
Volume **804**, 2024
https://doi.org/10.1090/conm/804/16111

Locally semialgebraic superspaces and Nash supermanifolds

Mahir Bilen Can

ABSTRACT. In this article, we present a novel theory of locally semialgebraic superspaces along with Nash supermanifolds. By adapting Batchelor's theorem to our framework, we show that all locally semialgebraic superspaces and affine Nash supermanifolds can be derived from appropriate vector bundles. Our analysis of Nash supermanifolds involves an investigation of a topology, where closed sets are defined by Nash subsets. Within this context, we establish the softness of the structure sheaf for an affine Nash manifold. Moreover, we establish that under this topology, the higher cohomology of quasi-coherent sheaves on an affine Nash manifold vanishes completely. These results open up new directions for cohomological studies on Nash manifolds.

1. Introduction

The main goal of our article is to introduce two new theories of $\mathbb{Z}_2$-graded objects. The prototypes of our spaces were developed in the ateliers of mathematical physicists Berezin and Leites [**BL75**], and the geometer Kostant [**Kos77**]. In his seminal work, Kostant expounded, among other things, the prequantizations of $\mathbb{Z}_2$-graded manifolds. In particular, he extended the orbit method to the representation theory of Lie supergroups. The primary objective of the research was to establish a robust mathematical foundation for the geometry of elementary particles. In quantum field theories, manifolds must possess the information of both bosonic and fermionic fields, represented by two anti-commuting variables. Consequently, the formal description of these manifolds necessitates the coordinate functions to assume values in a Grassmann algebra. Such manifolds are broadly referred to as *supermanifolds* [**DeW84, Rog80**]. The connection between these physical supermanifolds and the $\mathbb{Z}_2$-graded manifolds of Kostant was made precise by Batchelor. She proved in [**Bat80**] that the category of $\mathbb{Z}_2$-graded manifolds

2020 *Mathematics Subject Classification.* Primary 14P10, 58A07; Secondary 58C50.

Key words and phrases. Batchelor's theorem, locally semialgebraic spaces, Nash manifolds, supermanifolds, soft sheaf, flabby sheaf.

Final edits of this paper were completed while we were visiting the Okinawa Institute of Science and Technology (OIST) through the Theoretical Sciences Visiting Program (TSVP). We thank OIST-TSVP for providing us with a comfortable work environment. We gratefully acknowledge the support of the Louisiana Board of Regents (grant no. 090ENH-21) which partially funded this work. Additionally, we extend our gratitude to the anonymous referee for their meticulous review of our manuscript, offering invaluable feedback that greatly enhanced the quality of our article. Our thanks also go to Jörg Feldvoss and Naufil Sakran for their insightful review, which significantly contributed to the clarity of our writing.

is equivalent to the category of (DeWitt) supermanifolds. From this viewpoint, algebro-geometric versions of supermanifolds make natural appearance in physical theories as well, [**DM99**]. In fact, this is the underlying theme of Manin's influential book [**Man97**] on gauge field theory. While these exciting mathematical developments took place in the supergravity theory for almost half a century, potentially useful semi-algebro-geometric direction has not been pursued. Therefore, it is among the goals of our present article to initiate such a study for semialgebraic spaces. We achieve this objective by extending the work of Blattner and Rawnsley [**BR84**] to the setting of generalized topologies. The specific application we have in mind for our theory is a version of an important theorem of Batchelor [**Bat79**], which, roughly speaking, states that every $\mathbb{Z}_2$-graded smooth manifold comes from a vector bundle. In fact, we will prove this theorem for two related but distinct categories, that are,

(A) the category of locally semialgebraic spaces,
(B) the category of affine Nash manifolds.

We begin our discussion with (A). First, we will present a concise overview of the objects of the category of locally semialgebraic spaces, focusing on key concepts that may initially appear complex to the reader. However, we will leave out some details to keep our introduction brief and refer readers to the preliminaries section for more thorough elaboration.

Throughout our text, the set of positive integers is denoted by $\mathbb{Z}_+$. Let R be a real closed field. This means that R is an ordered field with no nontrivial algebraic extension $R_1 \supsetneq R$ such that R_1 is also an ordered field. Let $n \in \mathbb{Z}_+$. A *semialgebraic set in* R^n is a set that can be presented as a finite union of sets of the form

$$\{x \in R^n : \ f(x) = 0, \ g_1(x) > 0, \dots, g_k(x) > 0\},$$

where $f, g_1, \dots, g_k$ are some polynomials from $R[x_1, \dots, x_n]$. Let X be a semialgebraic set in R^n. A map $f : X \to R$ is called a *semialgebraic function* if the graph of f is a semialgebraic subset of R^{n+1}. Thanks to the ordering of the real closed field R, the affine space R^n is equipped with a natural topology for which the open balls form a basis of open subsets. We will call this topology the *ordinary topology* on R^n. We assume without further notice that all semialgebraic functions we consider in this paper are continuous with respect to the induced ordinary topology on the semialgebraic set under consideration. In fact, it is an important property of the semialgebraic functions that they remain continuous with respect to a weaker notion of a topology. Since this fact is an essential ingredient in our results, and since it further motivates our discussion, we will explain our remark here. The *restricted topology*, in the sense of [**DK81**, §7], is defined as follows. Let Y be a set. A family $\mathring{\mathfrak{S}}(Y)$ of subsets of Y is called a *restricted topology on* Y if the following two conditions are satisfied:

(1) $\{\varnothing, Y\} \subseteq \mathring{\mathfrak{S}}(Y)$,
(2) $\mathring{\mathfrak{S}}(Y)$ is closed under finite intersections and finite unions of its elements.

Then a function $f : X \to Y$ between two restricted topological spaces is called *continuous* if for every $U \in \mathring{\mathfrak{S}}(Y)$, the preimage $f^{-1}(U)$ is an element of $\mathring{\mathfrak{S}}(X)$.

In a semialgebraic set X, the open semialgebraic subsets form a restricted topology. In this case, the pair $(X, \mathscr{O}_X)$, where $\mathscr{O}_X$ is the sheaf of semialgebraic functions on X (with respect to the restricted topology) becomes a ringed space over R. An *affine semialgebraic space* is a ringed space over R that is isomorphic to the ringed space over R of a semialgebraic set in R^n, for some $n \in \mathbb{Z}_+$. Among the examples of affine semialgebraic spaces are the real algebraic varieties (projective or not). More generally, a *semialgebraic space* is a ringed space that has a finite open cover by affine semialgebraic subspaces.

Even though a restricted topology is not a true topology, it has proven to be valuable in resolving issues within semialgebraic geometry. Delfs and Knebusch have introduced in [**DK85**] a stricter concept, known as a *generalized topological space*, which leads to a substantial expansion of the theory of semialgebraic spaces. The precise definition of a generalized topology is more technical than the definition of a restricted topology. We postpone it to the preliminaries section. A *locally semialgebraic space* is a ringed space $(X, \mathscr{O}_X)$ over R, where X is a generalized topological space that is equipped with an admissible covering by open semialgebraic subspaces. Hereafter, if there is no danger for confusion, we denote a locally semialgebraic space $(X, \mathscr{O}_X)$ by X only. This paper will concentrate on locally semialgebraic spaces that are both regular and paracompact. To keep things brief in the introduction, we defer the precise definitions of these technical notions to the preliminaries section. It is important to remember that every semialgebraic space is inherently a locally semialgebraic space. Additionally, it should be noted that any semialgebraic subspace of a regular locally semialgebraic space is affine.

Let us write $\overline{0}$ (resp. $\overline{1}$) for the identity (resp. non-identity) element of $\mathbb{Z}_2$. By a *$\mathbb{Z}_2$-graded ring*, we mean an associative ring with unity, denoted A, together with a direct-sum decomposition of its additive group structure, $A = A_{\overline{0}} \oplus A_{\overline{1}}$, such that $A_m A_n \subset A_{m+n}$ for $\{m, n\} \subset \mathbb{Z}_2$. In this case, for $m \in \{\overline{0}, \overline{1}\}$, an element a from A_m is said to be a *homogeneous element of degree m*. As before, let R denote a real closed field. Let us assume that A has an R-algebra structure as well. If for every $a \in A_m$ and $b \in A_n$, where $\{m, n\} \subset \{\overline{0}, \overline{1}\}$, the equality $ab = (-1)^{m+n} ba$ holds, then A is said to be a *$\mathbb{Z}_2$-graded commutative R-algebra*. Let X be a locally semialgebraic space. Let $\mathscr{A}$ be a sheaf of $\mathbb{Z}_2$-graded commutative R-algebras on X. Let $s \in \mathbb{Z}_+$. We call the pair $(X, \mathscr{A})$ a *locally semialgebraic superspace of odd-dimension s* if the following two conditions hold:

(1) There is a short exact sequence of sheaves of $\mathbb{Z}_2$-graded commutative algebras of the form,
$$0 \to \mathscr{A}' \to \mathscr{A} \to \mathscr{O}_X \to 0,$$
where the $\mathbb{Z}_2$-grading on $\mathscr{O}_X$ is given by $(\mathscr{O}_X)_{\overline{0}} := \mathscr{O}_X$ and $(\mathscr{O}_X)_{\overline{1}} := 0$.

(2) There is an admissible cover by open semialgebraic spaces $\bigcup_{i \in I} U_i = X$, where, for each $i \in I$, the restricted sheaf $\mathscr{A}|_{U_i}$ is isomorphic to the restricted sheaf of $\mathbb{Z}_2$-graded commutative algebras, $(\mathscr{O}_X \otimes \bigwedge R^s)|_{U_i}$. Here, $\mathscr{O}_X$ is $\mathbb{Z}_2$-graded as in part 1, and $\bigwedge R^s$ is the total exterior algebra of R^s, whose $\mathbb{Z}_2$-grading is determined by the subspaces

$$(\bigwedge R^s)_{\overline{0}} := \bigoplus_{i\,:\,\text{even}} \bigwedge^i R^s \qquad \text{and} \qquad (\bigwedge R^s)_{\overline{1}} := \bigoplus_{i\,:\,\text{odd}} \bigwedge^i R^s. \tag{1.1}$$

It is not difficult to adapt the usual definition of a topological vector bundle to the setting of the locally semialgebraic spaces. In fact, later in this text we will work with even more general notion of a vector bundle that is defined on a generalized topological space. For now, we proceed with the assumption that $p : Y \to X$ is a locally semialgebraic vector bundle. In this case, the sum of all exterior powers of

Y, denoted $\bigwedge Y$, is a locally semialgebraic vector bundle also. Moreover, it is easily seen that the space of global sections of $\bigwedge Y$, denoted $\Gamma(X, \bigwedge Y)$, has the structure of a $\mathbb{Z}_2$-graded commutative algebra, where the $\mathbb{Z}_2$-grading is defined by

$$\Gamma(X, \bigwedge Y)_{\bar{0}} := \Gamma\left(Y \,,\, \bigoplus_{i\,:\,\text{even}} \bigwedge^i Y\right) \qquad \text{and} \qquad \Gamma(X, \bigwedge Y)_{\bar{1}} := \Gamma\left(Y \,,\, \bigoplus_{i\,:\,\text{odd}} \bigwedge^i Y\right).$$

In this notation, the thrust of our paper is the following result, which is analogous to the main theorem of [**Bat79**].

THEOREM 1.1. *Let X be a regular, paracompact locally semialgebraic space. If $(X, \mathscr{A})$ is a locally semialgebraic superspace of odd-dimension s, then there exists a locally semialgebraic vector bundle E such that $\mathscr{A}(X)$ is isomorphic to $\Gamma(X, \bigwedge E)$ as a $\mathbb{Z}_2$-graded commutative algebra.*

The proof of Theorem 1.1 is interesting by itself. Indeed, as we mentioned before, we introduce a metatheory. This theory yields immediately a description of the potential obstruction to the existence of a vector bundle E as in our Theorem 1.1. Then we show that, under our assumptions, there is no obstruction, namely the vanishing of higher cohomology, for certain sheaves of $\mathscr{O}_X$-modules that are related to our vector bundle.

We now proceed to introduce the ingredients of our second main result. To this end, we specialize our real closed field R to the field of real numbers, $\mathbb{R}$. Let $n \in \mathbb{Z}_+$. Roughly speaking, an *affine Nash submanifold in* $\mathbb{R}^n$ is a semialgebraic, analytic submanifold of $\mathbb{R}^n$. In classical algebraic geometry, an abstract variety is defined by gluing affine algebraic varieties. Similarly to the construction of an abstract algebraic variety, an abstract *Nash manifold* is obtained by gluing affine Nash manifolds. There are deep connections between Nash manifolds and real algebraic varieties. To give a general example let us consider an orbit of a real algebraic group action. The orbit is not necessarily a real algebraic variety but it always has the structure of a Nash manifold. We will give a more specific example of this phenomenon later in this text. It should be noted that when considering algebraic group actions defined over algebraically closed fields, there are no issues arising from objects outside the relevant algebraic category. In other words, every orbit resulting from an algebraic group action has a quasi-projective variety structure that is intrinsic to it.

Let M be a Nash manifold. There are several useful topologies on M. The first one is the topology that is inherited from the euclidean topology of $\mathbb{R}^n$. The second one is the restricted topology whose open sets are given by the open semialgebraic subsets of M. Now, for our second main result, we work with a weaker topology, which is similar to the Zariski topology on the prime spectrum of a ring. Let $f : U \to \mathbb{R}$ be a semialgebraic function defined on an open semialgebraic subset $U \subseteq M$. If, in addition, f is a real analytic function, then f is called a *Nash function* on U. It turns out that, under the usual point-wise addition and multiplication of functions, and the scaling action of $\mathbb{R}$ on them, the Nash functions on U form a noetherian $\mathbb{R}$-algebra. Our notation for this algebra is $\mathscr{N}_M(U)$.

We call a subset $V \subseteq M$ a *Nash subset* if V is given by the common zero-set of a finite collection of global Nash functions on M. The set of all Nash subsets of M is closed under arbitrary intersections and finite unions. This fact follows from [**BCR98**, Proposition 8.6.2]. Thus, the collection of all Nash subsets of M

defines a topology, which we call here the $\mathfrak{N}$-*topology of* M. When we want to emphasize this topology, we will write $(M, \mathfrak{N})$ instead of M only. The subtlety that we want to point out with this notion is that, in our next theorem, we work with the sheaves that are defined on the $\mathfrak{N}$-topology of M. We call the open subsets (resp. the closed subsets) of an $\mathfrak{N}$-topology the $\mathfrak{N}$-*open subsets* (resp. the $\mathfrak{N}$-closed subsets). While moving forward, our permanent assumption in this text will be that all of our $\mathfrak{N}$-topologies are *separated.* This means that the diagonal copy of M in $M \times M$ is an $\mathfrak{N}$-closed subset of $M \times M$. Note that the property of being separated does not necessarily imply the Hausdorff property, as the $\mathfrak{N}$-topology on $M \times M$ may not always coincide with the product topology.

Let $(M, \mathfrak{N})$ be an n-dimensional Nash manifold. We view $\mathscr{N}_M$ as a presheaf on $(M, \mathfrak{N})$. In fact, it is not difficult to check that $\mathscr{N}_M$ is a sheaf of R-algebras on $(M, \mathfrak{N})$. Let $s \in \mathbb{Z}_+$. We call a pair $(M, \mathscr{A})$, where $\mathscr{A}$ is a sheaf of $\mathbb{Z}_2$-graded commutative $\mathbb{R}$-algebras on $(M, \mathfrak{N})$, a *Nash supermanifold of odd-dimension* s if the following conditions hold:

(1) There is a short exact sequence of sheaves $\mathbb{Z}_2$-graded commutative algebras of the form,

$$0 \to \mathscr{A}' \to \mathscr{A} \to \mathscr{N}_M \to 0,$$

where, the $\mathbb{Z}_2$-grading on $\mathscr{N}_M$ is defined by setting

$$(\mathscr{N}_M)_{\bar{0}} := \mathscr{N}_M \quad \text{and} \quad (\mathscr{N}_M)_{\bar{1}} := 0.$$

(2) There is a finite open cover (with respect to the $\mathfrak{N}$-topology), $\bigcup_{i \in I} U_i = M$, where, for every $i \in I$, the restricted sheaf $\mathscr{A}|_{U_i}$ on U_i is isomorphic to the restricted sheaf of $\mathbb{Z}_2$-graded commutative algebras, $(\mathscr{N}_M \otimes \bigwedge \mathbb{R}^s)|_{U_i}$. Here, $\mathscr{N}_M$ is $\mathbb{Z}_2$-graded as in part 1, and $\bigwedge \mathbb{R}^s$ is the total exterior algebra of the euclidean space $\mathbb{R}^s$, whose grading is determined by the subspaces,

$$(\textstyle\bigwedge \mathbb{R}^s)_{\bar{0}} := \displaystyle\bigoplus_{i\,:\,\text{even}} \textstyle\bigwedge^i \mathbb{R}^s \quad \text{and} \quad (\textstyle\bigwedge \mathbb{R}^s)_{\bar{1}} := \displaystyle\bigoplus_{i\,:\,\text{odd}} \textstyle\bigwedge^i \mathbb{R}^s.$$

Let $\mathbb{R}[[X_1, \ldots, X_n]]_{\text{alg}}$ denote the trivially graded ring of power series which are algebraic over the polynomial ring $\mathbb{R}[X_1, \ldots, X_n]$. In a nutshell, the above definitions are saying that a Nash supermanifold of odd-dimension s is a Nash manifold together with a sheaf of $\mathbb{Z}_2$-graded commutative algebras that is locally isomorphic to a sheaf whose stalk at $x \in M$ is isomorphic to the $\mathbb{Z}_2$-graded commutative algebra,

$$\textstyle\bigwedge \mathbb{R}^s \otimes_{\mathbb{R}} \mathbb{R}[[X_1, \ldots, X_n]]_{\text{alg}}.$$

We proceed to fulfill the natural desire of determining the global structure of an affine Nash supermanifold.

As before, let $(M, \mathfrak{N})$ be an (affine) Nash manifold. A vector bundle $p : Y \to M$ is called an *(affine) Nash vector bundle* on M if Y is an (affine) Nash manifold such that 1) p is a Nash mapping, 2) for every $x \in M$, the addition and the scalar multiplication on the vector space $p^{-1}(x)$ are Nash mappings. Now let Y be an (affine) Nash vector bundle on M. Then the sum of all exterior powers of Y, denoted $\bigwedge Y$, is an (affine) Nash vector bundle on M as well. The space of global sections of $\bigwedge Y$ on $(M, \mathfrak{N})$ has the structure of a $\mathbb{Z}_2$-graded commutative algebra. In this notation, our second main result is the following statement.

THEOREM 1.2. *Let $(M, \mathscr{A})$ be an affine Nash supermanifold of odd-dimension s. Then there exists an affine Nash vector bundle E on M such that $\mathscr{A}(M)$ is isomorphic, as a $\mathbb{Z}_2$-graded commutative algebra, to $\Gamma(M, \bigwedge E)$.*

The main idea behind the proof of Theorem 1.2 is similar to the proof of Theorem 1.1. Nevertheless, for completing the proof, we utilized some additional facts about the sheaf of Nash functions. A sheaf $\mathscr{F}$ on a topological space X is called *flabby* (respectively, *soft*) if for every open (respectively, closed) subset $V \subseteq X$, the restriction map $\Gamma(X, \mathscr{F}) \to \Gamma(V, \mathscr{F})$, $s \mapsto s|_V$ is surjective. A flabby sheaf, also called a *flasque sheaf*, on a paracompact space is a soft sheaf, [**God58**, Theorem 3.3.1]. The proof of the next standard fact that we want to point out can be deduced from [**God58**, Theorem 4.4.3]: A soft sheaf on a paracompact topological space X is acyclic if the underlying topology of X is Hausdorff. It is important here to recall that while the $\mathfrak{N}$-topology on a Nash manifold is separated according to our convention, it is not Hausdorff unless M is a finite set. We should also mention here that every $\mathfrak{N}$-topology is paracompact. For the purpose of our discussion, we record the following assertion regarding the softness of $\mathcal{N}_M$.

THEOREM 1.3. *Let M be an affine Nash manifold. Then the sheaf $\mathscr{N}_M$ is a soft sheaf with respect to the $\mathfrak{N}$-topology on M.*

The proof of Theorem 1.3 is obtained through a straightforward application of the *Efroymson's extension theorem*, which is proven in the articles [**Efr82, Pec85**]. Although our sheaf of Nash functions $\mathscr{N}_M$ is soft, it fails to be a flabby sheaf. Fortunately, the category of sheaves of $\mathscr{N}_M$-modules that we are interested in exhibits the desired outcome of cohomology vanishing, which flabbiness could provide.

Let M be an affine Nash manifold. Let A denote the $\mathbb{R}$-algebra of all Nash functions on M. For $f \in A$, let $E(f)$ denote the $\mathfrak{N}$-open subset defined by

$$E(f) := \{x \in M : f(x) \neq 0\}.$$

A sheaf of $\mathscr{N}_M$-modules $\mathscr{F}$ is said to be a *quasi-coherent sheaf of $\mathscr{N}_M$-modules* if there exists an A-module K such that, for every $f \in A$, $\mathscr{F}(E(f))$ is the localization of the ring A at the element f. We will discuss these sheaves of modules in detail in Section 5. For now let us mention that the quasi-coherent sheaves of $\mathscr{N}_M$-modules on M give the quasi-coherent sheaves of $\mathscr{O}_X$-modules on the affine scheme $X := \operatorname{Spec}(A)$. The main technical result that we will use for proving Theorem 1.2 is the following vanishing result, which is well-known in the context of affine schemes.

THEOREM 1.4. *Let M be an affine Nash manifold. Let $\mathscr{F}$ be a quasi-coherent sheaf of $\mathscr{N}_M$-modules on M. Then we have $H^i(M, \mathscr{F}) = 0$ for all $i > 0$.*

We are now ready to describe the structure of our paper. The foundations of locally semialgebraic spaces, in the spirit of modern algebraic geometry, were laid by Delfs and Knebusch in several installments. Initially, they formulated the theory of semialgebraic spaces by embedding them within the real points of real algebraic varieties. Subsequently, they produced a significant manuscript that focused on locally semialgebraic spaces. After that Delfs continued to develop homological methods for the locally semialgebraic spaces. In the preliminaries section (Section 2) we make a brief introduction to the work of Delfs and Knebusch. Among the important results that we will review there are the *theorem of partition of unity* and the *extension theorem* of locally semialgebraic functions. These results will play a crucial role in proving our theorems. In [**BR84**], Blattner and Rawnsley describe a far reaching generalization of Bathchelor's theorem. The gist of their work is to reduce Batchelor's theorem to a vanishing statement for the first cohomology groups of certain sheaves. After we setup some terminology, in Section 3, we observe that the methods of Blattner and Rawnsley can be adapted in the general setting of a generalized topology. Although it looks somewhat abstract, the generality we achieve by this adaptation gives us enough space to prove Batchelor's Theorem for the regular locally semialgebraic spaces in Section 4. We give a proof of Theorem 1.2 in Section 5. As expected, the proof outline of this theorem shares similarities with that of Theorem 1.1. An integral element of the proof, namely Theorem 1.4 concerning the vanishing of higher cohomology sheaves, is also proven in Section 5. We close our paper by Section 6, where we mention our future work about the group objects in the category of Nash supermanifolds.

2. Preliminaries

In this article, we break new ground for certain noncommutative semialgebraic geometry. The key notion for our development is the notion of a semialgebraic space.

Not only in algebraic geometry but also in semialgebraic geometry, the finiteness of the coverings is a built-in property. In this sense, the notion of a restricted topology, which we introduced earlier, is a natural choice for our purposes. Recall that a family $\mathring{\mathfrak{S}}(X)$ of subsets of a set X is called a restricted topology on X if it satisfies the following conditions:

- $\{\emptyset, X\} \subseteq \mathring{\mathfrak{S}}(X)$,
- $\mathring{\mathfrak{S}}(X)$ is closed under finite intersections and finite unions.

In this case $(X, \mathring{\mathfrak{S}}(X))$ is called a *restricted topological space.* The elements of $\mathring{\mathfrak{S}}(X)$ are called the *open sets* of the restricted topology. A *continuous map* of restricted topological spaces is a function, $f : X \to Y$, such that for every $U \in \mathring{\mathfrak{S}}(Y)$, the set $f^{-1}(U)$ is open in X. From now on, for the sake of simplicity in notation, we will adopt our earlier convention: the restricted topological space $(X, \mathring{\mathfrak{S}}(X))$ will be referred to simply as X.

Let us now introduce a fundamental example of a restricted topological space for our intended purposes. Let $n \in \mathbb{Z}_+$. Let R be a real closed field. The collection of all semialgebraic subsets of R^n is the smallest family of subsets containing all sets of the form

$$\{x \in R^n : f(x) > 0\}, \qquad \text{where } f \in R[x_1, \ldots, x_n], \tag{2.1}$$

and closed under finite unions, finite intersections, and complements. We call a semialgebraic subset of R^n an *open semialgebraic subset* if it is open in the ordinary topology on R^n. A *closed semialgebraic subset of* R^n is the complement of an open semialgebraic subset in R^n.

The dimension of a semialgebraic subset $X \subseteq R^n$ can be defined in purely algebraic terms as follows. Let $\mathcal{I}(X)$ denote the ideal of polynomials $f \in R[x_1, \ldots, x_n]$ such that $f(x) = 0$ for every $x \in X$. Then the *dimension of* X is defined to be the Krull dimension of $R[x_1, \ldots, x_n]/\mathcal{I}(X)$. Note that the dimension of a semialgebraic subset $X \subseteq R^n$ is given by the dimension of the Zariski closure of X in R^n.

EXAMPLE 2.1. The semialgebraic subsets of $\mathbb{R}^1$ are exactly the finite unions of open intervals and points. Let us consider the following subsets of $\mathbb{R}^1$:

- $X_1 := \{x \in \mathbb{R}^1 : x \geqslant 0\}$,
- $X_2 := \{x \in \mathbb{R}^1 : 0 < x < 1\} \cup \{x \in \mathbb{R}^1 : 1 < x < \infty\}$,
- $X_3 := \mathbb{Z}$.

Then X_1 is a one dimensional closed semialgebraic subset of $\mathbb{R}^1$, and X_2 is a one dimensional open semialgebraic subset of $\mathbb{R}^1$. However, X_3 is not a semialgebraic subset of $\mathbb{R}^1$. Notice also that the Zariski closures of all of these subsets in $\mathbb{R}^1$ are equal to $\mathbb{R}^1$.

It is easy to check that the collection of all open semialgebraic subsets of $\mathbb{R}^n$, including the empty set, is a restricted topology on $\mathbb{R}^n$. Let $X \subseteq \mathbb{R}^n$. By intersecting the open semialgebraic subsets of $\mathbb{R}^n$ with X, we obtain the open sets of a restricted topology on X. We will refer to this restricted topology the *$\mathfrak{S}$-topology on X*.

Restricted topologies, as well as the upcoming discussion on generalized topologies, illustrate a more abstract concept of a Grothendieck site. However, in this paper, we will not extensively explore the theory of sheaves on Grothendieck sites as presented in [**Art62**] or [**Knu71**]. Nonetheless, we find a relevant resource in [**EP16**, Section 3.1], which provides a comprehensive overview of the current state of art concerning sheaves on the Grothendieck site of a locally semialgebraic space.

All presheaf operations on the categories of (pre)sheaves on restricted topological spaces are defined analogously to the presheaf operations on the categories of (pre)sheaves on the ordinary topological spaces. For example, if $\mathscr{F}$ and $\mathscr{G}$ are two presheaves on a restricted topology on X, then a presheaf map $\mathscr{F} \to \mathscr{G}$ is defined by the requirement that for every inclusion $U \hookrightarrow V$ in $\mathring{\mathfrak{S}}(X)$, there is a commutative diagram,

$$\begin{array}{ccc} \mathscr{F}(V) & \longrightarrow & \mathscr{G}(V) \\ \downarrow & & \downarrow \\ \mathscr{F}(U) & \longrightarrow & \mathscr{G}(U) \end{array}$$

Shortly, we will engage with sheaves of R-algebras on locally semialgebraic spaces, where R is an appropriately defined field. Such a space comes with a generalized topology (to be defined) on it. In preparation for explaining these notion, in our next definition, X will represent either a restricted topological space, a standard topological space, or a generalized topological space.

DEFINITION 2.2. A *ringed space over R on X* is a pair of the form $(X, \mathscr{A})$, where $\mathscr{A}$ is a sheaf of R-algebras on X. If every stalk of $\mathscr{A}$ is a local ring, then we call $(X, \mathscr{A})$ a *locally ringed space over R on X*.

Let $(X, \mathscr{A})$ and $(Y, \mathscr{B})$ be two ringed spaces over R. A morphism between them is a pair (φ, ϑ), where $\varphi : X \to Y$ is a continuous map, and $\vartheta : \mathscr{B} \to \varphi_* \mathscr{A}$ is a sheaf map. Here, $\varphi_* \mathscr{A}$ is the *direct image sheaf* on Y, defined by $(\varphi_* \mathscr{A})(U) := \mathscr{A}(\varphi^{-1}(U))$ where U is an object (that is, an open set) in Y.

The general theory of semialgebraic spaces is built on real closed fields. Recall that a *real closed field* is an ordered field that is maximal with respect to inclusions of ordered fields. Equivalently, by [**BCR98**, Theorem 1.2.2], a real closed field is a non-algebraically closed field R such that the degree two extension of R that is obtained by adding the solution of the equation $x^2 + 1 = 0$ to it is an algebraically closed field. We proceed with the assumption that R is a real closed field.

Recall also that an affine semialgebraic space is a ringed space over R that is isomorphic to $(X, \mathscr{O}_X)$, where X is a semialgebraic set in R^n for some $n \in \mathbb{Z}_+$, and $\mathscr{O}_X$ is the sheaf of semialgebraic functions on X.

A *semialgebraic subset* of an affine semialgebraic space $(X, \mathscr{O}_X)$ is a subset of X For such an affine semialgebraic space, a subset A is said to be *semialgebraic* if the image of A in R^n is isomorphic to a semialgebraic set in R^n also. A *semialgebraic space* is a ringed space $(X, \mathscr{O}_X)$ over R that has a finite open cover $X = \bigcup_{i\in I} X_i$ such that each ringed space $(X_i, \mathscr{O}_X|_{X_i})$, where $i \in I$, is an affine semialgebraic space. In particular, the sheaf $\mathscr{O}_X$ can be viewed as a subsheaf of the sheaf of all R-valued functions on X. Now, let $(X, \mathscr{O}_X)$ be a semialgebraic space over R. Let $X = \bigcup_{i\in I} X_i$ be a finite open cover of X, where the $(X_i, \mathscr{O}_X|_{X_i})$'s are affine semialgebraic spaces. Then, for each $i \in I$, the ringed space $(X_i, \mathscr{O}_X|_{X_i})$ is isomorphic to a semialgebraic set in R^n for some $n \in \mathbb{Z}_+$. A subset $A \subseteq X$ is called a *semialgebraic subset* if for every $i \in I$, the intersection $A \cap X_i$ is a semialgebraic subset of the affine semialgebraic space $(X_i, \mathscr{O}_X|_{X_i})$.

In his work [**Del91**], Delfs developed a sheaf cohomology theory for semialgebraic geometry. Before delving into Delfs' theorems, we will first provide a comprehensive introduction to the concept of a locally semialgebraic space.

Let X be a set. A *generalized topology* on X is a set $\mathring{\mathfrak{T}}(X)$ of subsets of X, called the *open subsets of* X, and a set Cov_X of families $\{U_i | i \in I\}$ in $\mathring{\mathfrak{T}}(X)$, called *admissible coverings*, such that the following properties hold:

(1) The empty set and X itself is an element of $\mathring{\mathfrak{T}}(X)$.
(2) All finite unions and intersections of the elements of $\mathring{\mathfrak{T}}(X)$ are also elements of $\mathring{\mathfrak{T}}(X)$.
(3) Every family $\{U_i | i \in I\}$, where I is a finite set, is an element of Cov_X.
(4) For every $\{U_i | i \in I\} \in \mathrm{Cov}_X$, the union $\bigcup_{i\in I} U_i$ is an element of $\mathring{\mathfrak{T}}(X)$.

For $U \in \mathring{\mathfrak{T}}(X)$, the set of all coverings $\{U_i | i \in I\}$ such that $\bigcup_{i\in I} U_i = U$ is denoted by $\mathrm{Cov}_X(U)$. The elements of $\mathrm{Cov}_X(U)$ are called the *admissible coverings of* U.

(5) Let $U \in \mathring{\mathfrak{T}}(X)$. If $\{U_i | i \in I\} \in \mathrm{Cov}_X(U)$ and $V \in \mathring{\mathfrak{T}}(X)$, then we have
$$\{V \cap U_i | i \in I\} \in \mathrm{Cov}_X(V).$$
(6) Let $U \in \mathring{\mathfrak{T}}(X)$. If $\{U_i | i \in I\} \in \mathrm{Cov}_X(U)$ and $\{V_{ij} | j \in I_i\} \in \mathrm{Cov}_X(U_i)$, then we have
$$\{V_{ij} | i \in I, j \in I_i\} \in \mathrm{Cov}_X(U).$$
(7) If $\{U_i | i \in I\}$ is a family in $\mathring{\mathfrak{T}}(X)$ such that $U := \bigcup_{i\in I} U_i$ is an element of $\mathring{\mathfrak{T}}(X)$, and if $\{V_j | j \in J\} \in \mathrm{Cox}_X(U)$ is a refinement of $\{U_i | i \in I\}$, that is to say, for every $j \in J$, there exists $i \in I$ such that $V_j \subseteq U_i$, then we have
$$\{U_i | i \in I\} \in \mathrm{Cov}_X(U).$$
(8) Let $U \in \mathring{\mathfrak{T}}(X)$. Let $\{U_i | i \in I\} \in \mathrm{Cov}_X(U)$. If V is a subset of U such that $U_i \cap V \in \mathring{\mathfrak{T}}(X)$ for every $i \in I$, then $V \in \mathring{\mathfrak{T}}(X)$. In particular, by 5., we have $\{U_i \cap V | i \in I\} \in \mathrm{Cov}_X(V)$.

DEFINITION 2.3. A *locally semialgebraic space* is a ringed space $(X, \mathscr{O}_X)$ over R, where X is a generalized topological space having an admissible covering $\{U_i | i \in$

$I\} \in \mathrm{Cov}_X$ such that the ringed subspace $(U_i, \mathscr{O}_X|_{U_i})$ over R is a semialgebraic space for every $i \in I$. A morphism of locally semialgebraic spaces is a morphism of the underlying ringed spaces.

REMARK 2.4.

(1) Clearly, if $(X, \mathscr{O}_X)$ is an ordinary semialgebraic space, then its restricted topology $\mathring{\mathfrak{S}}(X)$ satisfies all the axioms of a generalized topology. In other words, every semialgebraic space is automatically a locally semialgebraic space. More generally, let us consider a restricted topological space, $(X, \mathscr{O}_X)$. Let us define $\mathring{\mathfrak{T}}(X) := \mathring{\mathfrak{S}}(X)$. For the coverings of an element $U \in \mathring{\mathfrak{T}}(X)$, we consider the set of all families $\{U_i | i \in I\}$ in $\mathring{\mathfrak{S}}(X)$ such that $\bigcup_{i \in I} U_i = U$, and U is already covered by finitely many of the U_i's ($i \in I$). Then it is easy to check that with these coverings and open sets, $(X, \mathscr{O}_X)$ is equipped with a generalized topological space structure. In other words, every restricted topological space can be regarded as a generalized topological space. Thus, any general theorem that is applicable to the generalized topological spaces is applicable to the restricted topological spaces as well.

(2) The ringed space of a locally semialgebraic space is actually a *locally ringed space*, that is, for every $x \in X$, the stalk $\mathscr{O}_{X,x}$ is a local ring. In fact, its quotient ring $\mathscr{O}_{X,x}/\mathfrak{m}_x$, where $\mathfrak{m}_x$ is the unique maximal ideal of $\mathscr{O}_{X,x}$, is isomorphic to the real closed field, R.

A concise but very useful summary of the theory of locally semialgebraic spaces, up to year 1984, can be found in the article [**DK84**] of Delfs and Knebusch.

For our next definition, we fix a locally semialgebraic space M over R. This means that we fix an admissible covering $\{M_i | i \in I\} \in \mathrm{Cov}_M$, where each ringed subspace $(M_i, \mathscr{O}_M|_{M_i})$, $i \in I$, is a semialgebraic space.

DEFINITION 2.5. We maintain our notation from the previous paragraph. A *locally semialgebraic subset* of M is a subset $X \subseteq M$ such that for every element $W \in \mathring{\mathfrak{T}}(M)$ where $(W, \mathscr{O}_M|_W)$ is a semialgebraic space, the intersection $X \cap W$ is a semialgebraic subset of the semialgebraic space W. The set of all locally semialgebraic subsets of M is denoted by $\mathfrak{T}(M)$.

REMARK 2.6. To check that a subset $X \subseteq M$ is a locally semialgebraic subset, it suffices to check that $X \cap M_i$ is a semialgebraic subset of M_i for every $i \in I$.

EXAMPLE 2.7. Let $\mathring{\mathfrak{T}}(\mathbb{R}^2)$ denote the set $\{U_r : r \in \mathbb{N} \cup \{\infty\}\}$, where U_r is defined by

$$U_r := \begin{cases} \varnothing & \text{if } r = 0, \\ \{x \in \mathbb{R}^2 : |x| < r\} & \text{if } r \in \mathbb{Z}_+, \\ \mathbb{R}^2 & \text{if } r = \infty. \end{cases}$$

It is easy to check that the set of all families of the form $\{U_r : r \in I\}$, where $I \subseteq \mathbb{N} \cup \{\infty\}$, satisfies all axioms of being a set of admissible coverings. Let us denote this set of families by $\mathrm{Cov}_{\mathbb{R}^2}$. Hence, the pair $(\mathring{\mathfrak{T}}(\mathbb{R}^2), \mathrm{Cov}_{\mathbb{R}^2})$ gives a generalized topology on $\mathbb{R}^2$. For every $r \in \mathbb{Z}_+$, we will view U_r as an open semialgebraic subspace of the standard affine semialgebraic space $(\mathbb{R}^2, \mathscr{O}_{\mathbb{R}^2})$. Then we define $(X, \mathscr{O}_X)$ as the inductive limit $X := \varinjlim_{r \in \mathbb{N}} U_r$ in the category of locally semialgebraic spaces. The fact that this inductive limit exists in the category of locally semialgebraic spaces follows from [**DK84**, Lemma 1.2]. Notice that an admissible covering of X is given by $\{U_r : r \in \mathbb{N}\} \in \mathrm{Cov}_{\mathbb{R}^2}$.

While on a set-theoretical level, X can be readily identified with $\mathbb{R}^2$, as a locally semialgebraic space it is different from the standard affine semialgebraic space $(\mathbb{R}^2, \mathscr{O}_{\mathbb{R}^2})$. In fact, $(X, \mathscr{O}_X)$ does not even qualify as a semialgebraic space. This becomes evident when we consider the set of closed points in the generalized topology of $(X, \mathscr{O}_X)$, which is found to be empty. This follows from the fact that the basis of the generalized topology of X consists of the open sets U_r, where $r \in \mathbb{N} \cup \infty$. A closed semialgebraic subset Y of X can either be empty or unbounded. However, it is easy to see that every nonempty semialgebraic set possesses a closed point. Consequently, $(X, \mathscr{O}_X)$ is not a semialgebraic space.

Two other important properties of the (set of) locally semialgebraic subsets are as follows. The proofs of these properties are not difficult.

(1) If X is an element of $\mathfrak{T}(M)$, then so is $M \backslash X$.
(2) Both the union and the intersection of any locally finite family from $\mathfrak{T}(M)$ are contained in $\mathfrak{T}(M)$.

Let us explain what we mean by a locally finite family in this context.

Definition 2.8. Let M be a generalized topological space. A family $\{X_j | j \in J\}$ of subsets of M is called *locally finite* if any element $W \in \mathring{\mathfrak{T}}(M)$ meets only finitely many X_j's. Equivalently, for every $i \in I$, the open set M_i meets only finitely many X_j's. We note in passing that every locally finite family in $\mathring{\mathfrak{T}}(M)$ is an element of Cov_M.

So far, we dealt with locally semialgebraic subsets. Next, we want to explain the notion of a locally semialgebraic subspace.

Let M and $\mathfrak{T}(M)$ denote, as before, a locally semialgebraic space and the set of locally semialgebraic subsets of M, respectively. Let $\{M_i | i \in I\}$ be an admissible covering of M. We enlarge this admissible covering by adding to it all finite unions of its elements. Let $\{N_j | j \in J\}$ denote the resulting family of open semialgebraic subsets of M. It is easily checked that $\{N_j | j \in J\}$ is an admissible covering of M as well. Let X be an element of $\mathfrak{T}(M)$. Then every intersection $N_j \cap X$ $(j \in J)$ is a semialgebraic subspace of N_j.

Proposition-Definition 2.9. We maintain our notation from the previous paragraph. Let $(X, \mathscr{O}_X)$ denote the inductive limit of the directed system of semialgebraic spaces $\{N_j \cap X | j \in J\}$. Then $(X, \mathscr{O}_X)$ has a natural structure of a locally semialgebraic space. With this locally semialgebraic space structure, X is called a *locally semialgebraic subspace of* M.

The concept of a locally semialgebraic subspace provides a formal framework for precisely introducing the idea of a *semialgebraic subset* within a locally semialgebraic space.

Proposition-Definition 2.10. Let $(M, \mathscr{O}_M)$ be a locally semialgebraic space over R. A subset $X \subset M$ is called *semialgebraic* if X is a locally semialgebraic subset of M and the pair $(X, \mathscr{O}_X)$, where $\mathscr{O}_X$ is the restriction of $\mathscr{O}_M$ to X, is a semialgebraic space. If a locally semialgebraic subset $X \subseteq M$ is semialgebraic if and only if there exists a semialgebraic subset $Y \subseteq M$ such that $X \subset Y$.

EXAMPLE 2.11. We proceed the notation of Example 2.7. In Figure 1, we depict a curve $Z \subset \mathbb{R}^2$, defined by the graph of the piecewise linear function $f : \mathbb{R} \to \mathbb{R}$, where

$$f(x) = \frac{2}{\pi} \sin^{-1}(\sin(\pi x)), \quad x \in \mathbb{R}.$$

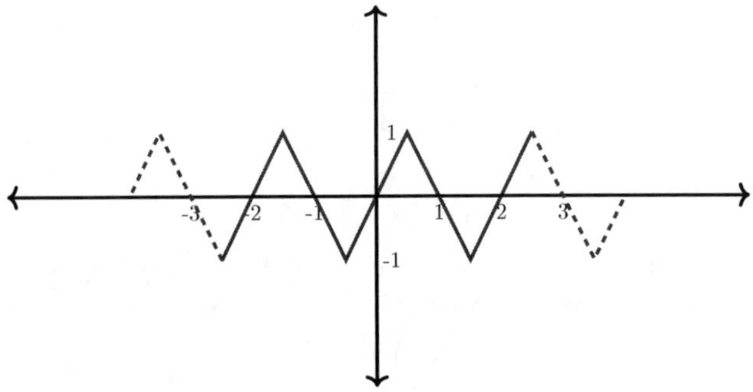

FIGURE 1. A (connected) locally semialgebraic subset $Z \subseteq \mathbb{R}^2$.

Clearly, for every $r \in \mathbb{N}$, the intersection $Z \cap U_r$ is a bounded semialgebraic subset of U_r. Indeed, $Z \cap U_r$ is a union of finitely many line segments. Hence, according to Remark 2.6, Z is a locally semialgebraic subset of $(X, \mathscr{O}_X)$. By a similar reasoning, we see that Z is a locally semialgebraic subset of $(\mathbb{R}^2, \mathscr{O}_{\mathbb{R}^2})$ as well. However, we will show that Z fails to qualify as a semialgebraic subset in either $(\mathbb{R}^2, \mathscr{O}_{\mathbb{R}^2})$ or $(X, \mathscr{O}_X)$.

Assume towards a contradiction that Z is a semialgebraic subset of $(\mathbb{R}^2, \mathscr{O}_{\mathbb{R}^2})$. Since intersection of two semialgebraic subsets is a semialgebraic subset, the intersection of Z with the line $y = 0$ in $\mathbb{R}^2$, denoted by Y, must be a semialgebraic subset as well. But Y consists of infinitely many isolated points in $\mathbb{R}^2$. Hence, Y cannot be expressed by finitely many polynomial equalities or inequalities. In particular, Y, hence Z, cannot a semialgebraic subset of $(\mathbb{R}^2, \mathscr{O}_{\mathbb{R}^2})$.

The fact that Z is not a semialgebraic subset of $(X, \mathscr{O}_X)$ follows from Definition 2.10. Indeed, if $(Z, \mathscr{O}_X|_Z)$ were a semialgebraic space, then it would have a closed point. However, a basis for the generalized topology of Z is given by $\{Z \cap U_r : r \in \mathbb{N} \cup \{\infty\}\}$. Since the complement of $Z \cap U_r$ $(r \in \mathbb{N} \cup \{\infty\})$ is either empty or unbounded, the generalized topology on Z does not possess any closed points. This contradiction shows that Z is not a semialgebraic subset of $(X, \mathscr{O}_X)$, also.

The preceding example generalizes when considering infinite locally semialgebraic spaces. In any such space M, there exists a locally semialgebraic subset $X \subsetneq M$ that does not qualify as semialgebraic. A proof of this fact mirrors the rationale employed in our previous example, but its details are omitted to avoid redundancy.

NOTATION 2.12. Let X be a locally semialgebraic space. Recall that $\mathring{\mathfrak{T}}(X)$ denotes the family of all open locally semialgebraic subsets of X. Recall also that by $\mathfrak{T}(X)$ we denoted the family of all locally semialgebraic subsets of X. Likewise, we will denote by $\mathfrak{S}(X)$ the family of all semialgebraic subsets of X. The family of all closed semialgebraic subsets of X will be denoted by $\bar{\mathfrak{S}}(X)$. Finally, let us remind that $\mathring{\mathfrak{S}}(X)$ stands for the family of all open semialgebraic subsets of X.

A (restricted) topological space X is said to be *separated* if the image of the diagonal mapping

$$\Delta : X \longrightarrow X \times X$$

is a closed subset. In algebraic geometry, separated schemes hold significant prominence. Notably, every affine scheme is separated (see [**Har77**, Proposition 4.1]). Furthermore, a widely accepted definition of an algebraic variety is that it is an integral separated scheme of finite type over an algebraically closed field (see [**Har77**, Example 4.0.2 and Proposition 4.10]). Serre proved in [**Ser56**] that a scheme defined over $\mathbb{C}$ is separated if and only if the associated analytic space is Hausdorff. These facts have direct consequences on a substantial family of semialgebraic spaces. Let X be an algebraic variety defined over R. Let M denote the set of R-rational points of X. Since all varieties are defined, at least locally, as the vanishing set of polynomials, M has the structure of a semialgebraic space over R. Since X is separated, M is a separated topological space as well. (Further discussion on such semialgebraic spaces can be found in [**DK81**].) For example, if X denotes the complex affine line as an algebraic variety, then its set of $\mathbb{R}$-rational points is given by $M := X(\mathbb{R}) = \mathbb{R}^2$. Then M is separated with respect to its Zariski topology. But every Zariski closed subset of $M \times M$ is a closed subset in the restricted topology of $M \times M$. It follows that M is separated with respect to its semialgebraic topology as well. For our first main theorem, we will utilize another separation property.

Definition 2.13. A locally semialgebraic space X is called *regular* if for every closed locally semialgebraic subset $A \subseteq X$ and every point $x \in X\backslash A$, there exist disjoint sets U and V from $\mathring{\mathfrak{T}}(X)$ such that $x \in U$, $A \subset V$.

An important theorem of Robson [**Rob83**] states that, a semialgebraic space X is regular if and only if X is affine. In particular, for every $n \in \mathbb{Z}_+$, the affine space $(R^n, \mathscr{O}_{R^n})$ is a regular semialgebraic space. However, a locally semialgebraic space need not be regular even if its underlying set is R^n for some $n \in \mathbb{Z}_+$.

Example 2.14. Let X denote the locally semialgebraic space that we constructed in Example 2.7. We will maintain the notation from that example. We claim that X is not a regular locally semialgebraic space. Let A denote the closed locally semialgebraic subset $X\backslash U_1$. Let x denote the origin of $\mathbb{R}^2$. Then $x \in X\backslash A$. Given the nested nature of the U_r sets, the set of all open subsets of X is given by $\mathring{\mathfrak{T}}(X) = \{U_r : r \in \mathbb{N}\} \cup \{\mathbb{R}^2\}$. Let $r, s \in \mathbb{Z}_+$. Then we have $U_r \cap U_s = U_{\min\{r,s\}} \neq \varnothing$. In other words, every two nonempty, open, locally semialgebraic subsets of X intersect nontrivially. This means that we cannot separate x from A by using the nonempty open subsets of X. Hence, X is not regular.

Another topological notion that is naturally a part of semialgebraic geometry is the following.

Definition 2.15. A locally semialgebraic space X is called *paracompact* if it possesses an admissible covering $\{X_i | i \in I\} \in \mathrm{Cov}_X$ such that $\{X_i | i \in I\}$ is a locally finite family, and every X_i $(i \in I)$ in this family is semialgebraic. (Recall that we introduced the notion of a locally finite family in Definition 2.8.)

Note that every semialgebraic space is paracompact.

Let X be a locally semialgebraic space. A *family of supports on* X is a family Φ of closed subsets of X such that

(a) every closed subset of a member of Φ is also an element of Φ;
(b) Φ is closed under finite unions.

If, in addition, the following conditions are satisfied, then we call Φ a *paracompactifying family of supports*:

(c) Every element of Φ is contained in some element of $\Phi \cap \mathfrak{T}(X)$.
(d) Every element of $\Phi \cap \mathfrak{T}(X)$ is regular and paracompact.
(e) Every element of Φ has a neighborhood in X which belongs to Φ.

EXAMPLE 2.16. Let X be an affine semialgebraic space. Then the family Φ of all closed subsets of X is a paracompactifying family of supports.

REMARK 2.17. The definition of a paracompactifying family of supports that we introduced above is the adaptation to the semialgebraic geometry of a similar notion from classical algebraic topology (see [**God58**, Section 3.2]). We will use the classical version in Section 5 when we focus on the sheaves on affine Nash manifolds.

Let X be a locally semialgebraic space. Let $\mathscr{F}$ be an *abelian sheaf* on X, that is, a sheaf on X with values in the category of abelian groups. Let $s \in \Gamma(X, \mathscr{F})$. For every $x \in X$, the section s defines an element in the stalk $\mathscr{F}_x$. The *support of* s, denoted $\operatorname{supp} s$, is the set $\{x \in X | s_x \neq 0\}$. Note that $\operatorname{supp} s$ is a closed subset of X. For any family Φ of supports on X, we set

$$\Gamma_\Phi(X, \mathscr{F}) := \{s \in \Gamma(X, \mathscr{F}) |\ \operatorname{supp} s \in \Phi\}.$$

Then it is easy to check that the functor $\mathscr{F} \mapsto \Gamma_\Phi(X, \mathscr{F})$ is left exact. Hence, by using its right derived functors, $R^q\Gamma_\Phi(X, -)$, we obtain the *sheaf cohomology groups with supports in* Φ. In more mathematical notation, we will write

$$H^q_\Phi(X, \mathscr{F}) := R^q\Gamma_\Phi(X, \mathscr{F})$$

for the q-th cohomology group with supports in Φ of the sheaf $\mathscr{F}$. A sheaf $\mathscr{F}$ on X is called Φ-acyclic if $H^q_\Phi(X, \mathscr{F}) = 0$ for every $q \in \mathbb{Z}_+$.

Let A be a closed subset of X. In our next definition, we will consider the restriction of a sheaf $\mathscr{F}$ onto A. For convenience, let us recall its definition. Then, the section ring $\Gamma(A, \mathscr{F})$ is defined as the inductive limit,

$$\Gamma(A, \mathscr{F}) := \varinjlim_{A \subset U \in \mathscr{F}} \mathscr{F}_M(U).$$

DEFINITION 2.18. Let Φ be a paracompactifying family of supports on a locally semialgebraic space X.

(1) A sheaf $\mathscr{F}$ on X is said to be Φ-*soft* if the restriction map

$$\Gamma(X, \mathscr{F}) \longrightarrow \Gamma(A, \mathscr{F})$$

is surjective for every $A \in \Phi$. If Φ consists of all closed subsets of X, then $\mathscr{F}$ is called *soft*.

(2) A sheaf $\mathscr{F}$ on X is called *sa-flabby* if the restriction map

$$\Gamma(X, \mathscr{F}) \longrightarrow \Gamma(U, \mathscr{F})$$

is surjective for every $U \in \mathring{\mathfrak{T}}(X)$.

We now state several useful propositions from Delfs work.

PROPOSITION 2.19. *Let X be a locally semialgebraic space. Let Φ be a paracompactifying family of supports on X. If $\mathscr{F}$ is Φ-soft, then $\mathscr{F}$ is Φ-acyclic.*

PROOF. This is a special case of the proof of [**Del91**, Proposition 4.17]. □

PROPOSITION 2.20. *Let X be a locally semialgebraic space. Let Φ be a paracompactifying family of supports on X. Let A be a principal ideal domain. Let $\mathscr{G}$ be a sheaf on X. If $\mathscr{F}$ is a flat and Φ-soft sheaf of A-modules on X, then the sheaf $\mathscr{F} \otimes \mathscr{G}$ is Φ-soft as well.*

PROOF. This is a restatement of [**Del91**, Proposition 4.22]. □

A locally semialgebraic space X is said to be *Lindelöf* if it possesses a countable admissible covering $\{X_i | i \in I\} \in \mathrm{Cov}_X(X)$. Clearly, every semialgebraic space is Lindelöf.

PROPOSITION 2.21. *Let X be a Lindelöf locally semialgebraic space. If $\mathscr{F}$ is an sa-flabby sheaf on X, then $\mathscr{F}$ is Φ-acyclic for every locally semialgebraic family Φ of supports on X.*

PROOF. The proof follows from [**Del91**, Theorem 4. 10]. □

We continue with the review of two more useful results about locally semialgebraic spaces. The first one is a partition of unity statement.

THEOREM 2.22. *([**DK85**, Theorem 4.12]) Let X be a regular paracompact locally semialgebraic space. Let $\{U_\lambda \to X | \lambda \in I\}$ be a locally finite covering of X by open semialgebraic subsets. Then there exists a family $\{\varphi_\lambda | \lambda \in I\}$ of locally semialgebraic functions $\varphi_\lambda : M \to [0,1]$ with $\mathrm{supp}(\varphi_\lambda) \subset U_\lambda$ for every $\lambda \in I$ and*

$$\sum_{\lambda \in I} \varphi_\lambda(x) = 1$$

for every $x \in X$.

The second result that we want to quote is a weak version of a locally semialgebraic analog of the well-known *Tietze's extension theorem* for continuous function. Its proof heavily depends on the previous partition of unity theorem.

THEOREM 2.23. *([**DK85**, Theorem 4.13]) Let X be a regular paracompact locally semialgebraic space. Let $f : A \to R$ be a locally semialgebraic function on a closed locally semialgebraic subset $A \subset M$. Then there exists a locally semialgebraic function $g : M \to R$ such that $g|_A = f$.*

REMARK 2.24. The special case of Theorem 2.23 for affine semialgebraic spaces is due to Efroymson [**Efr82**] and Pecker [**Pec85**].

We now proceed to introduce the group objects of the semialgebraic category. Let G be a group. If G has a locally semialgebraic space structure such that the group operations

$$\begin{aligned} m : G \times G &\longrightarrow G \\ (x,y) &\longmapsto xy \end{aligned}$$

and

$$i : G \longrightarrow G$$
$$x \longmapsto x^{-1}$$

are morphisms of locally semialgebraic spaces, then we call G a *locally semialgebraic group*. A *semialgebraic group* is a locally semialgebraic group whose underlying locally semialgebraic space is a semialgebraic space. An *affine semialgebraic group* is a semialgebraic group whose underlying semialgebraic space is isomorphic to an affine semialgebraic space.

EXAMPLE 2.25. Every real algebraic group is an affine semialgebraic group.

We know from the work [**Pil88**] of Pillay that every semialgebraic group has the unique structure of a Lie group but not conversely. Indeed, already the unit circle $\mathbb{S}^1$ can be endowed with infinitely many different semialgebraic group stuctures some of which are not even embeddable into any euclidean space. The first example of such a structure was found by Mazur. For a detailed discussion of the example of Mazur, see [**Shi87**, Chapter IV].

3. Ringed D-Spaces and Vector Bundles

Throughout this section, we fix the following data:

k	:	a field of characteristic zero,
D	:	a finite dimensional unital k-algebra,
$\varepsilon_0 : D \to k$	:	a k-algebra homomorphism,
J	:	the kernel of ε_0 (which will be assumed to be nilpotent),
X	:	a generalized topological space,
$(X, \mathscr{C})$	:	a locally ringed space over k.

As usual, if there is no danger for confusion, we will write X instead of $(X, \mathscr{C})$. We are now ready to introduce *ringed D-spaces*.

DEFINITION 3.1. A sheaf of k-algebras $\mathscr{A}$ on X is called a *ringed D-space structure on* X if there exist a map of sheaves of k-algebras $\varepsilon : \mathscr{A} \to \mathscr{C}$ and a covering $\{U_i | i \in I\}$ of X by open subsets, where, for each $i \in I$, there is a sheaf isomorphism

$$\tau_i : \mathscr{A}|_{U_i} \longrightarrow (\mathscr{C} \otimes D)|_{U_i}$$

such that

$$(id \otimes \varepsilon_0) \circ \tau_i = \varepsilon|_{U_i}.$$

We will show that the notion of a ringed D-space is closely related to a notion of a vector bundle over X with values in D. We proceed with the assumption that X is connected in an appropriate sense in its Grothendieck topology. This assumption is not essential but simplifies our discussion. Also, temporarily, we consider only the vector space structure on D.

DEFINITION 3.2. A *vector bundle over X of class $\mathscr{C}$* is a sheaf $\mathscr{E}$ of $\mathscr{C}$-modules on X for which there is an open covering $\{X_\alpha \to X | \alpha \in I\}$ satisfying the following conditions:

(1) For every $\alpha \in I$, there is an isomorphism of sheaves of $\mathscr{C}$-modules,

$$\tau_\alpha : \mathscr{E}|_{X_\alpha} \longrightarrow \mathscr{C}|_{X_\alpha} \otimes D. \tag{3.1}$$

(2) For $\{\alpha, \beta\} \subset I$, the map of sheaves of $\mathscr{C}$-modules,

$$g_{\alpha\beta} := \tau_\alpha \circ \tau_\beta^{-1} : \mathscr{C}|_{X_\alpha \cap X_\beta} \otimes D \longrightarrow \mathscr{C}|_{X_\alpha \cap X_\beta} \otimes D, \tag{3.2}$$

called the *transition function of $\mathscr{E}$ on $X_\alpha \cap X_\beta$*, is an isomorphism of $\mathscr{C}$-modules.

In this case, the general linear group $\mathrm{GL}(D)$ is said to be the *structure group of $\mathscr{E}$*.

We follow the notation of Definition 3.2. For every $x \in X_\alpha \cap X_\beta$, the transition function $g_{\alpha,\beta}$ in (3.2) gives a $\mathscr{C}_x$-linear automorphism $(g_{\alpha,\beta})_x$ of $\mathscr{C}_x \otimes D$. Since $\mathscr{C}_x$ is a local k-algebra, and since D is a k-vector space, such an automorphism is given by an element of the general linear group $\mathrm{GL}(D)$. In fact, the data of transition functions of $\mathscr{E}$ is equivalent to a data of maps,

$$\{X_\alpha \cap X_\beta \to \mathrm{GL}(D) \mid \{\alpha, \beta\} \subset I\},$$

satisfying the *Čech cocycle conditions*, $(g_{\alpha,\alpha})_x = \mathrm{id}_D$ and $(g_{\beta,\gamma})_x (g_{\alpha,\beta})_x = (g_{\alpha,\gamma})_x$ for every $\{\alpha, \beta, \gamma\} \subset I$.

EXAMPLE 3.3. In this example, we will find out that the ordinary topological vector bundles are among the examples of our Definition 3.2. Let X be a topological space. Let $k := \mathbb{R}$. Let $\mathscr{C}^0$ denote the sheaf of continuous k-valued functions on X. Let $\mathscr{E}$ be a locally free sheaf of $\mathscr{C}^0$-modules on X such that there is an open cover $\{X_\alpha | \alpha \in I\}$ of X satisfying the properties in (3.1) and (3.2). After fixing a basis for D, we may assume that D is isomorphic to $\mathbb{R}^r$ for some $r \in \mathbb{Z}_+$. Then we have the isomorphism,

$$\mathscr{E}|_{X_\alpha} \cong \underbrace{\mathscr{C}^0|_{X_\alpha} \oplus \cdots \oplus \mathscr{C}^0|_{X_\alpha}}_{r\text{-copies}} \qquad (\alpha \in I).$$

Thus, locally, the transition functions are given by the elements of $\mathrm{GL}_r(\mathbb{R})$. They satisfy the Čech cocycle conditions. Therefore, we recover the well-known definition of a topological vector bundle on X.

The structure group of a vector bundle over X of class $\mathscr{C}$ is a group object in the category that X belongs to. This is forced upon us by the transition functions of the vector bundle. In this sense, the structure group of our previous example, that is $\mathrm{GL}_r(\mathbb{R})$, is viewed as a topological group. Of course, $\mathrm{GL}_r(\mathbb{R})$ is a real algebraic group. Hence, it is also a locally semialgebraic group. In this regard, for us, a locally semialgebraic vector bundle on a locally semialgebraic space X is a vector bundle $\mathscr{A}$ over X of type $\mathscr{O}_X$ whose structure group is $\mathrm{GL}_r(\mathbb{R})$.

We proceed with the assumption that D is a k-algebra. Let $\mathrm{Aut}(D)$ denote the group of all k-algebra automorphisms of D. Then $\mathrm{Aut}(D)$ is an algebraic subgroup of $\mathrm{GL}(D)$. Let $\mathscr{E}$ be a vector bundle over X of class $\mathscr{C}$. In the extreme case that (the images in the stalks of) every transition function of E as in (3.2) is contained in $\mathrm{Aut}(D)$, the sheaf $\mathscr{E}$ can be viewed as a sheaf of $\mathscr{C}$-algebras so that the map (3.1) is an isomorphism of sheaves of $\mathscr{C}$-algebras. Let $\varepsilon_0 : D \to k$ be a k-algebra homomorphism. The automorphisms of D that preserve ε_0 form a subgroup of $\mathrm{Aut}(D)$, denoted by $\mathrm{Aut}(D, \varepsilon_0)$. Let us assume that $\mathscr{E}$ is a vector bundle over X of class $\mathscr{C}$ such that all transition functions of $\mathscr{E}$ preserve the map $\mathrm{id} \otimes \varepsilon_0$.

DEFINITION 3.4. If $\mathscr{E}$ is a vector bundle over X of class $\mathscr{C}$ such that all transition functions of $\mathscr{E}$ preserve the map $\mathrm{id} \otimes \varepsilon_0$, then we say that $\mathscr{E}$ *has the structure group* $\mathrm{Aut}(D, \varepsilon_0)$.

Let $\mathscr{E}$ be a vector bundle over X of class $\mathscr{C}$ with structure group $\mathrm{Aut}(D, \varepsilon_0)$. Then we have a morphism $\varepsilon : \mathscr{E} \to \mathscr{C}$ such that locally on U_α it is given by

$$\varepsilon = (\mathrm{id} \otimes \varepsilon_0) \circ \tau_\alpha.$$

Let us put these definitions together in the form of a lemma.

LEMMA 3.5. *Let $\mathscr{E}$ be a vector bundle over X of class $\mathscr{C}$. If all transition functions of $\mathscr{E}$ are contained in* $\mathrm{Aut}(D, \varepsilon_0)$, *then $\mathscr{E}$ has the structure of a ringed D-space on X.*

Conversely, a ringed D-space $\mathscr{E}$ on X is a vector bundle of class $\mathscr{C}$ if $\mathscr{E}$ is locally free as a sheaf of $\mathscr{C}$-modules.

Let $\mathscr{A}$ be a sheaf of k-algebras on X. Let $\{U_\alpha | \alpha \in I\}$ be an open cover of X. Then the assignment $U_\alpha \mapsto Z(\mathscr{A}(U_\alpha))$, where $Z(\cdot)$ stands for the center, gives us a presheaf of k-algebras on X. We denote the sheafification of this presheaf by $Z(\mathscr{A})$. We note in passing that, in general, the existence of the sheafification of a presheaf with values in the category of k-algebras is a nontrivial fact; see [**Sta22**].

In our next proposition, a criterion for checking when a ringed D-space is locally free is presented.

PROPOSITION 3.6. *Let $\varepsilon : \mathscr{A} \to \mathscr{C}$ be a ringed D-space structure on X. If $\varphi : \mathscr{C} \to Z(\mathscr{A})$ is a map of sheaves of k-algebras such that $\varepsilon \circ \varphi = \mathrm{id}_{\mathscr{C}}$, then $\mathscr{A}$ is a locally free sheaf of $\mathscr{C}$-modules. In this case, the structure group of $\mathscr{A}$ is contained in* $\mathrm{Aut}(D, \varepsilon_0)$.

The proof of [**BR84**, Proposition 1] relies on the following useful lemma.

LEMMA 3.7. ([**BR84**, Lemma 1]) *Let $\theta : D \to D$ be a unipotent endomorphism of D. Then $\log \theta := \sum_{p=1}^{\infty} (-1)^{p-1} (\theta - \mathrm{id})^p / p$ is a nilpotent derivation of D. If δ is a nilpotent derivation of D, then $\exp \delta := \mathrm{id} + \sum_{p=1}^{\infty} \delta^p / p!$ is a unipotent automorphism of D. The operations* log *and* exp *are inverses of each other. Furthermore,* log *is a bijection between the set of unipotent automorphisms of D and the set of nilpotent derivations of D.*

PROOF OF PROPOSITION 3.6. This proof is identical to the proof of [**BR84**, Proposition 1]. We record here since it is instructive.

Let us define an endomorphism $\theta_\alpha : \mathscr{C}|_{U_\alpha} \otimes D \to \mathscr{C}|_{U_\alpha} \otimes D$ by setting locally $\theta_\alpha(c \otimes d) := \tau_\alpha(\varphi(c)) 1 \otimes d$, where τ_α is as in (3.1). Then we have

$$\theta_\alpha(c \otimes d) - c \otimes d = (\tau_\alpha(\varphi(c)) - c \otimes 1) \cdot (1 \otimes d).$$

Notice that

$$\mathrm{id} \otimes \varepsilon_0(\tau_\alpha(\varphi(c)) - c \otimes 1) = (\varepsilon \circ \varphi)(c) - c = 0.$$

In other words, $\tau_\alpha(\varphi(c)) - c \otimes 1$ is an element of $\mathscr{C}|_{U_\alpha} \otimes J$. It follows that $\theta_\alpha - \mathrm{id}$ maps $\mathscr{C}|_{U_\alpha} \otimes J^p$ into $\mathscr{C}|_{U_\alpha} \otimes J^{p+1}$. Hence, we have $(\theta_\alpha - \mathrm{id})^{n+1} = 0$ for some

$n \in \mathbb{Z}_+$. This means that θ_α is a unipotent endomorphism of sheaves of algebras. Then by Lemma 3.7, θ_α is of the form $\exp \delta_\alpha$ for some nilpotent endomorphism δ_α. Hence, θ_α is a sheaf automorphism. Let $\sigma_\alpha : \mathscr{A}|_{U_\alpha} \to C|_{U_\alpha} \otimes D$ denote the map defined by

$$\sigma_\alpha := \theta_\alpha^{-1} \circ \tau_\alpha.$$

It is easy to see that σ_α is an isomorphism of sheaves of $\mathscr{C}$-modules. Thus, $\mathscr{A}$ is locally free with transition functions $\sigma_\alpha \circ \sigma_\beta^{-1}$. These transition functions have values in $\mathrm{Aut}(D)$. The following calculation shows that θ_α^{-1}, hence, θ_α preserves $\mathrm{id} \otimes \varepsilon_0$:

$$\begin{aligned} \mathrm{id} \otimes \varepsilon_0(\theta_\alpha(c \otimes d)) &= \mathrm{id} \otimes \varepsilon_0(\tau_\alpha(\varphi(c))1 \otimes d) \\ &= (\varepsilon \circ \varphi)(c)\varepsilon_0(d) \\ &= c\varepsilon_0(d) \\ &= \mathrm{id} \otimes \varepsilon_0(c \otimes d). \end{aligned}$$

It follows that $\sigma_\alpha \circ \sigma_\beta^{-1}$ has in fact values in $\mathrm{Aut}(D, \varepsilon_0)$. This finishes the proof of the proposition. □

We see from Proposition 3.6 that, to show that a ringed D-space $\mathscr{A}$ is a vector bundle, it suffices to show that the map $\varepsilon : Z(\mathscr{A}) \to \mathscr{C}$ is split as a surjective map of sheaves of $\mathscr{C}$-algebras. Indeed, this is the main strategy of the article of Blattner and Rawnsley, [**BR84**] for proving the Batchelor's theorem in a general setup. We will adapt their strategy into our context.

Recall our notation that J is the kernel of a k-algebra homomorphism, $\varepsilon_0 : D \to k$. We fix some additional notation. First, we introduce the notation corresponding to the ideal J in the center of D, that is,

$$J_0 := J \cap Z(D).$$

If $\varepsilon : \mathscr{A} \to \mathscr{C}$ is a ringed D-space structure on X, then we denote by $\mathscr{I}$ the ideal sheaf

$$\mathscr{I} := \ker \varepsilon.$$

Then we have an ideal sheaf defined by

$$\mathscr{I}_0 := \mathscr{I} \cap Z(\mathscr{A}).$$

Finally, for $p \in \mathbb{Z}_+$, we denote by $\mathrm{Der}(\mathscr{C}, \mathscr{I}_0^p/\mathscr{I}_0^{p+1}))$ the sheafification of the abelian presheaf of "derivations of $\mathscr{C}$ into $\mathscr{I}_0^p/\mathscr{I}_0^{p+1}$." Here, by a *derivation of $\mathscr{C}$ into $\mathscr{I}_0^p/\mathscr{I}_0^{p+1}$*, we mean a map $\mathscr{C} \to \mathscr{I}_0^p/\mathscr{I}_0^{p+1}$ of sheaves of $\mathscr{C}$-algebras that has, locally, the well-known Leibniz property. Expressed using this notation, the principal theorem from [**BR84**] takes the form of the following assertion for the ringed D-spaces associated with a generalized topological space.

THEOREM 3.8. *We maintain our previous data and notation. In addition, we assume that J is a nilpotent ideal. If $\mathscr{A}$ is a ringed D-space structure on X such that for every $p \in \mathbb{Z}_+$ the cohomology group*

$$H^1(X, \mathrm{Der}(\mathscr{C}, \mathscr{I}_0^p/\mathscr{I}_0^{p+1}))$$

is trivial, then there is a surjective map $\varphi : \mathscr{C} \to Z(\mathscr{A})$ of sheaves of algebras on X such that $\varepsilon \circ \varphi = \mathrm{id}_{\mathscr{C}}$. In this case, as a sheaf of $\mathscr{C}$-modules, $\mathscr{A}$ is a vector bundle over X of type $\mathscr{C}$ with structure group in $\mathrm{Aut}(D, \varepsilon_0)$.

The proof of Theorem 3.8 develops in the same way as the proof of the corresponding theorem of [**BR84**] does. The only difference is that we consider generalized topologies instead of topological spaces. To avoid an excessive amount of repetition, for the details of the proof, we refer the reader to the article [**BR84**].

4. Proof of Theorem 1.1

In this section we will provide a proof of our first result, Theorem 1.1.

Racall that a $\mathbb{Z}_2$-graded commutative R-algebra is a $\mathbb{Z}_2$-graded R-algebra A such that for every $a \in A_m$ and $b \in A_n$, where $\{m,n\} \subseteq \{\overline{0},\overline{1}\}$, the following anti-commutation relation holds:

$$ab = (-1)^{m+n} ba. \tag{4.1}$$

For our purposes, the most important example is the full exterior algebra $\bigwedge R^s$. Let us denote it by D. Then we have

$$D := R^s \oplus \bigwedge^1 R^s \oplus \bigwedge^2 R^s \oplus \cdots \oplus \bigwedge^s R^s.$$

The $\mathbb{Z}_2$-grading on D is given by the decomposition $D = D_{\overline{0}} \oplus D_{\overline{1}}$, where

$$D_{\overline{0}} := (\bigwedge R^s)_{\overline{0}} = \bigoplus_{i\ :\ \text{even}} \bigwedge R^i \qquad \text{and} \qquad D_{\overline{1}} := (\bigwedge R^s)_{\overline{1}} = \bigoplus_{i\ :\ \text{odd}} \bigwedge R^i.$$

It is easy to check that, under this $\mathbb{Z}_2$-grading, the anti-commutation rule (4.1) holds for D. In other words, D is a $\mathbb{Z}_2$-graded commutative R-algebra. At the same time, it is also easy to check from (4.1) that the center of D is given by $D_{\overline{0}}$.

Let $e_1, \ldots, e_s$ denote the standard basis vectors for R^s. Then as an R-algebra, D is generated by $1, e_1, \ldots, e_s$. In this notation, there is a unique R-algebra homomorphism $\varepsilon_0 : D \to R$ such that $\varepsilon_0(e_i) = 0$ and $\varepsilon_0(1) = 1$. Let J denote the kernel, $\ker \varepsilon_0$. Evidently, J is a nilpotent ideal of D.

We proceed to show that the hypothesis of Theorem 3.8 is satisfied for these choices.

LEMMA 4.1. *Let $\varepsilon : \mathscr{A} \to \mathscr{O}_X$ be a ringed D-space structure on a regular, paracompact locally semialgebraic space X. Let $\mathscr{I}$ denote the ideal sheaf defined by*

$$\mathscr{I} := \ker \varepsilon.$$

Let $\mathscr{I}_0$ denote the corresponding ideal sheaf in $Z(\mathscr{A})$,

$$\mathscr{I}_0 := \mathscr{I} \cap Z(\mathscr{A}).$$

Then for every $p \in \mathbb{Z}_+$, the first cohomology of $\mathrm{Der}(\mathscr{O}_X, \mathscr{I}_0^p/\mathscr{I}_0^{p+1})$, *that is,*

$$H^1(X, \mathrm{Der}(\mathscr{O}_X, \mathscr{I}_0^p/\mathscr{I}_0^{p+1}))$$

is trivial.

PROOF. We already know from Theorem 2.23 that the Tietze's extension theorem holds for our locally semialgebraic spaces. This theorem readily implies that the sheaf $\mathscr{O}_X$ is a soft sheaf. Note that, since $\mathscr{O}_X$ is a sheaf of R-algebras, it has a natural structure of a sheaf of $\mathbb{Z}$-modules, where we view the elements of $\mathbb{Z}$ as (locally) constant functions on X. Since the sheaf $\mathrm{Der}(\mathscr{O}_X, \mathscr{I}_0^p/\mathscr{I}_0^{p+1})$ is a sheaf of $\mathscr{O}_X$-modules, we see by Proposition 2.20 that $\mathrm{Der}(\mathscr{O}_X, \mathscr{I}_0^p/\mathscr{I}_0^{p+1})$ is a soft sheaf on X. In particular, it is an acyclic sheaf. This finishes the proof of our assertion. □

We are now ready to finish the proof of our Theorem 1.1. Let us recall its statement for convenience:

Let X be a regular, paracompact locally semialgebraic space. If $(X, \mathscr{A})$ is a $\mathbb{Z}_2$-graded semialgebraic space of odd-dimension s on X, then there exists a locally semialgebraic vector bundle E on X such that $\mathscr{A}(X)$ is isomorphic, as a $\mathbb{Z}_2$-graded commutative algebra, to $\Gamma(X, \bigwedge E)$.

Proof of Theorem 1.1. Recall that D is the $\mathbb{Z}_2$-graded commutative algebra $\bigwedge R^s$. Let us denote by $\mathrm{Aut}_0(D, \varepsilon_0)$ the subgroup of $\mathrm{Aut}(D, \varepsilon_0)$ consisting of its $\mathbb{Z}_2$-grading preserving elements. Since $\mathrm{Aut}_0(D, \varepsilon_0)$ is an algebraic subgroup of $\mathrm{Aut}(D, \varepsilon_0)$, which is defined over R, it is also an affine semialgebraic subgroup. Now, following the general idea of the proof of the Batchelor's theorem in [**BR84**, Remark 2], we use the cohomology vanishing result. To this end, we apply Lemma 4.1. Then Theorem 3.8 shows that our sheaf of $\mathbb{Z}_2$-graded commutative algebras $\mathscr{A}$ on X can be viewed as the sheaf of sections of a vector bundle over X of type $\mathscr{O}_X$ with structure group in $\mathrm{Aut}_0(D, \varepsilon_0)$. To finish off the proof, we still need to show that the structure group of $\mathscr{A}$ can be reduced to the $\mathbb{Z}$-grading preserving subgroup of $\mathrm{Aut}(D, \varepsilon_0)$, which is $\mathrm{GL}_s(R)$. Of course, this is also an affine semialgebraic subgroup. Also, the reasoning presented in the second half of [**BR84**, Remark 2] applies verbatim to the semialgebraic groups $\mathrm{Aut}_0(D, \varepsilon_0), \mathrm{Aut}(D, \varepsilon_0)$, and $\mathrm{GL}_s(R)$. Indeed, the quotients $\mathrm{Aut}(D, \varepsilon_0)/\mathrm{GL}_s(R)$ and $\mathrm{Aut}_0(D, \varepsilon_0)/\mathrm{GL}_s(R)$ are affine spaces. Therefore, the former quotient can be be contracted by a semialgebraic homotopy to the latter quotient. Thus, we see that the structure group of $\mathscr{A}$ reduces to $\mathrm{GL}_s(R)$ in the semialgebraic category as well. This finishes the proof of our first main theorem. □

5. Nash Manifolds and Their $\mathfrak{N}$-topologies

In this section, we prove our second main result, Theorem 1.2. Towards this end, we will review, in more detail, the theory of affine Nash manifolds. Central to our development in this section is a class of functions on the affine Nash manifolds that behaves like the class of polynomial functions on an affine algebraic variety. We begin with a discussion of the relevant functions for semialgebraic spaces.

Recall that previously we defined the morphisms of locally semialgebraic spaces as the morphisms of the underlying locally ringed spaces. Let us make this definition more explicit in the case of affine semialgebraic spaces.

Definition 5.1. Let $n \in \mathbb{Z}_+$. Let M be a semialgebraic subset of $\mathbb{R}^n$. If, in addition, M is a smooth submanifold of $\mathbb{R}^n$, then M is called a *Nash submanifold of* $\mathbb{R}^n$.

Let A (resp. B) be a semialgebraic subset in $\mathbb{R}^m$ (resp. in $\mathbb{R}^n$). A map $f : A \to B$ is called a *semialgebraic map* if the graph of f is a semialgebraic subset in $\mathbb{R}^m \times \mathbb{R}^n$. Let $f : A \to B$ be a semialgebraic map, as defined above. Also, we assume that A and B are open sets. If, in addition, f is a smooth map, then f is called a *Nash map*. A *Nash diffeomorphism* is a bijective Nash map $f : A \to B$ whose inverse $f^{-1} : B \to A$ is a Nash map. Notice that these definitions regarding Nash maps did not use Nash submanifolds of $\mathbb{R}^n$. To clarify this picture, let us reintroduce the Nash submanifolds of $\mathbb{R}^n$ by using Nash maps.

A semialgebraic subset $M \subset \mathbb{R}^n$ is said to be a Nash submanifold of $\mathbb{R}^n$ of dimension d if, for every $x \in M$, there exists a Nash diffeomorphism φ from an open semialgebraic neighborhood Ω of the origin $0 \in \mathbb{R}^n$ onto an open semialgebraic neighborhood Ω' of x in $\mathbb{R}^n$ such that $\varphi(0) = x$ and

$$\varphi(\mathbb{R}^d \times \{0\} \cap \Omega) = M \cap \Omega'.$$

Clearly, every open semialgebraic subset X of a Nash submanifold of $\mathbb{R}^n$ of dimension d is a Nash submanifold of $\mathbb{R}^n$ of dimension d.

REMARK 5.2. We would like to highlight a characteristic of affine Nash manifolds that resembles a property found in every affine algebraic variety defined over an algebraically closed field. It is well-known that an affine algebraic variety $X \subset \mathbb{C}^n$ need not be closed in $\mathbb{C}^n$. For example, the set $X := \mathbb{C}\backslash\{0\}$ is a Zariski open subset of $\mathbb{C}$. But X is isomorphic to the "hyperbola" $V := \{(a,b) \in \mathbb{C}^2 : ab = 1\}$ in $\mathbb{C}^2$ via the natural morphism

$$\begin{aligned} X &\longrightarrow V, \\ x &\longmapsto (x, 1/x). \end{aligned}$$

In other words, although X is not a closed subset of $\mathbb{C}$, it is isomorphic to a Zariski closed subvariety in a higher dimensional affine space. Likewise, a Nash submanifold of $\mathbb{R}^n$ $(n \in \mathbb{Z}_+)$ need not be closed in $\mathbb{R}^n$. Nonetheless, every Nash submanifold of $\mathbb{R}^n$ is Nash diffeomorphic to a closed Nash submanifold of some $\mathbb{R}^m$ $(m \in \mathbb{Z}_+)$. A proof of this fact is given in [**Shi87**, Corollary I.4.3]. We note also that, since every open semialgebraic subset X of a Nash submanifold $M \subseteq \mathbb{R}^n$ has the structure of a Nash submanifold of $\mathbb{R}^n$, X is isomorphic to a closed Nash submanifold of some $\mathbb{R}^m$ $(m \in \mathbb{Z}_+)$.

Let M be a Nash submanifold of $\mathbb{R}^n$ for some n. A *Nash function* on an open semialgebraic subset $U \subseteq M$ is a smooth semialgebraic function. In fact, by a theorem of Malgrange [**Mal67**], every such map is a real analytic map. Thus, our definition of a Nash function here agrees with the one that is given in the introduction section. The Nash functions defined on an open semialgebraic subset $U \subseteq M$, denoted by $\mathscr{N}(U)$, form a subring of the ring of all functions defined on U. In fact, $\mathscr{N}(U)$ is a noetherian ring. This fact was originally proved independently by Efroymson in [**Efr74**] and by Riseler in [**Ris73**] for globally defined Nash functions on Nash submanifolds of $\mathbb{R}^n$. It follows from our Remark 5.2 that it holds for every open semialgebraic subset of a Nash submanifold M of $\mathbb{R}^n$. Also, it is not difficult to check that the assignment $U \mapsto \mathscr{N}(U)$, $U \in \mathring{\mathfrak{S}}(M)$ defines a sheaf on M, denoted by $\mathscr{N}_M$. We call it the *sheaf of Nash functions on* M.

DEFINITION 5.3. Let M be a Nash submanifold of $\mathbb{R}^n$. A *Nash subset of* M is a semialgebraic subset of the form

$$\mathcal{Z}_M(f_1, \ldots, f_r) := \{x \in M | f_1(x) = \cdots = f_r(x) = 0\},$$

where $f_1, \ldots, f_r$ are Nash functions from M to $\mathbb{R}$.

We continue with the assumption that M is a Nash submanifold of $\mathbb{R}^n$ for some n. Since $\mathscr{N}(M)$ is a noetherian ring, every Nash subset of M is defined by an ideal of $\mathscr{N}(M)$. Recall from the introduction section that the $\mathfrak{N}$-topology on M is the restricted topology whose closed subsets are the Nash subsets of M. Hence, we know that every $\mathfrak{N}$-closed subset of M is a zero-set of the form

$$\mathcal{Z}_M(I) := \{x \in M | f(x) = 0 \text{ for every } f \in I\},$$

where I is an ideal of $\mathscr{N}(M)$.

Since the intersection of a family of closed sets in a restricted topology may not be a closed set in that restricted topology, the closure operation may not exist. Fortunately, in the restricted topology of semialgebraic subsets, every semialgebraic subset has a semialgebraic closure. A proof of this fact can be found in [**AG08**, Corollary 2.2.7]. Of course, in the case of the $\mathfrak{N}$-topology, we are not concerned about this remark since $\mathfrak{N}$-topology is indeed a topology.

NOTATION 5.4. The set of all open (resp. closed) subsets of the $\mathfrak{N}$-topology on M will be denoted by $\mathring{\mathfrak{N}}(M)$ (resp. by $\bar{\mathfrak{N}}(M)$). By abuse of notation, we denote the sheaf obtained by restricting the sheaf of Nash functions $\mathscr{N}_M$ to the $\mathring{\mathfrak{N}}(M)$-topology by $\mathscr{N}_M$. Likewise, we denote the sheaf obtained by restricting the sheaf of continuous semialgebraic functions $\mathscr{O}_M$ to the $\mathring{\mathfrak{N}}(M)$-topology by $\mathscr{O}_M$ as well.

Since every Nash function is a continuous semialgebraic function, every $\mathfrak{N}$-closed subset of M is an $\mathfrak{S}$-closed subset as well. However, as we demonstrate it in the following example, the containment $\bar{\mathfrak{N}}(M) \subset \bar{\mathfrak{S}}(M)$ can be a strict inclusion.

EXAMPLE 5.5. Let M denote $\mathbb{R}^1$ with its unique affine Nash manifold structure. Let U denote the interval $(0, \infty)$ in $\mathbb{R}^1$. The semialgebraic closure of U is given by the half-closed interval,

$$[0, \infty) := \{x \in \mathbb{R}^1 |\; x \geqslant 0\}.$$

We claim that $[0, \infty)$ cannot be an $\mathfrak{N}$-closed subset of M. Let us assume towards a contradiction that $[0, \infty) \in \bar{\mathfrak{N}}(M)$ holds. Then we find finitely many Nash functions $f_1, \dots, f_r$ on M such that $[0, \infty) = \bigcap_{i=1}^r f_i^{-1}(0)$. In particular, since $[0, \infty)$ is an unbounded closed subset of M, there is a nonempty open subinterval $I \in \mathring{\mathfrak{N}}(M)$ such that $I \subset f_i^{-1}(0)$ for each $i \in \{1, \dots, r\}$. But since M is semialebraically connected, we know the following implication from [**BCR98**, Proposition 8.1.9]:

$$f_i|_I = 0 \implies f_i = 0 \text{ on } M \text{ for every } i \in \{1, \dots, r\}.$$

This means that $\bigcap_{i=1}^r f_i^{-1}(0) = M$, which is absurd. Therefore, the $\mathfrak{N}$-topology and the $\mathfrak{S}$-topology on M are different topologies.

In [**AG08**, Definition 3.2.4], an $\mathbb{R}$-*space* is defined as a ringed space $(M, \mathscr{O}_M)$, where M is a restricted topological space, and $\mathscr{O}_M$ is a sheaf of $\mathbb{R}$-algebras over M which is a subsheaf of the sheaf of all continuous real-valued functions on M. In [**AG08**, Example 3.1.1], a ringed space of the form $(M, \mathscr{N}_M)$, where M is a Nash submanifold of $\mathbb{R}^n$ ($n \in \mathbb{Z}_+$) and $\mathscr{N}_M$ is the sheaf of Nash functions defined on the restricted topology of M is called the $\mathbb{R}$-*space* of the Nash submanifold M of $\mathbb{R}^n$. Then an *affine Nash manifold* is defined as an $\mathbb{R}$-space which is isomorphic to the $\mathbb{R}$-space of a Nash submanifold of $\mathbb{R}^n$ for some $n \in \mathbb{Z}_+$.

REMARK 5.6. In the preliminaries section we introduced the affine semialgebraic groups as the group objects in the category of affine semialgebraic spaces. An *affine Nash group* can be defined similarly, as a group object in the category of affine Nash manifolds.

DEFINITION 5.7. Let M be a Nash submanifold of $\mathbb{R}^n$. We call $(M, \mathscr{N}_M)$ the $\mathbb{R}$-*space of* M *in the* $\mathfrak{N}$-*topology*. An *affine Nash manifold* is a ringed space that is isomorphic to the $\mathbb{R}$-*space in the* $\mathfrak{N}$-*topology* of a Nash submanifold of $\mathbb{R}^n$ for some $n \in \mathbb{Z}_+$.

EXAMPLE 5.8. Every smooth real algebraic variety is an affine Nash manifold. The curve C in $\mathbb{R}^2$ defined by the polynomial equation $y^2 = x^3$ is an affine semialgebraic space but not an affine Nash manifold since it has a singularity at the origin.

We proceed with a review of some useful facts about the semialgebraic coverings of M.

Let F be a continuous semialgebraic function on M. Following [**AG08**, Notation A.3.1] we denote by M_F the set $\{x \in M : F(x) \neq 0\}$. If the restriction of F to M_F is a positive Nash function, then F is said to be a *basic semialgebraic function.* Let $\{F_1, \dots, F_m\}$ be a set of basic semialgebraic functions on M such that, for every $i \in \{1, \dots, m\}$, F_i is continuous (w.r.t. the $\mathfrak{S}(M)$-topology). We call $\{F_1, \dots, F_m\}$ a *basic collection* if for every $x \in M$ at least one of the functions $F_1, \dots, F_m$ is greater than 1 at x. An open cover $M = \bigcup_{j=1}^m V_j$ is said to be a *refinement* of another open cover $M = \bigcup_{i=1}^n U_i$ if for any $j \in \{1, \dots, m\}$ there exists $i \in \{1, \dots, n\}$ such that $V_j \subset U_i$. We now state a result of Aizenbud and Gourevitch as a lemma so that we can refer to it easily.

LEMMA 5.9. *([**AG08**, Proposition A.3.4]) If $M = \bigcup_{i=1}^n U_i$ is a finite open semialgebraic cover of a Nash manifold M, then there exists a basic collection $\{F_1, \dots, F_m\}$ such that $M = \bigcup_{i=j}^m M_{F_j}$ is a refinement of $M = \bigcup_{i=1}^n U_i$.*

We will apply this lemma to the stalks of the sheaf $\mathcal{N}_M$ with respect to the $\mathfrak{N}$-topology. The ring of sections of $\mathcal{N}_M$ on a closed subset $V \subset M$ is given by the ring of global sections of the restricted sheaf, $i^{-1}\mathcal{N}_M$, where $i : V \hookrightarrow M$ is the closed embedding of V into M. Equivalently, it is defined by the direct limit,

$$\mathcal{N}_M(V) := \varinjlim_{V \subset U \in \mathring{\mathfrak{N}}(M)} \mathcal{N}_M(U). \tag{5.1}$$

This means that, for $s \in \mathcal{N}_M(V)$, there exists a pair $(U, \tilde{s})$, where U is an $\mathfrak{N}$-open set such that $V \subset U$, and $\tilde{s}$ is an element of $\mathcal{N}_M(U)$ such that $s(x) = \tilde{s}(x)$ for every $x \in V$. Two such pairs $(U, \tilde{s})$ and $(U', \tilde{s}')$ represent the same section $s \in \mathcal{N}_M(V)$ if and only if there exists a third $\mathfrak{N}$-open set U'' such that $V \subset U'' \subseteq U \cap U'$ and $\tilde{s}(x) = \tilde{s}'(x)$ for every $x \in U''$. By specializing (5.1) to the closed sets of the form, $V = \{x\}$, where x is a point in M, we recover the definition of the *stalk of $\mathcal{N}_M$ at* x, which we denote by $\mathcal{N}_{M,x}$.

PROPOSITION 5.10. *Let M be an affine Nash manifold. Then the pair $(M, \mathcal{N}_M)$ with respect to the $\mathfrak{N}$-topology of M is a locally ringed space over $\mathbb{R}$.*

PROOF. Since any open cover in the $\mathfrak{N}$-topology is an open cover in the $\mathfrak{S}$-topology on M, all sheaf axioms hold automatically for $\mathcal{N}_M$ with respect to the $\mathfrak{N}$-topology. It follows that $(M, \mathcal{N}_M)$ is indeed a ringed space over $\mathbb{R}$. We proceed to show that the stalk $\mathcal{N}_{M,x}$ is a local ring for every $x \in M$. Let U be an open semialgebraic neighborhood of x in M. We view U as an open subset of a semialgebraic cover of M. Then Lemma 5.9 shows that x is contained in an open subset $M_F \subseteq U$, where F is a basic semialgebraic function on M. This argument shows that every open neighborhood of x can be shrinked to an element of $\mathring{\mathfrak{N}}(M)$. Hence, we have the following identification of direct limits:

$$\varinjlim_{V \subset U \in \mathring{\mathfrak{S}}(M)} \mathcal{N}_M(U) = \varinjlim_{V \subset U \in \mathring{\mathfrak{N}}(M)} \mathcal{N}_M(U).$$

But it is well-known that [**BCR98**, Corollary 8.1.6] the stalk $\varinjlim_{V\subset U\in\mathring{\mathfrak{S}}(M)} \mathscr{N}_M(U)$ is a local ring. This finishes the proof of our assertion. $\square$

We will now review a significant theorem that is very useful for the cohomological analysis of Nash manifolds. This theorem is known as "Efroymson's extension theorem."

LEMMA 5.11. *([**Efr82, Pec85**])(Efroymson's extension theorem) Let M be an affine Nash manifold. Let $\varphi : M \to \mathbb{R}$ be a Nash function on M. Let g be a Nash function that is defined on a semialgebraic neighborhood U of $\varphi^{-1}(0)$. Then there exists a Nash function $h : M \to \mathbb{R}$ such that $g - h|_U$ is given by the product of $\varphi|_U$ and a Nash function on U. In partigular, g and h coincide on the closed Nash set $\varphi^{-1}(0)$.*

We will use Efroymson's extension theorem in the proof of Theorem1.3. We recall its statement for convenience.

If M is an affine Nash manifold, then the sheaf $\mathscr{N}_M$ is a soft sheaf with respect to the $\mathfrak{N}$-topology on M.

PROOF OF THEOREM 1.3. Let V be an $\mathfrak{N}$-closed subset of M. Then there are finitely many Nash functions $f_1, \dots, f_m$ on M such that $V = \{x \in M|\ f_1(x) = \cdots = f_m(x) = 0\}$. We define a Nash function $\varphi : M \to \mathbb{R}$ by the formula, $\varphi := f_1^2 + \cdots + f_m^2$. Clearly, for $x \in M$, we have

$$\varphi(x) = 0 \iff f_i(x) = 0 \text{ for all } i \in \{1, \dots, m\}.$$

In other words, V is the zero set of φ, that is, $V = \varphi^{-1}(0)$.

We are now ready to show that the following natural map

$$\begin{aligned} \mathrm{res}_V : \Gamma(M, \mathscr{N}_M) &\longrightarrow \Gamma(V, \mathscr{N}_M) \\ h &\longmapsto h|_V \end{aligned}$$

is surjective. To this end, let s be a section of $\Gamma(V, \mathscr{N}_M)$. Then there exists an $\mathfrak{N}$-open neighborhood $V \subset U$ and a section $\tilde{s} \in \mathscr{N}_M(U)$ such that $\tilde{s}(x) = s(x)$ for every $x \in V$. By Lemma 5.11, we know that there exists a Nash function $h : M \to \mathbb{R}$ such that $h - \tilde{s}$ is a product of $\varphi|_U$ and a Nash function on U. It follows that $h|_V = s$ on V. In other words, we have $\mathrm{res}_V(h) = s$. Hence, we showed that the map res_V is surjective. This finishes the proof of our theorem. $\square$

Let X be a topological space having a paracompactifying family of supports. As we mentioned it in the introductory section, every flasque (flabby) sheaf on X is a soft sheaf. Unfortunately, our sheaf of Nash functions $\mathscr{N}_M$ with respect to the $\mathfrak{N}$-topology is not flabby unless M is finite.

EXAMPLE 5.12. Let M denote $\mathbb{R}^1$ with its standard Nash manifold structure. Let U denote the $\mathfrak{N}$-open subset, $U := M\backslash\{0\}$. It is easy to check that the rational map $f : U \to \mathbb{R}$, $x \mapsto 1/x$ gives a section of $\mathscr{N}_M$ on U. However, no globally defined Nash function $s : M \to \mathbb{R}$ has the property that $s|_U(x) = f(x)$ for every $x \in U$. In other words, the restriction map $\mathrm{res}_V : \Gamma(M, \mathscr{N}_M) \to \Gamma(U, \mathscr{N}_M)$ is not surjective.

However, by utilizing standard techniques from algebraic geometry, we are able to establish the anticipated cohomological properties of our sheaves. To this end, we attach an affine scheme to every affine Nash manifold.

DEFINITION 5.13. Let M be an affine Nash manifold. Let A denote the ring of globally defined Nash functions on M, that is, $A := \mathcal{N}_M(M)$. Then we define the *Nash scheme of* M, denoted $\mathrm{Nsp}(M)$, as the affine scheme $\operatorname{Spec} A$.

Since the $\mathbb{R}$-algebra of global Nash functions on an affine Nash manifold M is a noetherian ring, the associated Nash scheme is a scheme of finite type over $\mathbb{R}$. In particular, the set of closed points is dense in $\mathrm{Nsp}(M)$ (see [**Har77**, Ch II, Exercise 3.14]). We will show that the closed points of $\mathrm{Nsp}(M)$ are in bijection with the points of M.

LEMMA 5.14. *Let M be an affine Nash manifold. Let A denote the ring $\mathcal{N}_M(M)$. If x is a point of M and I is the ideal defined by $I := \{f \in A : f(x) = 0\}$, then I is a maximal ideal. Conversely, if I is a maximal ideal of A, and V is the vanishing set defined by $V := \{x \in M : f(x) = 0$ for every $f \in I\}$, then $V = \{x\}$ for some $x \in M$.*

PROOF. Let B denote the stalk of $\mathcal{N}_M$ at x. Then we have a surjective ring homomorphism $\varphi : A \to B$ defined by sending a global Nash function s to its germ at x. We know from Proposition 5.10 that B is a local ring. It is easy to see that the unique maximal ideal $\tilde{I}$ of B consists of pairs (U, s), where U is an open neighborhood (w.r.t. $\mathfrak{N}$-topology) of x and $s \in \mathcal{N}_M(U)$ is such that $s(x) = 0$. Since φ is surjective, $\varphi^{-1}(\tilde{I})$ is a maximal ideal of A. But $\varphi^{-1}(\tilde{I})$ consists of $f \in A$ such that $f(x) = 0$. In other words, we have $I = \varphi^{-1}(\tilde{I})$. Hence, the proof of our first assertion follows.

Our second assertion is stated in [**BCR98**, Corollary 8.6.3]. Hence, the proof of our lemma is finished. □

Lemma 5.14 shows that the Nash scheme of M is packed with much information about the geometry of M.

LEMMA 5.15. *Let M be an affine Nash manifold. Let $A := \mathcal{N}_M(M)$. Let $\beta : M \to \mathrm{Nsp}(M)$ be the map defined by $\beta(x) := I_x$, where I_x is the maximal ideal of A corresponding to $x \in M$. Then β is a homeomorphism onto the subspace of closed points of* $\mathrm{Nsp}(M)$.

PROOF. Lemma 5.14 shows that β is a bijection of M onto the set of closed points of $\mathrm{Nsp}(M)$. If V is a closed subset of $\mathrm{Nsp}(M)$, then there exists a prime ideal $P \subset A$ such that $V = \{Q : Q$ is a prime ideal of A containing $P\}$. Then we have

$$\begin{aligned}\beta^{-1}(V) &= \{x \in M : P \subseteq I_x\} \\ &= \{x \in M : f(x) = 0 \text{ for every } f \in P\}.\end{aligned}$$

Since A is noetherian, P is finitely generated. Hence, we can write $\beta^{-1}(V)$ as the vanishing set of a single Nash function by using the argument in the first paragraph of the proof of Theorem 1.3. This means that $\beta^{-1}(V)$ is a closed subset of M. Hence, β is a continuous map. Conversely, if W is a Nash closed subset of M, then we can write it as the zero set of a single Nash function $f : M \to \mathbb{R}$. In particular, the image of W under β consists of all maximal ideals Q of A such that $f \in Q$. Since the set of all prime ideals of A that contains f is a closed subset, the intersection of this closed subset with $\beta(M)$ is precisely the subset $t(W)$. Hence, $t(W)$ is a closed subset of the subspace of closed points of $\mathrm{Nsp}(M)$. This finishes the proof of the fact that β is a bijective map such that both β and its inverse are continuous. Therefore, β is a homeomorphism. $\square$

As a consequence of the proof of Lemma 5.15, we note that the structure sheaf of the Nash scheme of an affine Nash manifold M restricts to give the structure sheaf of M. We will make this more precise in our next theorem. First, we recall the definition of a quasi-coherent sheaf of $\mathcal{N}_M$-modules.

Let V be an $\mathfrak{N}$-closed subset of an affine Nash manifold M. Let A denote the ring of Nash functions, $A := \mathcal{N}_M(M)$. As in the first paragraph of the proof of Theorem 1.3, we write V as the vanishing set of a single Nash function, $V = \{x \in M : f(x) = 0\}$ for some $f \in A$. Then we denote the complement of V in M by $E(f)$. By using Lemma 5.15, we can identify $E(f)$ with the intersection of the space of maximal ideals of A with a basic Zariski-open subset,

$$\beta(E(f)) = \beta(M) \cap D(f), \qquad \text{where} \qquad D(f) := \{y \in \mathrm{Nsp}(M) : f(y) \neq 0\}.$$

In particular, $\{E(f) : f \in A\}$ gives a basis for the $\mathfrak{N}$-topology on M. Hence, every sheaf on M is characterized by its values on the sets $E(f)$, $f \in A$.

DEFINITION 5.16. A sheaf of $\mathcal{N}_M$-modules $\mathcal{G}$ is called a quasi-coherent sheaf of $\mathcal{N}_M$-modules if there exists an A-module K such that for every $f \in A$ we have

$$\mathcal{G}(E(f)) = A_f := \{g/f^m : g \in A \text{ and } m \in \mathbb{N}\}.$$

THEOREM 5.17. *We maintain the notation from the above paragraphs. In particular, $\beta : M \to \mathrm{Nsp}(M)$ denotes the homeomorphism that is defined in Lemma 5.15. Then the direct image functor β_* gives an equivalence between the category of quasi-coherent $\mathcal{N}_M$-modules and the category of quasi-coherent* $\mathrm{Nsp}(M)$*-modules.*

PROOF. Since $\beta(M)$ is precisely the (dense) subset of closed points of $\mathrm{Nsp}(M)$, the Zariski open subsets of $\mathrm{Nsp}(M)$ are uniquely determined by the open Nash subsets of M. For every pair of open subsets $U \subset V$ in $\mathrm{Nsp}(M)$, we have a commuting diagram of maps as in Figure 2, where the vertical maps are restrictions, and the vertical maps are given by β. Hence, β defines a map of locally ringed

$$\begin{array}{ccc}
\mathcal{N}_M(\beta^{-1}(V)) & \longrightarrow & \mathcal{O}_{\mathrm{Nsp}(M)}(V) \\
\downarrow & & \downarrow \\
\mathcal{N}_M(\beta^{-1}(U)) & \longrightarrow & \mathcal{O}_{\mathrm{Nsp}(M)}(U)
\end{array}$$

FIGURE 2

spaces. Let $x \in M$. The stalk of $\mathcal{N}_M$ at x and the stalk of $\mathcal{O}_{\mathrm{Nsp}(M)}$ at $\beta(x)$ are the same. Indeed, the stalk $\mathcal{N}_{M,x}$ is a local ring and the ideal I_x consisting of Nash functions vanishing at x is a maximal ideal by Lemma 5.14. Therefore, the stalk $\mathcal{N}_{M,x}$ is the localization of $A := \mathcal{N}_M(M)$ at I_x. But this is precisely the stalk of the structure sheaf $\mathcal{O}_{\mathrm{Nsp}(M)}$ of the affine scheme $\mathrm{Spec}(A)$ at the point I_x. After this observation, the fact that $(\beta, \beta^\#) : (M, \mathcal{N}_M) \to (\beta(M), \mathcal{O}_{\mathrm{Nsp}(M)}|_{\beta(M)})$ are isomorphic as locally ringed spaces is clear. Thus, β_* defines a functor between the category of sheaves on M and the category of sheaves on $\mathrm{Nsp}(M)$.

Let K be an A-module. Then K defines a quasi-coherent sheaf of $\mathcal{O}_{\mathrm{Nsp}(M)}$-modules $\mathcal{F}$ as well as a quasi-coherent sheaf of $\mathcal{N}_M$-modules $\mathcal{G}$. The stalk of the direct image $\beta_*\mathscr{G}$ at $\beta(x)$, where $x \in M$, is isomorphic to the stalk of the restriction of $\mathscr{F}$ at $\beta(x)$. This is a consequence of the following facts: 1) there is a 1-1 correspondence between the open subsets of $\beta(M)$ and $\mathrm{Nsp}(M)$, 2) the spaces of global sections of $\mathscr{F}$ and $\mathscr{G}$ on the basic open subsets $D(f)$ and $E(f)$, where $f \in A$, are the same A-module, A_f. Hence, we see that $\beta_*\mathscr{G} = \mathscr{F}$. This means that β_* restricts to give a functor from the category of quasi-coherent sheaves of $\mathscr{N}_M$-modules to the category of quasi-coherent sheaves of $\mathscr{O}_{\mathrm{Nsp}(M)}$-modules. By abuse of notation, we will denote this subfunctor by β_*, also. We proceed to show that this restricted functor is an equivalence of categories. In other words, we will show that it is a full, faithful, and essentially surjective functor.

The essentially surjectivity of β_* readily follows from our arguments in the first paragraph of this proof. To prove that β_* is full and faithful, we will show that it maps bijectively the short exact sequences of quasi-coherent $\mathscr{N}_M$-modules to the short exact sequences of quasi-coherent $\mathscr{O}_{\mathrm{Nsp}(M)}$-modules. It follows from the first paragraph of this proof that every exact sequence of sheaves of $\mathscr{N}_M$-modules of the form

$$0 \to \mathscr{F}' \to \mathscr{F} \to \mathscr{F}'' \to 0 \tag{5.2}$$

gives rise to a short exact sequence of $\mathscr{O}_{\mathrm{Nsp}(M)}|_{\beta(M)}$-modules of the form

$$0 \to \beta_*\mathscr{F}'|_{\beta(M)} \to \beta_*\mathscr{F}|_{\beta(M)} \to \beta_*\mathscr{F}''|_{\beta(M)} \to 0. \tag{5.3}$$

Assuming that (5.3) is a short exact sequence of quasi-coherent $\mathscr{O}_{\mathrm{Nsp}(M)}|_{\beta(M)}$-modules, we proceed to show that

$$0 \to \beta_*\mathscr{F}' \to \beta_*\mathscr{F} \to \beta_*\mathscr{F}'' \to 0 \tag{5.4}$$

is a short exact sequence of quasi-coherent $\mathscr{O}_{\mathrm{Nsp}(M)}$-modules. Equivalently, we will show that for every point $P \in \mathrm{Nsp}(M)$ the corresponding sequence of stalks,

$$0 \to (\beta_*\mathscr{F}')_P \to (\beta_*\mathscr{F})_P \to (\beta_*\mathscr{F}'')_P \to 0,$$

is exact. Recall that two modules are isomorphic if and only if their localizations at every maximal ideal are isomorphic. Hence, we have to check that

$$0 \to (\beta_*\mathscr{F}')_I \to (\beta_*\mathscr{F})_I \to (\beta_*\mathscr{F}'')_I \to 0 \tag{5.5}$$

is exact for every maximal ideal $I \subseteq A$. But this was our initial assumption in (5.3). Hence, (5.4) is a short exact sequence of quasi-coherent $\mathscr{O}_{\mathrm{Nsp}(M)}$-modules. Clearly, these implications are reversible. In other words, if we start with a short exact sequence of quasi-coherent sheaves of $\mathscr{O}_{\mathrm{Nsp}_M}$-modules, then its restriction to $\beta(M)$ gives a short exact sequence of quasi-coherent $\beta_*\mathscr{N}_M$-modules. Since $(\beta, \beta^\sharp)$ is an isomorphism, we obtain a short exact sequence of quasi-coherent $\mathscr{N}_M$-modules. Hence, the proof of our theorem is complete. □

We are now ready to finish the proof our Theorem 1.4 from the introduction. We record it here as a corollary of Theorem 5.17.

COROLLARY 5.18. *(Theorem 1.4) Let M be an affine Nash manifold. If $\mathscr{G}$ is a quasi-coherent sheaf of $\mathscr{N}_M$-modules, then we have $H^i(M, \mathscr{F}) = 0$ for every $i > 0$.*

PROOF. Let $\mathfrak{Qcn}(M)$ (resp. $\mathfrak{Qco}(\mathrm{Nsp}(M))$) denote the category of quasi-coherent sheaves of $\mathscr{N}_M$-modules (resp. the category of quasi-coherent sheaves of

$\mathscr{O}_{\mathrm{Nsp}(M)}$-modules). By Theorem 5.17, $\mathscr{G}$ gives a quasi-coherent sheaf of $\mathscr{O}_{\mathrm{Nsp}(M)}$-modules, $\mathscr{F} := \beta_*\mathscr{G}$, where $\beta_* : \mathfrak{Qcn}(M) \to \mathfrak{Qco}(\mathrm{Nsp}(M))$ is the equivalence of categories. We know from [**Har77**, Ch III, Theorem 3.5] that

$$H^i(\mathrm{Nsp}(M), \mathscr{F}) = 0 \quad \text{for every } i > 0.$$

Since $\mathfrak{Qco}(\mathrm{Nsp}(M))$ is equivalent to $\mathfrak{Qcn}(M)$, the vanishing of the higher cohomology of $\mathscr{F}$ implies the vanishing of the higher cohomology of $\mathscr{G}$. This finishes the proof of our assertion. □

Before revisiting the definition of an affine Nash vector bundle, let us briefly discuss Nash groups. In the preliminaries section, we introduced affine semialgebraic groups as the group objects within the category of affine semialgebraic spaces. Similarly, we establish a parallel definition for Nash groups.

DEFINITION 5.19. A *Nash group* is a group which is at the same time a Nash manifold such that all group operations are Nash maps. A *Nash homomorphism* is a group homomorphism between two Nash groups which is at the same time a Nash map. If the underlying Nash manifold of a Nash group G is an affine Nash manifold, then we call G an *affine Nash group.*

In other words, (affine) Nash groups are precisely the group objects in the category of (affine) Nash manifolds. A general introduction to Nash groups can be found in Sun's article [**Sun15**].

Let K be a real algebraic group. Then the underlying real algebraic variety of K is a smooth manifold. Furthermore, the group operations of K are smooth maps. It follows that every real algebraic group is naturally a Nash manifold. Here is an example of an affine Nash group which is not a real algebraic group.

EXAMPLE 5.20. Let H denote the group of positive real numbers under multiplication. Then H is an affine Nash group. We will show that H is not a real algebraic group. To this end, we consider the $\mathbb{R}$-algebra of regular functions on H, denoted by $\mathbb{R}[H]$. It is clear that $\mathbb{R}[H]$ contains all rational functions of the form $\frac{1}{x+a}$, where $a \in H$. These rational functions are all algebraically independent over $\mathbb{R}[x]$, which implies that $\mathbb{R}[H]$ is not a finitely generated $\mathbb{R}$-algebra. Therefore, H is not a real algebraic variety. In particular, it follows that H is not a real algebraic group.

We will use our previous example to illustrate several subtle distinctions between the theory of real algebraic groups and the theory of complex algebraic groups.

First, we will show that the connected component of the identity element in a real algebraic group need not be a real algebraic group. This is in contrast to the case of complex algebraic groups.

Let G denote the set $G := \{(x,y) \in \mathbb{R}^2 : xy = 1\}$, which is a real algebraic group with respect to entry-wise multiplication. Geometrically, G is a hyperbola in the real plane. The defining ideal of G is the prime ideal $(xy-1)$ in $\mathbb{R}[x,y]$. Therefore, G is irreducible as a real algebraic variety. However, it has two connected components in the ordinary topology. The connected component containing the identity element $(1,1) \in G$ is Nash isomorphic to H via the Nash imbedding

$$\begin{aligned} f : H &\longrightarrow G, \\ x &\longmapsto (x, x^{-1}). \end{aligned}$$

We conclude from this example that the connected component of the identity element in a real algebraic group need not be a real algebraic subgroup.

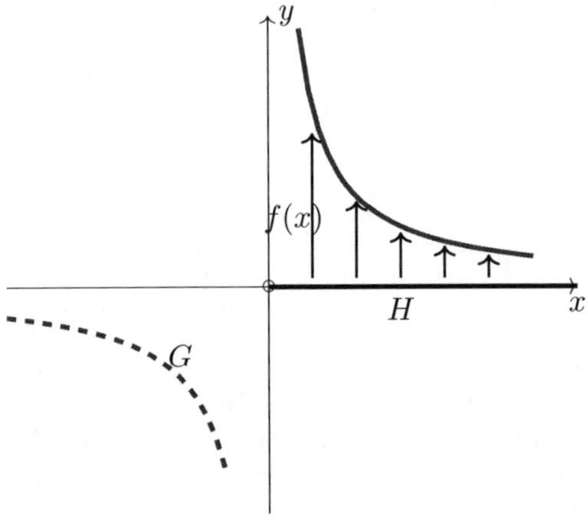

Next, we will show that even a homomorphic image of a real algebraic group under a real algebraic group homomorphism need not be a real algebraic group. This is in contrast to the case of complex algebraic groups. Once again, we focus on our previous example.

EXAMPLE 5.21. Let G denote the multiplicative group of nonzero real numbers, $(\mathbb{R}^*, \times)$. Then G is isomorphic to the hyperbola of the previous example. In particular, G is a real algebraic group. Since G is abelian, the squaring map, $sq : G \to G$, $g \mapsto g^2$ is an algebraic group homomorphism. The image of sq is canonically isomorphic to the group $(\mathbb{R}^+, \times)$, which we denoted by H in Example 5.20. As we showed in that example, H is not a real algebraic group but an affine Nash group.

An important result of Hrushovsky and Pillay [**HP94**] states that for every connected affine Nash group H, there exists a complex algebraic group G, defined over $\mathbb{R}$, that admits a finite surjective Nash group homomorphism $H \to G(\mathbb{R})^0$, where $G(\mathbb{R})^0$ is the connected component of the $\mathbb{R}$-rational points of G such that $1 \in G(\mathbb{R})^0$. In other words, affine Nash groups are precisely the finite covers of real algebraic groups.

We go back to our discussion about the $\mathfrak{N}$-topology. We will consider the affine Nash vector bundles in our topology. A locally free sheaf of $\mathscr{N}_M$-modules $\mathscr{E}$ on M is said to be an affine Nash vector bundle of class $\mathscr{N}_M$ if there is a finite open cover $M = \bigcup_{i=1}^m U_i$ in the $\mathfrak{N}$-topology such that for every $i \in \{1, \dots, m\}$ there is an isomorphism $\tau_i : \mathcal{E}|_{U_i} \to \mathscr{N}_M|_{U_i} \otimes_{\mathbb{R}} D$, and the transition functions $g_{i,j} := \tau_i \circ \tau_j^{-1}$ $(i, j \in \{1, \dots, m\})$ are isomorphisms of $\mathscr{N}_M$-modules. We note that the maps induced by the transition functions on the stalks $\mathscr{N}_{M,x} \otimes D$ are contained in $\mathrm{GL}(D)$. The structure group of an affine Nash vector bundle is naturally an affine Nash group.

Let us recall the Nash analog of a ringed D-space structure. A sheaf of k-algebras $\mathscr{A}$ on an affine Nash manifold M is called a *ringed D-space structure on M* if there exists a map of sheaves of k-algebras $\varepsilon : \mathscr{A} \to \mathscr{N}_M$, and a covering

$\{U_i \to X | i \in I\}$ in the $\mathfrak{N}$-topology equipped with the maps of sheaves of $\mathcal{N}_M$-algebras,

$$\tau_i : \mathscr{A}|_{U_i} \longrightarrow (\mathcal{N}_M \otimes D)|_{U_i} \qquad (i \in I)$$

such that $(id \otimes \varepsilon_0) \circ \tau_i = \varepsilon$. The hypothesis of our next proposition is a specialization of the hypothesis of Theorem 3.8. Hence we have the corresponding conclusion. We omit the details of its proof.

LEMMA 5.22. *Let $(M, \mathcal{N}_M)$ be an affine Nash manifold. Let $\varepsilon_\circ : D \to \mathbb{R}$ be a surjective homomorphism of finite dimensional $\mathbb{R}$-algebras such that $J := \ker \varepsilon_\circ$ is a nilpotent ideal. Let $\varepsilon : \mathscr{A} \to \mathcal{N}_M$ be an affine Nash D-manifold structure on M. Let $\mathscr{I}$ denote the kernel of ε. If the sheaf defined by $\mathscr{I}_0 := \mathscr{I} \cap Z(\mathscr{A})$ satisfies the conditions*

$$H^1(M, \mathrm{Der}(\mathcal{N}_M, \mathscr{I}_0^p / \mathscr{I}_0^{p+1})) = 0$$

for every $p \in \mathbb{Z}_+$, then there is a map of sheaves of $\mathcal{N}_M$-algebras $\varphi : \mathcal{N}_M \to Z(\mathscr{A})$ splitting ε. In this case, as a sheaf of $\mathcal{N}_M$-modules, $\mathscr{A}$ is an affine Nash vector bundle over M of class $\mathcal{N}_M$ with structure group in $\mathrm{Aut}(D, \varepsilon_0)$.

For the rest of this section, we will denote by D an exterior algebra $\bigwedge \mathbb{R}^s$ $(s \in \mathbb{Z}_+)$ with grading as described at the beginning of Section 4. Let us recall the statement of Theorem 1.2 from the introduction section.

Let $(M, \mathscr{A})$ be an affine Nash supermanifold of odd-dimension s. Then there exists a Nash vector bundle E on M and a finite cover $M = \bigcup_{i \in I} U_i$ by $\mathfrak{N}$-open sets such that, for each $i \in I$, there is an isomorphism of $\mathbb{Z}_2$-graded algebras, $\mathscr{A}(U_i) \cong \Gamma(U_i, \bigwedge E)$.

PROOF OF THEOREM 1.2. We begin with showing the vanishing of all higher cohomologies of the sheaves $\mathrm{Der}(\mathcal{N}_M, \mathscr{I}_0^p / \mathscr{I}_0^{p+1})$ for $p \in \mathbb{Z}_+$. To this end, we notice that $\mathrm{Der}(\mathcal{N}_M, \mathscr{I}_0^p / \mathscr{I}_0^{p+1})$ is a quasi-coherent sheaf of $\mathcal{N}_M$-modules. Hence, by Theorem 1.4, we have

$$H^i(M, \mathrm{Der}(\mathcal{N}_M, \mathscr{I}_0^p / \mathscr{I}_0^{p+1})) = 0 \quad \text{for every } \{i, p\} \subset \mathbb{Z}_+.$$

Now, our Lemma 5.22 implies that, for every D-manifold structure $\varepsilon : \mathscr{A} \to \mathcal{N}_M$, there is a splitting homomorphism $\varphi : \mathcal{N}_M \to Z(\mathscr{A})$. Hence, we can view our sheaf of $\mathbb{Z}_2$-graded commutative algebras $\mathscr{A}$ on M as a vector bundle over M of class $\mathcal{N}_M$ with structure group in $\mathrm{Aut}_0(D, \varepsilon_0)$. Note here that $\mathrm{Aut}_0(D, \varepsilon_0)$ is the subgroup of $\mathrm{Aut}(D, \varepsilon_0)$ consisting of $\mathbb{Z}_2$-grading preserving elements. Since both of these groups are real algebraic groups, we can view them as affine Nash groups. To finish the proof, we repeat the reasoning of the second half of the proof of our Theorem 1.1. Since the arguments are almost identical except that we now have affine Nash subgroups instead of affine semialgebraic groups, we conclude that the structure group of $\mathscr{A}$ can be reduced to $\mathrm{GL}_s(\mathbb{R})$ in the category of affine Nash groups. This finishes the proof of our second main result. □

6. Final Remarks

Every Nash group G is automatically a real Lie group since the underlying space of G is a smooth manifold and the group operations are $\mathscr{C}^\infty$ maps. As we mentioned earlier, a theorem of Malgrange shows that these maps are analytic maps. The classification of one-dimensional (real) Lie groups is a straightforward process. Indeed, a connected one-dimensional Lie group is isomorphic to either $(\mathbb{R}, +)$ or $(\mathbb{S}, \cdot)$, where $\mathbb{S} := \{z \in \mathbb{C}^* : |z| = 1\}$, and the group operation of $\mathbb{S}$ is the multiplication of complex numbers. However, the classification of connected one-dimensional Nash groups turned out to be substantially more intricate. This task was successfully accomplished by Madden and Stanton in their work [**MS92**]. A readily accessible exposition of the results of Madden and Stanton can be found in the comprehensive article [**Shi96**].

For the classification of connected one-dimensional Nash groups, an essential ingredient is the notion of a *locally Nash group*. Such a structure can be defined as a group object in the category of *locally Nash manifolds*. Here, a locally Nash manifold is defined as an analytic manifold with a possibly infinite system of Nash coordinate neighborhoods. Equivalently, a *locally Nash manifold* M is a group object in the category of locally semialgebraic spaces such that the underlying space of M is an analytic manifold, and the group operations of M are analytic maps.

EXAMPLE 6.1. Let us consider the unit circle $\mathbb{S}$ (with its group structure inherited from $\mathbb{C}^*$) as an affine Nash group. Let $\widetilde{\mathbb{S}}$ denote the universal covering space of $\mathbb{S}$. Then $\widetilde{\mathbb{S}}$ has the structure of a locally Nash manifold and a compatible group structure on it. In other words, $\widetilde{\mathbb{S}}$ is locally Nash group. Then $\widetilde{\mathbb{S}}$ is not a Nash group. To see this, consider the covering homomorphism $p : \widetilde{\mathbb{S}} \to \mathbb{S}$. If $\widetilde{\mathbb{S}}$ was a Nash group, then p would be a Nash group homomorphism. However, the kernel of p is isomorphic to $\mathbb{Z}$, which is not a Nash group but a locally Nash group. It follows that $\widetilde{\mathbb{S}}$ is not a Nash group.

We will now provide a very brief overview of the development of the classification of connected one-dimensional Nash groups.

To begin with, it can be shown that the universal covering group of a locally Nash group is also a locally Nash group. This observation leads to a separation of the classification problem into two distinct parts:

(1) The classification of isomorphism classes of simply connected locally Nash groups.
(2) The classification of isomorphism classes of the quotients of a simply connected locally Nash group.

Let G be a one-dimensional, connected, and simply connected locally Nash group. Then the underlying locally Nash manifold is locally Nash isomorphic to $\mathbb{R}$ with an appropriate (possibly infinite) system of locally Nash coordinate charts. In particular, there is an analytic Nash coordinate chart $\psi : U \to \mathbb{R}$ such that $0 \in U$, and the Nash group operations on every sufficiently small neighborhood of 0 are transferred via ψ from $(\mathbb{R}, +)$. Furthermore, this coordinate chart ψ determines, up to a locally Nash group isomorphism, the locally Nash group structure on the whole of G. Therefore, it makes sense to denote the locally Nash group G by the pair $(\mathbb{R}, \psi)$. The gist of the classification of the pairs $(\mathbb{R}, \psi)$ is the *addition theorem of Weierstrass*. All of this is explained in more detail in the original article [**MS92**] as well as in the Shiota's article [**Shi96**, pages 107–108].

In the present article, we have introduced the categories of Nash supermanifolds and locally semialgebraic superspaces implicitly. Therefore, the following definition now makes sense.

DEFINITION 6.2. A *Nash supergroup* (or a *$\mathbb{Z}_2$-graded Nash group*) is a group object in the category of Nash supermanifolds. Likewise, a *locally semialgebraic supergroup* (or a *$\mathbb{Z}_2$-graded locally semialgebraic group*) is a group object in the category of locally semialgebraic supergroups.

Continuing our work in this paper and building upon the research of Madden and Stanton, we have classified the connected one-dimensional affine locally Nash supergroups. The details of this work will appear elsewhere.

References

[AG08] A. Aizenbud and D. Gourevitch, *Schwartz functions on Nash manifolds*, Int. Math. Res. Not. IMRN **5** (2008), Art. ID rnm 155, 37 pp., DOI 10.1093/imrn/rnm155. MR2418286

[Art62] Michael Artin, *Grothendieck topologies: notes on a seminar*, Harvard University, Dept. of Mathematics, 1962.

[Bat79] M. Batchelor, *The structure of supermanifolds*, Trans. Amer. Math. Soc. **253** (1979), 329–338, DOI 10.2307/1998201. MR536951

[Bat80] M. Batchelor, *Two approaches to supermanifolds*, Trans. Amer. Math. Soc. **258** (1980), no. 1, 257–270, DOI 10.2307/1998294. MR554332

[BCR98] J. Bochnak, M. Coste, and M.-F. Roy, *Real algebraic geometry*, Ergebnisse der Mathematik und ihrer Grenzgebiete (3) [Results in Mathematics and Related Areas (3)], vol. 36, Springer-Verlag, Berlin, 1998. Translated from the 1987 French original; Revised by the authors, DOI 10.1007/978-3-662-03718-8. MR1659509

[BL75] F. A. Berezin and D. A. Leĭtes, *Supermanifolds* (Russian), Dokl. Akad. Nauk SSSR **224** (1975), no. 3, 505–508. MR402795

[BR84] R. J. Blattner and J. H. Rawnsley, *Remarks on Batchelor's theorem*, Mathematical aspects of superspace (Hamburg, 1983), NATO Adv. Sci. Inst. Ser. C: Math. Phys. Sci., vol. 132, Reidel, Dordrecht, 1984, pp. 161–171. MR773083

[Del91] H. Delfs, *Homology of locally semialgebraic spaces*, Lecture Notes in Mathematics, vol. 1484, Springer-Verlag, Berlin, 1991, DOI 10.1007/BFb0093939. MR1176311

[DeW84] B. DeWitt, *Supermanifolds*, Cambridge Monographs on Mathematical Physics, Cambridge University Press, Cambridge, 1984. MR778559

[DK81] H. Delfs and M. Knebusch, *Semialgebraic topology over a real closed field. II. Basic theory of semialgebraic spaces*, Math. Z. **178** (1981), no. 2, 175–213, DOI 10.1007/BF01262039. MR631628

[DK84] H. Delfs and M. Knebusch, *An introduction to locally semialgebraic spaces*, Rocky Mountain J. Math. **14** (1984), no. 4, 945–963, DOI 10.1216/RMJ-1984-14-4-945. Ordered fields and real algebraic geometry (Boulder, Colo., 1983). MR773141

[DK85] H. Delfs and M. Knebusch, *Locally semialgebraic spaces*, Lecture Notes in Mathematics, vol. 1173, Springer-Verlag, Berlin, 1985, DOI 10.1007/BFb0074551. MR819737

[DM99] P. Deligne and J. W. Morgan, *Notes on supersymmetry (following Joseph Bernstein)*, Quantum fields and strings: a course for mathematicians, Vol. 1, 2 (Princeton, NJ, 1996/1997), Amer. Math. Soc., Providence, RI, 1999, pp. 41–97. MR1701597

[Efr74] G. A. Efroymson, *A Nullstellensatz for Nash rings*, Pacific J. Math. **54** (1974), 101–112. MR360576

[Efr82] G. A. Efroymson, *The extension theorem for Nash functions*, Real algebraic geometry and quadratic forms (Rennes, 1981), Lecture Notes in Math., vol. 959, Springer, Berlin-New York, 1982, pp. 343–357. MR683141

[EP16] M. J. Edmundo and L. Prelli, *Sheaves on $\mathcal{T}$-topologies*, J. Math. Soc. Japan **68** (2016), no. 1, 347–381, DOI 10.2969/jmsj/06810347. MR3454562

[God58] R. Godement, *Topologie algébrique et théorie des faisceaux* (French), Publications de l'Institut de Mathématiques de l'Université de Strasbourg [Publications of the Mathematical Institute of the University of Strasbourg], No. 13, Hermann, Paris, 1958. Actualités Scientifiques et Industrielles [Current Scientific and Industrial Topics], No. 1252. MR102797

[Har77] R. Hartshorne, *Algebraic geometry*, Graduate Texts in Mathematics, No. 52, Springer-Verlag, New York-Heidelberg, 1977. MR463157

[HP94] E. Hrushovski and A. Pillay, *Groups definable in local fields and pseudo-finite fields*, Israel J. Math. **85** (1994), no. 1-3, 203–262, DOI 10.1007/BF02758643. MR1264346

[Knu71] D. Knutson, *Algebraic spaces*, Lecture Notes in Mathematics, Vol. 203, Springer-Verlag, Berlin-New York, 1971. MR302647

[Kos77] B. Kostant, *Graded manifolds, graded Lie theory, and prequantization*, Differential geometrical methods in mathematical physics (Proc. Sympos., Univ. Bonn, Bonn, 1975), Lecture Notes in Math., Vol. 570, Springer, Berlin-New York, 1977, pp. 177–306. MR580292

[Mal67] B. Malgrange, *Ideals of differentiable functions*, Tata Institute of Fundamental Research Studies in Mathematics, vol. 3, Tata Institute of Fundamental Research, Bombay; Oxford University Press, London, 1967. MR212575

[Man97] Y. I. Manin, *Gauge field theory and complex geometry*, 2nd ed., Grundlehren der mathematischen Wissenschaften [Fundamental Principles of Mathematical Sciences], vol. 289, Springer-Verlag, Berlin, 1997. Translated from the 1984 Russian original by N. Koblitz and J. R. King; With an appendix by Sergei Merkulov, DOI 10.1007/978-3-662-07386-5. MR1632008

[MS92] J. J. Madden and C. M. Stanton, *One-dimensional Nash groups*, Pacific J. Math. **154** (1992), no. 2, 331–344. MR1159515

[Pec85] D. Pecker, *On Efroymson's extension theorem for Nash functions*, J. Pure Appl. Algebra **37** (1985), no. 2, 193–203, DOI 10.1016/0022-4049(85)90097-0. MR796409

[Pil88] A. Pillay, *On groups and fields definable in o-minimal structures*, J. Pure Appl. Algebra **53** (1988), no. 3, 239–255, DOI 10.1016/0022-4049(88)90125-9. MR961362

[Ris73] J.-J. Risler, *Sur l'anneau des fonctions de Nash globales* (French), C. R. Acad. Sci. Paris Sér. A-B **276** (1973), A1513–A1516. MR318510

[Rob83] R. Robson, *Embedding semi-algebraic spaces*, Math. Z. **183** (1983), no. 3, 365–370, DOI 10.1007/BF01176477. MR706394

[Rog80] A. Rogers, *A global theory of supermanifolds*, J. Math. Phys. **21** (1980), no. 6, 1352–1365, DOI 10.1063/1.524585. MR574696

[Ser56] J.-P. Serre, *Géométrie algébrique et géométrie analytique* (French), Ann. Inst. Fourier (Grenoble) **6** (1955/56), 1–42. MR82175

[Shi87] M. Shiota, *Nash manifolds*, Lecture Notes in Mathematics, vol. 1269, Springer-Verlag, Berlin, 1987, DOI 10.1007/BFb0078571. MR904479

[Shi96] M. Shiota, *Nash functions and manifolds*, Lectures in real geometry (Madrid, 1994), De Gruyter Exp. Math., vol. 23, de Gruyter, Berlin, 1996, pp. 69–112. MR1440210

[Sta22] The Stacks project authors, *The stacks project*, `https://stacks.math.columbia.edu`, 2022.

[Sun15] B. Sun, *Almost linear Nash groups*, Chinese Ann. Math. Ser. B **36** (2015), no. 3, 355–400, DOI 10.1007/s11401-015-0915-7. MR3341164

Department of Mathematics, Tulane University, Louisiana 70118

Current address: Department of Mathematics, Tulane University, New Orleans, Louisiana 70118

Email address: `mahirbilencan@gmail.com`

Contemporary Mathematics
Volume **804**, 2024
https://doi.org/10.1090/conm/804/16112

A survey of support theories for Lie superalgebras and finite supergroup schemes

Christopher M. Drupieski and Jonathan R. Kujawa

ABSTRACT. We survey the current state of various support variety theories for Lie superalgebras and finite supergroup schemes. We pay particular attention to the theory in characteristic zero developed by Boe, Kujawa, and Nakano using relative Lie superalgebra cohomology, and to the theory developed in positive characteristic in our previous work.

CONTENTS

1. Introduction
2. Background on support varieties and Lie superalgebra cohomology

Part 1. Characteristic zero
3. Support varieties via relative Lie superalgebra cohomology
4. Support varieties for finite supergroup schemes in characteristic zero

Part 2. Positive characteristic
5. Support varieties via ordinary Lie superalgebra cohomology
6. Finite group schemes: recollections from the non-super theory
7. Infinitesimal supergroup schemes
Acknowledgments
References

1. Introduction

1.1. Overview. Let A be a Hopf algebra or similar algebraic structure over an algebraically closed field k. Support varieties are tools that help relate the representation theory of A to the geometry encoded by the spectrum $|A|$ of the cohomology ring $\mathrm{H}^\bullet(A,k) = \mathrm{Ext}_A^\bullet(k,k)$. The enveloping algebra $U(\mathfrak{g})$ of a finite-dimensional Lie algebra $\mathfrak{g}$ has finite global dimension $d = \dim(\mathfrak{g})$, so the Lie algebra

2020 *Mathematics Subject Classification.* Primary 17B56; Secondary 20G10, 17B50, 17B55.
Key words and phrases. Support varieties, Lie superalgebras, finite supergroup schemes.
The first author was supported in part by Simons Collaboration Grant for Mathematicians No. 4269055.
The second author was supported in part by Simons Collaboration Grant for Mathematicians No. 525043.

cohomology ring $\mathrm{H}^\bullet(\mathfrak{g},k) = \mathrm{H}^\bullet(U(\mathfrak{g}),k)$ vanishes in all large degrees, and hence its spectrum does not provide an interesting geometry. On the other hand, in positive characteristic the cohomology of finite-dimensional *restricted* Lie algebras—or more generally, of finite group schemes—frequently gives rise to rich algebro-geometric structures. Support variety theories in these contexts have been developed and explored by Friedlander and Parshall [**25**], Suslin, Friedlander, and Bendel [**49**], and Friedlander and Pevtsova [**27**], and many others.

The aim of this paper is to give an overview of what is currently known about cohomological support varieties for Lie *super*algebras and finite *super*group schemes. For a finite-dimensional Lie superalgebra $\mathfrak{g}$, the enveloping superalgebra $U(\mathfrak{g})$ may or may not have finite global dimension, depending on whether or not $\mathfrak{g}$ contains any odd elements such that $[x,x] = 0$. In characteristic zero, even if a Lie superalgebra $\mathfrak{g}$ has infinite global dimension, the ordinary Lie superalgebra cohomology ring $\mathrm{H}^\bullet(\mathfrak{g},k)$ may still be finite, and hence may not give rise to any interesting ambient geometry; see for example the calculations in [**29**, §2.6]. But replacing the ordinary Lie superalgebra cohomology ring with the relative cohomology ring for the pair $(\mathfrak{g},\mathfrak{g}_{\overline{0}})$, one gets a rich support variety theory in many cases. In positive characteristic, the Lie superalgebra cohomology ring $\mathrm{H}^\bullet(\mathfrak{g},k)$ gives rise to a nontrivial geometry whenever the enveloping superalgebra $U(\mathfrak{g})$ is not of finite global dimension. In general, support varieties for finite-dimensional restricted Lie superalgebras (and more generally, for finite supergroup schemes) are not completely understood, with the fullest picture available only for p-nilpotent restricted Lie superalgebras (and more generally, for unipotent finite supergroup schemes). What is known at present parallels the classical, non-super theory in many ways, but with interesting twists arising from new super phenomena.

The paper is organized as follows: We begin in Section 2 by recalling some of the basic definitions and properties for support varieties and Lie superalgebra cohomology. Part 1 of the paper is devoted to what is known in characteristic zero. In Section 3 we summarize the support variety theory for Lie superalgebras developed in a series of papers by Boe, Kujawa, Nakano, and collaborators using relative Lie superalgebra cohomology. In Section 4 we give a brief account of support varieties for finite supergroup schemes in characteristic zero. Part 2 of the paper is devoted to what is known in (odd) positive characteristic. In Section 5 we describe the support variety theory one gets for a finite-dimensional Lie superalgebra $\mathfrak{g}$ from considering its ordinary Lie superalgebra cohomology ring $\mathrm{H}^\bullet(\mathfrak{g},k)$. Then, to aid the reader in comparing the super theory with its classical (non-super) counterpart, in Section 6 we summarize the classical results for restricted Lie algebras and infinitesimal group schemes (as developed by Friedlander and Parshall and by Suslin, Friedlander, and Bendel), before discussing in Section 7 our work to date related to restricted Lie superalgebras and infinitesimal supergroup schemes. As a complement to our own work on infinitesimal supergroups, we also recommend to the reader the recent survey article by Benson, Iyengar, Krause, and Pevtsova [**4**].

1.2. Conventions. Throughout, k will denote a ground field of characteristic $p \neq 2$. For simplicity, we will assume that k is algebraically closed, though it may be possible to formulate some results without this general assumption. Except when specified, all vector spaces will be k-vector spaces, all algebras will be (associative, unital) k-algebras, and all unadorned tensor products will be tensor products over k. Given a k-vector space V, let $V^* = \mathrm{Hom}_k(V,k)$ be its k-linear dual.

Set $\mathbb{Z}_2 = \mathbb{Z}/2\mathbb{Z} = \{\overline{0}, \overline{1}\}$. Following the literature, we use the prefix 'super' to indicate that an object is $\mathbb{Z}_2$-graded, and we assume that the reader is generally familiar with the standard conventions of 'supermathematics'. In particular, we denote the decomposition of a vector superspace into its $\mathbb{Z}_2$-homogeneous components by $V = V_{\overline{0}} \oplus V_{\overline{1}}$, calling $V_{\overline{0}}$ and $V_{\overline{1}}$ the even and odd subspaces of V, respectively, and writing $\overline{v} \in \mathbb{Z}_2$ to denote the superdegree of a homogeneous element $v \in V_{\overline{0}} \cup V_{\overline{1}}$. When written without additional adornment, we consider the ground field to be a superspace concentrated in even superdegree. Whenever we state a formula in which homogeneous degrees of elements are specified, we mean that the formula is true as written for homogeneous elements and that it extends by linearity to non-homogeneous elements.

We use the adjective *graded* to indicate that an object admits an additional $\mathbb{Z}$-grading that is compatible with its underlying structure. Thus a *graded superspace* is a $(\mathbb{Z} \times \mathbb{Z}_2)$-graded vector space, a *graded superalgebra* is a $(\mathbb{Z} \times \mathbb{Z}_2)$-graded algebra, etc. Given a graded superspace V and a homogeneous element $v \in V$ of bidegree $(s,t) \in \mathbb{Z} \times \mathbb{Z}_2$, we write $\deg(v) = s$ and $\overline{v} = t$ for the $\mathbb{Z}$-degree and the superdegree of v, respectively. If A is a graded superalgebra, we say that A is *graded-commutative* provided that for all homogeneous elements $a, b \in A$, one has

$$ab = (-1)^{\deg(a) \cdot \deg(b) + \overline{a} \cdot \overline{b}} ba. \tag{1.2.1}$$

The notion of *graded-cocommutativity* for a graded super-coalgebra is defined similarly. If A and B are graded superalgebras, then $A \otimes B$ is a graded superalgebra, with $\deg(a \otimes b) = \deg(a) + \deg(b)$, and with the product defined by $(a \otimes b)(c \otimes d) = (-1)^{\deg(b) \cdot \deg(c) + \overline{b} \cdot \overline{c}} ac \otimes bd$; we denote this graded superalgebra by $A \,{}^g\!\otimes B$, and call it the *graded tensor product of superalgebras*.

Let $\mathbb{N} = \{0, 1, 2, 3, \ldots\}$ be the set of non-negative integers.

2. Background on support varieties and Lie superalgebra cohomology

2.1. Basic notions for support varieties. In this section we recall some of the basic definitions concerning (cohomological) support varieties for supermodules over a Hopf superalgebra. For further details, the reader can consult [**15**, §2].

Let k be a (for simplicity, algebraically closed) field of characteristic $p \neq 2$, and let A be a Hopf superalgebra over k. The cohomology ring $\mathrm{H}^\bullet(A,k) = \mathrm{Ext}^\bullet_A(k,k)$ is naturally a graded superalgebra: the $\mathbb{Z}$-component of the grading is the cohomological grading, and the $\mathbb{Z}_2$-component is the 'internal' grading coming from the superalgebra structure on A. The cup product makes $\mathrm{H}^\bullet(A,k)$ into a graded-commutative superalgebra in the sense of (1.2.1). In particular, the subspaces $\mathrm{H}^{\mathrm{ev}}(A,k)_{\overline{1}}$ and $\mathrm{H}^{\mathrm{odd}}(A,k)_{\overline{0}}$ of $\mathrm{H}^\bullet(A,k)$ consist of nilpotent elements. Then modulo its nilradical, $\mathrm{H}^\bullet(A,k)$ is a commutative ring in the ordinary sense, with the superdegree of a homogeneous element equal simply to the reduction modulo 2 of its $\mathbb{Z}$-degree.

Given A-supermodules M and N, the cohomology ring $\mathrm{H}^\bullet(A,k)$ acts on the space of extensions $\mathrm{Ext}^\bullet_A(M,N)$ via the cup product. In all of the examples we will consider, the following properties will hold:

(FG1) $\mathrm{H}^\bullet(A,k)$ is a finitely-generated k-algebra

(FG2) If M and N are finite-dimensional, then $\mathrm{Ext}^\bullet_A(M,N)$ is a finite $\mathrm{H}^\bullet(A,k)$-module.

Let $I_A(M)$ be the annihilator ideal for the left cup product action of $\mathrm{H}^\bullet(A,k)$ on $\mathrm{Ext}_A^\bullet(M,M)$:

$$I_A(M) = \mathrm{ann}_{\mathrm{H}^\bullet(A,k)} \mathrm{Ext}_A^\bullet(M,M).$$

Equivalently, $I_A(M)$ is the kernel of the algebra map

$$\Phi_M : \mathrm{Ext}_A^\bullet(k,k) \to \mathrm{Ext}_A^\bullet(M,M)$$

induced by the functor $- \otimes M$. Now the cohomological spectrum of A is

$$|A| = \mathrm{MaxSpec}\, \mathrm{H}^\bullet(A,k),$$

the maximal ideal spectrum of $\mathrm{H}^\bullet(A,k)$, and the support variety of M, denoted $|A|_M$, is the Zariski closed subset of $|A|$ defined by $I_A(M)$:

$$|A|_M = \mathrm{MaxSpec}\left(\mathrm{H}^\bullet(A,k)/I_A(M)\right).$$

Support varieties for supermodules over a Hopf superalgebra satisfy a list of standard properties; see for example [**15**, §2.3] or [**51**, §8.3]. In particular, if A is a finite-dimensional (hence self-injective) Hopf superalgebra, then under the 'finite-generation' hypotheses (FG1) and (FG2), one gets for a finite-dimensional A-supermodule M that the geometric dimension of the support variety $|A|_M$ is equal to $\mathrm{cx}_A(M)$, the complexity of M as an A-supermodule, and the support variety $|A|_M$ is zero (i.e., is equal to a single point) if and only if M is projective; see [**15**, Proposition 2.3.13].

In some contexts we may instead consider support varieties defined via either the prime ideal spectrum $\mathrm{Spec}\,\mathrm{H}^\bullet(A,k)$ or the projective spectrum $\mathrm{Proj}\,\mathrm{H}^\bullet(A,k)$. When confusion is unlikely, we may write

$$|A| = \mathrm{Spec}\,\mathrm{H}^\bullet(A,k) \quad \text{and} \quad |A|_M = \mathrm{Spec}\left(\mathrm{H}^\bullet(A,k)/I_A(M)\right).$$

In Part 1 we will consider support varieties that are defined via relative cohomology groups. Aside from the switch to relative cohomology, the basic definitions of the cohomological spectrum and of support varieties are the same. Support varieties defined via relative cohomology may not satisfy the full list of 'standard' properties as those defined via non-relative cohomology.

2.2. Generalities of Lie superalgebra cohomology. In this section we give an overview of how ordinary and relative Lie superalgebra cohomology can be calculated in terms of the ordinary and relative Koszul complexes.

Let $\mathfrak{g}$ be a Lie superalgebra over the field k, and let $U(\mathfrak{g})$ be the universal enveloping superalgebra of $\mathfrak{g}$. By construction, a $\mathfrak{g}$-supermodule is the same thing as a $U(\mathfrak{g})$-supermodule. The field k is an $\mathfrak{g}$-supermodule via the augmentation map $\varepsilon : U(\mathfrak{g}) \to k$. Given $\mathfrak{g}$-supermodules M and N, set

$$\mathrm{Ext}_{\mathfrak{g}}^\bullet(M,N) = \mathrm{Ext}_{U(\mathfrak{g})}^\bullet(M,N) \quad \text{and} \quad \mathrm{H}^\bullet(\mathfrak{g},M) = \mathrm{Ext}_{\mathfrak{g}}^\bullet(k,M).$$

By definition, the differentials in a $U(\mathfrak{g})$-supermodule resolution are required to be even linear maps, i.e., we only consider resolutions in $(\mathsf{sMod})_{\mathrm{ev}}$, the 'underlying even subcategory' of the category $\mathsf{sMod}_{U(\mathfrak{g})}$ of all $U(\mathfrak{g})$-supermodules. Then $\mathrm{Ext}_{\mathfrak{g}}^\bullet(M,N)$ can be computed as the cohomology of the complex $\mathrm{Hom}_{U(\mathfrak{g})}(P_\bullet,N)$, or as the cohomology of the complex $\mathrm{Hom}_{U(\mathfrak{g})}(M,Q^\bullet)$, for any projective $U(\mathfrak{g})$-supermodule resolution $P_\bullet \to M$, or any injective $U(\mathfrak{g})$-supermodule resolution $N \to Q^\bullet$.

An explicit $U(\mathfrak{g})$-free resolution of the ground field is provided by the Koszul resolution, denoted $Y(\mathfrak{g})$. As a left $U(\mathfrak{g})$-supermodule, $Y(\mathfrak{g}) = U(\mathfrak{g}) \otimes \boldsymbol{A}(\mathfrak{g})$, where $\boldsymbol{A}(\mathfrak{g}) = \bigoplus_{n\in\mathbb{N}} \boldsymbol{A}^n(\mathfrak{g})$ denotes the alternating power superalgebra on $\mathfrak{g}$, as defined for example in [**13**, §2.3.7]. As a graded superalgebra, $\boldsymbol{A}(\mathfrak{g}) \cong \Lambda(\mathfrak{g}_{\overline{0}}) \,{}^g\!\otimes\, \Gamma(\mathfrak{g}_{\overline{1}})$. Here $\Lambda(\mathfrak{g}_{\overline{0}}) = \bigoplus_{n \in \mathbb{N}} \Lambda^n(\mathfrak{g}_{\overline{0}})$ is the ordinary exterior algebra on $\mathfrak{g}_{\overline{0}}$ (considered as a graded superalgebra concentrated in superdegree $\overline{0}$), and $\Gamma(\mathfrak{g}_{\overline{1}}) = \bigoplus_{n \in \mathbb{N}} \Gamma^n(\mathfrak{g}_{\overline{1}})$ is the ordinary divided power algebra on $\mathfrak{g}_{\overline{1}}$ (considered as a graded superalgebra with $\Gamma^n(\mathfrak{g}_{\overline{1}})$ concentrated in superdegree $\overline{n}$). One can define a graded super-bialgebra structure on $Y(\mathfrak{g})$ such that the Koszul differential makes $Y(\mathfrak{g})$ into a differential graded super-bialgebra, and such that the coproduct $Y(\mathfrak{g}) \to Y(\mathfrak{g}) \,{}^g\!\otimes Y(\mathfrak{g})$ restricts to the usual coproducts on $U(\mathfrak{g})$, $\Lambda(\mathfrak{g}_{\overline{0}})$, and $\Gamma(\mathfrak{g}_{\overline{1}})$. For more details, see [**12**, §3.1]. In particular, an explicit formula for the Koszul differential $d : Y_n(\mathfrak{g}) \to Y_{n-1}(\mathfrak{g})$ is given in [**12**, Remark 3.1.4].

REMARK 2.2.1. By definition, $\boldsymbol{A}^n(\mathfrak{g})$ is a subspace of $\mathfrak{g}^{\otimes n}$. Then via the canonical maps $\boldsymbol{A}^n(\mathfrak{g}) \subseteq \mathfrak{g}^{\otimes n} \subseteq U(\mathfrak{g})^{\otimes n}$, one can show that $Y(\mathfrak{g})$ is a subcomplex of the left bar complex for $U(\mathfrak{g})$.

In the special case that k is a field of characteristic zero, one has $\gamma_a(y) = \frac{1}{a!}[\gamma_1(y)]^a$ in $\Gamma(\mathfrak{g}_{\overline{1}})$, and $\Gamma(\mathfrak{g}_{\overline{1}})$ is isomorphic to the symmetric algebra $S(\mathfrak{g}_{\overline{1}})$. Thus in characteristic zero, the underlying space for the Koszul resolution can be realized as $Y(\mathfrak{g}) = U(\mathfrak{g}) \otimes \boldsymbol{\Lambda}(\mathfrak{g})$, where $\boldsymbol{\Lambda}(\mathfrak{g})$ denotes the superexterior algebra on the superspace $\mathfrak{g}$, as discussed for example in [**13**, §2.3.5].

Now given a $\mathfrak{g}$-supermodule M, $\mathrm{H}^\bullet(\mathfrak{g}, M)$ can be computed as the cohomology of the cochain complex $C^\bullet(\mathfrak{g}, M) := \mathrm{Hom}_{U(\mathfrak{g})}(Y_\bullet(\mathfrak{g}), M)$. Denote the differential on $C^\bullet(\mathfrak{g}, M)$ by ∂. In the case $M = k$, the coproduct on $Y(\mathfrak{g})$ induces an algebra structure on $C^\bullet(\mathfrak{g}, k)$, which in turn induces the cup product on the cohomology ring $\mathrm{H}^\bullet(\mathfrak{g}, k)$. Explicitly, $C^\bullet(\mathfrak{g}, k)$ is isomorphic as a graded superalgebra to the superexterior algebra $\boldsymbol{\Lambda}(\mathfrak{g}^*)$. The graded superalgebra isomorphism

$$C^\bullet(\mathfrak{g}, k) = \mathrm{Hom}_{U(\mathfrak{g})}(U(\mathfrak{g}) \otimes \boldsymbol{A}(\mathfrak{g}), k) \cong \mathrm{Hom}_k(\boldsymbol{A}(\mathfrak{g}), k) \cong \boldsymbol{\Lambda}(\mathfrak{g}^*)$$

corresponds to the duality of strict polynomial superfunctors $\boldsymbol{A}^\# \cong \boldsymbol{\Lambda}$ discussed in [**13**, §2.6]. As a graded superalgebra, $\boldsymbol{\Lambda}(\mathfrak{g}^*) \cong \Lambda(\mathfrak{g}_{\overline{0}}^*) \,{}^g\!\otimes\, S(\mathfrak{g}_{\overline{1}}^*)$, where $S(\mathfrak{g}_{\overline{1}}^*) = \bigoplus_{n \in \mathbb{N}} S^n(\mathfrak{g}_{\overline{1}}^*)$ denotes the ordinary symmetric algebra on $\mathfrak{g}_{\overline{1}}^*$ (considered as a graded superalgebra with $S^n(\mathfrak{g}_{\overline{1}}^*)$ concentrated in superdegree $\overline{n}$). The differential ∂ makes $C^\bullet(\mathfrak{g}, k)$ into a differential graded superalgebra.

The left adjoint action of $\mathfrak{g}$ on itself extends by superderivations to an action of $\mathfrak{g}$ on $\boldsymbol{A}(\mathfrak{g})$. Considering $\boldsymbol{A}^n(\mathfrak{g})$ as a subspace of $\mathfrak{g}^{\otimes n}$, this is just the restriction of the usual adjoint action on tensor space. On a monomial $\gamma_a(y) \in \Gamma(\mathfrak{g}_{\overline{1}})$, the adjoint action of $u \in \mathfrak{g}$ is given by $u.\gamma_a(y) = s([u, y])\gamma_{a-1}(y)$, where $s : \mathfrak{g} \to \boldsymbol{A}^1(\mathfrak{g})$ is the canonical map.

Now let $\mathfrak{s}$ be a Lie superalgebra over k, and let $\mathfrak{t}$ be a Lie sub-superalgebra of $\mathfrak{s}$. The adjoint action of $\mathfrak{s}$ on $\boldsymbol{A}(\mathfrak{s})$ descends to an action of $\mathfrak{t}$ on $\boldsymbol{A}(\mathfrak{s}/\mathfrak{t})$. Also, $U(\mathfrak{s})$ becomes a right $\mathfrak{t}$-supermodule via right multiplication. Set

$$Y(\mathfrak{s}, \mathfrak{t}) = U(\mathfrak{s}) \otimes_{U(\mathfrak{t})} \boldsymbol{A}(\mathfrak{s}/\mathfrak{t}).$$

The Koszul differential on $Y(\mathfrak{s})$ induces a differential on $Y(\mathfrak{s}, \mathfrak{t})$, which together with the induced augmentation map $U(\mathfrak{s}) \otimes_{U(\mathfrak{t})} \boldsymbol{A}(\mathfrak{s}/\mathfrak{t}) \to U(\mathfrak{s}) \otimes_{U(\mathfrak{t})} k \to k$ makes $Y(\mathfrak{s}, \mathfrak{t})$ into a resolution of the field k. We call this complex the *relative Koszul resolution for the pair* $(\mathfrak{s}, \mathfrak{t})$.

The relative Lie superalgebra cohomology of the pair $(\mathfrak{s}, \mathfrak{t})$ with coefficients in an $\mathfrak{s}$-supermodule M, denoted $\mathrm{H}^\bullet(\mathfrak{s}, \mathfrak{t}; M)$, is by definition the cohomology of the cochain complex $\mathrm{Hom}_{U(\mathfrak{s})}(Y(\mathfrak{s}, \mathfrak{t}), M)$. If k is a field of characteristic zero (e.g., if

$k = \mathbb{C}$), and if $\mathfrak{s}$ is finitely-semisimple under the adjoint action of $\mathfrak{t}$, then $Y(\mathfrak{s},\mathfrak{t})$ is a $(U(\mathfrak{s}),U(\mathfrak{t}))$-projective resolution of the field k, and hence $\mathrm{H}^\bullet(\mathfrak{s},\mathfrak{t};M)$ is equal to the relative Ext-group $\mathrm{Ext}^\bullet_{(U(\mathfrak{s}),U(\mathfrak{t}))}(k,M)$ as defined for example by Hochschild [**33**]. For more details on relative homological algebra, see for example [**39**, Chapter III].

Part 1. Characteristic zero

In this part, let $k = \mathbb{C}$.

3. Support varieties via relative Lie superalgebra cohomology

In this section we give an overview of the support variety theory for classical complex Lie superalgebras developed by Boe, Kujawa, and Nakano using relative Lie superalgebra cohomology. See [**42**] for further information about complex Lie superalgebras.

3.1. Relative cohomology and support varieties. A finite-dimensional complex Lie superalgebra $\mathfrak{g} = \mathfrak{g}_{\overline{0}} \oplus \mathfrak{g}_{\overline{1}}$ is called *classical* if there is a connected reductive algebraic group $G_{\overline{0}}$ such that $\mathrm{Lie}(G_{\overline{0}}) = \mathfrak{g}_{\overline{0}}$ and if there is an action of $G_{\overline{0}}$ on $\mathfrak{g}_{\overline{1}}$ that differentiates to the adjoint action of $\mathfrak{g}_{\overline{0}}$ on $\mathfrak{g}_{\overline{1}}$. In this section we will assume $\mathfrak{g}$ is simple and classical, or a closely related Lie superalgebra such as $\mathfrak{gl}(m|n)$, $\mathfrak{sl}(n|n)$, or $\mathfrak{p}(n)$. Our running example will be $\mathfrak{g} = \mathfrak{gl}(m|n)$. Recall that

$$\mathfrak{gl}(m|n) = \left\{ \left(\begin{array}{c|c} A & B \\ \hline C & D \end{array} \right) \right\}, \tag{3.1.1}$$

where A is a $m \times m$ matrix, B is a $m \times n$ matrix, C is a $n \times m$ matrix, and D is a $n \times n$ matrix. The $\mathbb{Z}_2$-grading is defined so that $\mathfrak{g}_{\overline{0}}$ consists of those matrices with $B = 0$ and $C = 0$, $\mathfrak{g}_{\overline{1}}$ is those matrices with $A = 0$ and $D = 0$, and the Lie bracket is the supercommutator. Note that $G_{\overline{0}} = GL_m \times GL_n$ acts on $\mathfrak{g}_{\overline{1}}$ by matrix conjugation, and $\mathfrak{gl}(m|n)$ is classical.

Let $\mathcal{F} = \mathcal{F}(\mathfrak{g},\mathfrak{g}_{\overline{0}})$ be the category of finite-dimensional $\mathfrak{g}$-supermodules that are completely reducible when restricted to $\mathfrak{g}_{\overline{0}}$. It is common that $\mathfrak{g}_{\overline{0}}$ is a semisimple Lie algebra, in which case $\mathcal{F}$ is the category of all finite-dimensional $\mathfrak{g}$-supermodules. One can use the relative cohomology introduced in Section 2.2 to compute cohomology in $\mathcal{F}$. Namely, given supermodules M and N in $\mathcal{F}$, [**7**, Theorem 2.5] states that for all $d \geq 0$,

$$\mathrm{Ext}^d_{\mathcal{F}}(M,N) \cong \mathrm{H}^d(\mathfrak{g},\mathfrak{g}_{\overline{0}}; M^* \otimes N)$$

as superspaces. In the particular case when $M = N = \mathbb{C}$, it is easy to see that the differentials in the relative Koszul complex are identically zero and hence the cohomology ring is a ring of invariants:

$$\mathrm{Ext}^\bullet_{\mathcal{F}}(\mathbb{C},\mathbb{C}) \cong \mathrm{H}^\bullet(\mathfrak{g},\mathfrak{g}_{\overline{0}};\mathbb{C}) \cong S^\bullet(\mathfrak{g}_{\overline{1}}^*)^{\mathfrak{g}_{\overline{0}}} = S^\bullet(\mathfrak{g}_{\overline{1}}^*)^{G_{\overline{0}}}.$$

Since $\mathfrak{g}$ is classical, $\mathrm{H}^\bullet(\mathfrak{g},\mathfrak{g}_{\overline{0}};\mathbb{C})$ is a finitely-generated algebra and $\mathrm{Ext}^\bullet_{\mathcal{F}}(M,M)$ is a finitely-generated $\mathrm{H}^\bullet(\mathfrak{g},\mathfrak{g}_{\overline{0}};\mathbb{C})$-module. Thus conditions (FG1) and (FG2) are satisfied.

Using notions from invariant theory allows us to compute $\mathrm{H}^\bullet(\mathfrak{g},\mathfrak{g}_{\overline{0}};\mathbb{C})$ in many cases. If there exists an element $x_0 \in \mathfrak{g}_{\overline{1}}$ such that the orbit $G_{\overline{0}} \cdot x_0$ is both closed and has maximal dimension among $G_{\bar{0}}$-orbits, then the action of $G_{\bar{0}}$ on $\mathfrak{g}_{\bar{1}}$ is said to be *stable*. Set

$$\mathrm{Stab}_{G_{\bar{0}}}(x_0) = \{g \in G_{\bar{0}} : g \cdot x_0 = x_0\}.$$

When the action of $G_{\bar{0}}$ on $\mathfrak{g}_{\bar{1}}$ is stable one can define the detecting subalgebra $\mathfrak{f} = \mathfrak{f}_{\bar{0}} \oplus \mathfrak{f}_{\bar{1}}$ via

$$\mathfrak{f}_{\bar{1}} = \mathfrak{g}_{\bar{1}}^{\mathrm{Stab}_{G_{\bar{0}}}(x_0)} \qquad \text{and} \qquad \mathfrak{f}_{\bar{0}} = [\mathfrak{f}_{\bar{1}}, \mathfrak{f}_{\bar{1}}].$$

For short, we say that $\mathfrak{g}$ is *stable* if the action of $G_{\bar{0}}$ on $\mathfrak{g}_{\bar{1}}$ is stable. Set

$$N = \mathrm{Norm}_{G_{\bar{0}}}(\mathfrak{f}_{\bar{1}}) = \{g \in G_{\bar{0}} : g \cdot \mathfrak{f}_{\bar{1}} \subseteq \mathfrak{f}_{\bar{1}}\}.$$

If there exists a vector $v \in \mathfrak{g}_{\bar{1}}$ such that

$$\dim_k \{x \in \mathfrak{g}_{\bar{1}} : \mathfrak{g}_{\bar{0}}.x \subseteq \mathfrak{g}_{\bar{0}}.v\} = \dim S^\bullet(\mathfrak{g}_{\bar{1}}^*)^{G_{\bar{0}}},$$

then the action of $G_{\bar{0}}$ on $\mathfrak{g}_{\bar{1}}$ is said to be *polar*. In this case, after fixing a choice of such a $v \in \mathfrak{g}_{\bar{1}}$ one can define another detecting subalgebra $\mathfrak{e} = \mathfrak{e}_{\bar{0}} \oplus \mathfrak{e}_{\bar{1}}$ via

$$\mathfrak{e}_{\bar{1}} = \{x \in \mathfrak{g}_{\bar{1}} : \mathfrak{g}_{\bar{0}}.x \subseteq \mathfrak{g}_{\bar{0}}.v\} \qquad \text{and} \qquad \mathfrak{e}_{\bar{0}} = [\mathfrak{e}_{\bar{1}}, \mathfrak{e}_{\bar{1}}].$$

Set

$$W = \mathrm{Norm}_{G_{\bar{0}}}(\mathfrak{e}_{\bar{1}}) / \mathrm{Stab}_{G_{\bar{0}}}(\mathfrak{e}_{\bar{1}}).$$

It is known for polar actions that W will always be a finite pseudoreflection group. For short, we say that $\mathfrak{g}$ is *polar* if the action of $G_{\bar{0}}$ on $\mathfrak{g}_{\bar{1}}$ is polar. Note that both $\mathfrak{f}$ and $\mathfrak{e}$ will again be classical Lie superalgebras and, in the case when the action is both polar and stable, one can choose $x_0 = v$ and then $\mathfrak{e} \subseteq \mathfrak{f}$, and we will always do so.

By a case-by-case check of the classification of simple classical Lie superalgebras [**36**], one sees each is either stable or polar, and most are both [**7**, Table 5]. The Lie superalgebra $\mathfrak{gl}(m|n)$ is both stable and polar. In this case, the detecting subalgebras can be chosen so that $\mathfrak{f}$ is all matrices of the form given in (3.1.1) where A, B, C, and D are diagonal with the additional requirement that the (i,i) and $(m+i, m+i)$ entries are equal for all $1 \le i \le m$. The subalgebra $\mathfrak{e} \subseteq \mathfrak{f}$ is those matrices with the further condition that $(i, m+i)$ and $(m+i, i)$ are equal for all $1 \le i \le m$.

Both $\mathfrak{f}$ and $\mathfrak{e}$ are significantly smaller than $\mathfrak{g}$. For example, $\mathfrak{e}$ is nearly abelian (namely, $[\mathfrak{e}_{\bar{0}}, \mathfrak{e}] = 0$) and $\mathfrak{f}$ is only a little more complicated. In many respects, a detecting subalgebra simultaneously plays the role of the Cartan subalgebra in Lie theory and the role of the elementary abelian subgroups of a finite group in support variety theory. The following theorem illustrates this philosophy. Note that while the cited result is for classical Lie superalgebras that are stable *and* polar, the proof handles each case separately and so remains valid for those that are stable *or* polar.

THEOREM 3.1.1 ([**7**, Theorem 4.1]). *Assume $\mathfrak{g}$ is stable, or polar, or both. Then the restriction maps induced by the inclusions $\mathfrak{f} \hookrightarrow \mathfrak{g}$ and $\mathfrak{e} \hookrightarrow \mathfrak{g}$*

$$\mathrm{res} : \mathrm{H}^\bullet(\mathfrak{g}, \mathfrak{g}_{\bar{0}}; \mathbb{C}) \to \mathrm{H}^\bullet(\mathfrak{f}, \mathfrak{f}_{\bar{0}}; \mathbb{C}), \qquad \textit{and}$$
$$\mathrm{res} : \mathrm{H}^\bullet(\mathfrak{g}, \mathfrak{g}_{\bar{0}}; \mathbb{C}) \to \mathrm{H}^\bullet(\mathfrak{e}, \mathfrak{e}_{\bar{0}}; \mathbb{C})$$

define algebra isomorphisms

$$\mathrm{H}^\bullet(\mathfrak{g}, \mathfrak{g}_{\bar{0}}; \mathbb{C}) \cong \mathrm{H}^\bullet(\mathfrak{f}, \mathfrak{f}_{\bar{0}}; \mathbb{C})^N = S^\bullet(\mathfrak{f}_{\bar{1}}^*)^N, \quad \textit{and}$$
$$\mathrm{H}^\bullet(\mathfrak{g}, \mathfrak{g}_{\bar{0}}; \mathbb{C}) \cong \mathrm{H}^\bullet(\mathfrak{e}, \mathfrak{e}_{\bar{0}}; \mathbb{C})^W = S^\bullet(\mathfrak{e}_{\bar{1}}^*)^W.$$

If $\mathfrak{g}$ is polar, then the previous theorem describes $\mathrm{H}^\bullet(\mathfrak{g}, \mathfrak{g}_{\bar{0}}; \mathbb{C})$ as the invariant ring of a finite pseudoreflection group, and hence $\mathrm{H}^\bullet(\mathfrak{g}, \mathfrak{g}_{\bar{0}}; \mathbb{C})$ is a polynomial ring. A case-by-case check verifies that this remains true for the non-polar simple classical Lie superalgebras; see [**7**, Appendix].

Given a finite-dimensional Lie superalgebra $\mathfrak{a}$ and a finite-dimensional supermodule $M \in \mathcal{F}(\mathfrak{a}, \mathfrak{a}_{\bar{0}})$, we write $\mathcal{V}_{(\mathfrak{a},\mathfrak{a}_{\bar{0}})}(M)$ for the *relative support variety* defined via the action of the relative cohomology ring $\mathrm{H}^\bullet(\mathfrak{a}, \mathfrak{a}_{\bar{0}}; \mathbb{C})$ on the relative cohomology group $\mathrm{Ext}^\bullet_{\mathcal{F}(\mathfrak{a},\mathfrak{a}_{\bar{0}})}(M, M) \cong \mathrm{H}^\bullet(\mathfrak{a}, \mathfrak{a}_{\bar{0}}; M^* \otimes M)$, i.e.,

$$\mathcal{V}_{(\mathfrak{a},\mathfrak{a}_{\bar{0}})}(M) = \operatorname{MaxSpec}\left(\mathrm{H}^\bullet(\mathfrak{a}, \mathfrak{a}_{\bar{0}}; \mathbb{C})/\operatorname{ann}_{\mathrm{H}^\bullet(\mathfrak{a},\mathfrak{a}_{\bar{0}};\mathbb{C})} \mathrm{H}^\bullet(\mathfrak{a}, \mathfrak{a}_{\bar{0}}; M^* \otimes M)\right).$$

Importantly for both computations and for theoretical results, the elementary structure of $\mathfrak{e}$ allows one to prove that its support varieties have a rank variety description:

THEOREM 3.1.2 ([**7**, Theorem 6.4]). *Given an $\mathfrak{e}$-supermodule M, set*

$$\mathcal{V}_{\mathfrak{e}}^{\mathrm{rank}}(M) = \{x \in \mathfrak{e}_{\bar{1}} : M \text{ is not projective as an } \langle x\rangle\text{-supermodule}\} \cup \{0\}. \tag{3.1.2}$$

Then there is an isomorphism

$$\mathcal{V}_{\mathfrak{e}}^{\mathrm{rank}}(\mathbb{C}) \cong \mathcal{V}_{(\mathfrak{e},\mathfrak{e}_{\bar{0}})}(\mathbb{C}),$$

which restricts for each finite-dimensional $\mathfrak{e}$-supermodule M to an isomorphism

$$\mathcal{V}_{\mathfrak{e}}^{\mathrm{rank}}(M) \cong \mathcal{V}_{(\mathfrak{e},\mathfrak{e}_{\bar{0}})}(M)$$

Let M be a supermodule in $\mathcal{F}(\mathfrak{g}, \mathfrak{g}_{\bar{0}})$. If the action of $G_{\bar{0}}$ on $\mathfrak{g}_{\bar{1}}$ is stable and/or polar, then there are morphisms of varieties induced by restriction,

$$\mathcal{V}_{(\mathfrak{f},\mathfrak{f}_{\bar{0}})}(M) \to \mathcal{V}_{(\mathfrak{g},\mathfrak{g}_{\bar{0}})}(M),$$
$$\mathcal{V}_{(\mathfrak{e},\mathfrak{e}_{\bar{0}})}(M) \to \mathcal{V}_{(\mathfrak{g},\mathfrak{g}_{\bar{0}})}(M).$$

It was conjectured in [**7**] that these maps induce isomorphisms of varieties

$$\mathcal{V}_{(\mathfrak{f},\mathfrak{f}_{\bar{0}})}(M)/N \cong \mathcal{V}_{(\mathfrak{g},\mathfrak{g}_{\bar{0}})}(M), \tag{3.1.3}$$

$$\mathcal{V}_{(\mathfrak{e},\mathfrak{e}_{\bar{0}})}(M)/W \cong \mathcal{V}_{(\mathfrak{g},\mathfrak{g}_{\bar{0}})}(M). \tag{3.1.4}$$

This conjecture was proven for the simple classical Lie superalgebras other than type P in [**32**] using BBW parabolic subalgebras. One important consequence of this identification is that the rank variety description for $\mathfrak{e}$-support varieties can be transported to $\mathfrak{f}$- and $\mathfrak{g}$-support varieties. In particular, this identification implies that support varieties for these Lie superalgebras satisfy the Tensor Product Property:

$$\mathcal{V}_{(\mathfrak{g},\mathfrak{g}_{\bar{0}})}(M \otimes N) = \mathcal{V}_{(\mathfrak{g},\mathfrak{g}_{\bar{0}})}(M) \cap \mathcal{V}_{(\mathfrak{g},\mathfrak{g}_{\bar{0}})}(N).$$

Support varieties for simple supermodules and other interesting supermodules have been computed in some cases; see [**6**].

REMARK 3.1.3. An alternative proof of the conjectured isomorphisms (3.1.3) and (3.1.4) was given in [**40**] for the Type I classical Lie superalgebras, that is, for the classical Lie superalgebras that have a compatible $\mathbb{Z}$-grading concentrated in degrees -1, 0, and 1. However, as the following example shows, there is a gap in the proof of [**40**, Theorem 3.2.1].

Let $\mathfrak{g} = \mathfrak{gl}(m|n) = \mathfrak{g}_{-1} \oplus \mathfrak{g}_0 \oplus \mathfrak{g}_1$ be the standard $\mathbb{Z}$-grading. Let $N = \mathfrak{sl}(\mathfrak{g}_1) \oplus \mathfrak{g}_1$ viewed as a superspace by setting $N_{\bar{0}} = \mathfrak{sl}(\mathfrak{g}_1)$ and $N_{\bar{1}} = \mathfrak{g}_1$. Let $\rho : \mathfrak{g}_0 \to \mathrm{End}(\mathfrak{g}_1)$

be the structure map defining the action of $\mathfrak{g}_0$ on $\mathfrak{g}_1$. Define an action of $\mathfrak{g}^+ = \mathfrak{g}_0 \oplus \mathfrak{g}_1$ on N by

(1) $x.f = \rho(x) \circ f - f \circ \rho(x) \in \mathfrak{sl}(\mathfrak{g}_1) = N_{\bar{0}}$ for all $x \in \mathfrak{g}_0$ and $f \in \mathfrak{sl}(\mathfrak{g}_1) = N_{\bar{0}}$;
(2) $x.y = [x, y] \in \mathfrak{g}_1 = N_{\bar{1}}$ for all $x \in \mathfrak{g}_0$ and $y \in \mathfrak{g}_1 = N_{\bar{1}}$;
(3) $y.f = f(y) \in \mathfrak{sl}(\mathfrak{g}_1) = N_{\bar{0}}$ for all $y \in \mathfrak{g}_1$ and $f \in \mathfrak{sl}(\mathfrak{g}_1) = N_{\bar{1}}$;
(4) $y.y' = [y, y'] = 0 \in \mathfrak{g}_1 = N_{\bar{1}}$ for all $y \in \mathfrak{g}_1$ and $y' \in \mathfrak{g}_1 = N_{\bar{1}}$.

Now let $M = U(\mathfrak{g}) \otimes_{U(\mathfrak{g}^+)} N$, and let $\varphi : \mathfrak{g}_1 \to M$ be the even linear map that identifies $\mathfrak{g}_1$ with $M_{\bar{1}}$. Fix an $\mathfrak{f}_0$-module decomposition $\mathfrak{g}_1 = \mathfrak{f}_1 \oplus W$, and set $r = \dim_{\mathbb{C}}(\mathfrak{f}_1)/\dim_{\mathbb{C}}(W)$. Let $\psi = \mathrm{id}_{\mathfrak{f}_1} - (r \cdot \mathrm{id}_W) \in \mathfrak{sl}(\mathfrak{g}_1) = M_{\bar{1}}$. Then one can check via direct calculation that φ defines a nonzero cohomology class $[\varphi] \in \mathrm{H}^1(\mathfrak{g}^+, \mathfrak{g}_0; M)$, that ψ defines a 0-cochain for the pair $(\mathfrak{f}^+, \mathfrak{f}_0)$, and the image of $[\varphi]$ under the restriction map

$$\mathrm{res} : \mathrm{H}^1(\mathfrak{g}^+, \mathfrak{g}_0; M) \to \mathrm{H}^1(\mathfrak{f}^+, \mathfrak{f}_0, M)$$

is equal to the coboundary of ψ. In particular, the restriction map fails to be injective in this case.

We expect that restriction defines an injective map

$$\mathrm{res} : \mathrm{H}^d(\mathfrak{g}, \mathfrak{g}_0; W) \to \mathrm{H}^d(\mathfrak{f}, \mathfrak{f}_0, W) \tag{3.1.5}$$

for all finite-dimensional $\mathfrak{g}$-supermodules W and all $d \geq 0$. Indeed, [**45, 46**] give related results for supergroups that imply an injectivity statement of this type. However, the reader should note that the Lie superalgebras of the splitting subgroups given therein for types P and Q are not quite the detecting subalgebras discussed here. In any case, given the injectivity of (3.1.5), the proof of (3.1.3) and (3.1.4) given in [**40**] is validated.

PROBLEM 3.1.4. Compute the support varieties of supermodules in other interesting cases.

QUESTION 3.1.5. A related question is which subvarieties can be realized as the support variety of a $\mathfrak{g}$-supermodule. As an ingredient in describing the Balmer spectrum of the stable category for $\mathcal{F}$ (see Section 3.3), in [**9**] it was shown for $\mathfrak{gl}(m|n)$ that every N-stable closed subset of $\mathcal{V}_{(\mathfrak{f}, \mathfrak{f}_{\bar{0}})}(\mathbb{C})$ can be realized as the support variety of a $\mathfrak{g}$-supermodule. Is this realization statement true for other stable Lie superalgebras?

3.2. Defect, atypicality, and complexity. Based on other settings, it is reasonable to hope that the dimensions of relative support varieties should encode interesting representation-theoretic invariants. In many classical settings, the dimension of a module's support variety is equal to the module's *complexity*; that is, the rate of growth of the module's minimal projective resolution. As we next explain, for classical Lie superalgebras something different occurs.

Fix a Cartan subalgebra $\mathfrak{h}_{\bar{0}} \subseteq \mathfrak{g}_{\bar{0}}$ for $\mathfrak{g}_{\bar{0}}$. By definition, the Cartan subalgebra, $\mathfrak{h}$, of $\mathfrak{g}$ is the centralizer of $\mathfrak{h}_{\bar{0}}$. The roots Φ of $\mathfrak{g}$ are the roots of the adjoint action of $\mathfrak{h}_{\bar{0}}$ on $\mathfrak{g}$. In particular, we write $\Phi_{\bar{0}}$ and $\Phi_{\bar{1}}$ for the roots that have a root vector in $\mathfrak{g}_{\bar{0}}$ and $\mathfrak{g}_{\bar{1}}$, respectively. Fix a choice of postive (resp., negative) roots Φ^+ (resp., Φ^-). Let $\mathfrak{b} = \mathfrak{h} \oplus_{\alpha \in \Phi^+} \mathfrak{g}_\alpha$ be the corresponding Borel subalgebra. For example, for $\mathfrak{gl}(m|n)$, $\mathfrak{h} = \mathfrak{h}_{\bar{0}}$ can be taken to be the diagonal matrices and $\mathfrak{b}$ can be taken to be the upper triangular matrices. For $r \in \mathbb{Z}_2$, set $\Phi_r^\pm = \Phi^\pm \cap \Phi_r$. Put

$$\rho = \frac{1}{2} \sum_{\alpha \in \Phi_{\overline{0}}^+} \alpha - \frac{1}{2} \sum_{\alpha \in \Phi_{\overline{1}}^+} \alpha.$$

A classical Lie superalgebra is called *basic* if it admits a nondegenerate, invariant, supersymmetric, even bilinear form $(-,-)$. For example, the supertrace defines such a bilinear form on $\mathfrak{gl}(m|n)$ by the formula $(X,Y) = \operatorname{str}(XY)$. By the usual identifications, a basic classical Lie superalgebra then has a nondegenerate bilinear form on $\mathfrak{h}_{\overline{0}}^*$, which we also write as $(-,-)$. By definition, the *defect* of $\mathfrak{g}$, denoted $\operatorname{def}(\mathfrak{g})$, is the cardinality of a maximal set of mutually orthogonal, isotropic elements of $\Phi_{\overline{1}}^+$ with respect to this bilinear form on $\mathfrak{h}_{\overline{0}}^*$. For example, the defect of $\mathfrak{gl}(m|n)$ is the minimum of m and n. We will assume $\mathfrak{g}$ is basic classical for the remainder of this section.

As discussed after Theorem 3.1.1, for a simple basic classical Lie superalgebra the relative cohomology ring $\mathrm{H}^\bullet(\mathfrak{g}, \mathfrak{g}_{\overline{0}}; \mathbb{C})$ is a polynomial ring in some number of variables. A case-by-case check verifies that the number of variables is precisely the defect of $\mathfrak{g}$:

$$\operatorname{def}(\mathfrak{g}) = \dim \mathrm{H}^\bullet(\mathfrak{g}, \mathfrak{g}_{\overline{0}}; \mathbb{C}) = \dim \mathcal{V}_{(\mathfrak{g},\mathfrak{g}_{\overline{0}})}(\mathbb{C}).$$

Thus the dimension of the ambient space for support varieties encodes the defect of $\mathfrak{g}$. This raises the question of whether root system combinatorics can also be used to compute the dimension of support varieties for interesting supermodules (e.g., the simple supermodules). As we next explain, the answer appears to be yes.

Given $\lambda \in \mathfrak{h}_{\overline{0}}^*$, the *atypicality* of λ, denoted $\operatorname{atyp}(\lambda)$, is the cardinality of a maximal set of mutually orthogonal, isotropic elements of $\Phi_{\overline{1}}^+$ that are orthogonal to $\lambda + \rho$. Clearly,

$$\operatorname{atyp}(\lambda) \leq \operatorname{def}(\mathfrak{g}).$$

As for ordinary Lie algebras, the simple $\mathfrak{g}$-supermodules in $\mathcal{F}$ can be labeled by their highest weight with respect to our choice of Cartan and Borel subalgebras. Let $X^+ \subseteq \mathfrak{h}_{\overline{0}}^*$ be the set of weights which index the simple supermodules in $\mathcal{F}$. Let

$$\{L(\lambda) : \lambda \in X^+ \subseteq \mathfrak{h}_{\overline{0}}^*\}$$

be a complete, irredundant set of simple $\mathfrak{g}$-supermodules in $\mathcal{F}$ according to this labeling. It is well-known that the atypicality of the highest weight encodes important representation-theoretic information about the simple supermodule it labels. For example, Kac and Wakimoto [**37**] conjectured that the superdimension of a simple supermodule,

$$\operatorname{sdim}_{\mathbb{C}} L(\lambda) := \dim_{\mathbb{C}} L(\lambda)_{\overline{0}} - \dim_{\mathbb{C}} L(\lambda)_{\overline{1}},$$

should be nonzero if and only if $\operatorname{atyp}(\lambda) = \operatorname{def}(\mathfrak{g})$. In [**30**], the second author along with Geer and Paturau-Mirand formulated a generalization of this conjecture for all degrees of atypicality by relating it to the nonvanishing of certain modified dimension functions. The Kac–Wakimoto conjecture is known for the finite-type symmetrizable Kac-Moody superalgebras and its generalization is known to be true for $\mathfrak{gl}(m|n)$ and $\mathfrak{osp}(m|2n)$; see [**23, 24, 31, 38, 44**].

It is easy to see from the rank variety description given by (3.1.2) and (3.1.4) that if $\operatorname{sdim}_k L(\lambda) \neq 0$, then $\mathcal{V}_{(\mathfrak{g},\mathfrak{g}_{\overline{0}})}(L(\lambda)) = \mathcal{V}_{(\mathfrak{g},\mathfrak{g}_{\overline{0}})}(\mathbb{C})$. That is, the nonvanishing of the superdimension is related to the support variety being as large as possible, and hence having dimension equal to the defect of $\mathfrak{g}$. In light of the generalized Kac–Wakimoto conjecture, one could imagine that atypicality is related to the dimension of the support variety more generally. The next result confirms this speculation.

THEOREM 3.2.1 ([**6**, Theorem 4.7], [**38**, Theorem 4.3]). *Let $\mathfrak{g}$ be $\mathfrak{gl}(m|n)$ or $\mathfrak{osp}(m|2n)$, and let $L(\lambda)$ be a simple $\mathfrak{g}$-supermodule in $\mathcal{F}$. Then,*

$$\dim \mathcal{V}_{(\mathfrak{g},\mathfrak{g}_{\bar{0}})}(L(\lambda)) = \operatorname{atyp}(\lambda).$$

This gives a cohomological/geometric interpretation of atypicality. Moreover, it naturally suggests to define the atypicality of an arbitrary supermodule in $\mathcal{F}$ as the dimension of its support variety.

QUESTION 3.2.2. Suppose $\mathfrak{g}$ is a classical Lie superalgebra that is not basic, e.g., a Lie superalgebra of type P or type Q. Can one give a combinatorial method of computing the dimension of a simple $\mathfrak{g}$-supermodule's support variety?

The dimension of a relative support variety does not directly equal the complexity of a supermodule, but we can ask if they are nevertheless related. It turns out that there are two relevant notions of complexity. As mentioned earlier, the *complexity* of a supermodule $M \in \mathcal{F}$ is the rate of growth of the dimensions of the terms in a minimal projective resolution $P_\bullet$ in $\mathcal{F}$ of M. That is, the complexity of M is the minimal integer c such that there is a constant $K > 0$ for which $\dim_{\mathbb{C}} P_d \leq K d^{c-1}$ for all $d \geq 0$ (declaring the complexity to be infinite if no such c exists). A second invariant is the *z-complexity* of M, defined to be the rate of growth of the number of indecomposable summands of the minimal resolution. Importantly, unlike complexity, z-complexity is invariant under category equivalences. Let us write $c_{\mathcal{F}}(M)$ and $z_{\mathcal{F}}(M)$ for the complexity and z-complexity of M, respectively.

The *odd nullcone* of a Lie superalgebra $\mathfrak{g}$ is the set

$$\mathcal{N}_{\text{odd}}(\mathfrak{g}) = \mathcal{X}_{\mathfrak{g}}(\mathbb{C}) = \{x \in \mathfrak{g}_{\bar{1}} : [x,x] = 0\}.$$

Given M in $\mathcal{F}$, one has the *associated variety* introduced by Duflo and Serganova [**20**]:

$$\mathcal{X}_{\mathfrak{g}}(M) = \left\{x \in \mathcal{N}_{\text{odd}}(L) : M|_{\langle x\rangle} \text{ is not free}\right\} \cup \{0\}. \tag{3.2.1}$$

If $\mathfrak{g}$ is $\mathfrak{gl}(m|n)$, $\mathfrak{osp}(2|2n)$, $\mathfrak{osp}(k|2)$, $D(2,1;\alpha)$, $G(3)$, or $F(4)$, then it is known that the complexity and z-complexity of a simple supermodule each have a geometric interpretation.

THEOREM 3.2.3 ([**8, 21, 22**]). *Let $\mathfrak{g}$ be a Lie superalgebra from the above list and let L be a simple supermodule in $\mathcal{F}$. Then*

$$\begin{aligned} c_{\mathcal{F}}(L) &= \dim \mathcal{V}_{(\mathfrak{g},\mathfrak{g}_{\bar{0}})}(L) + \dim \mathcal{X}_{\mathfrak{g}}(L), \quad \textit{and} \\ z_{\mathcal{F}}(L) &= \dim \mathcal{V}_{(\mathfrak{f},\mathfrak{f}_{\bar{0}})}(L). \end{aligned}$$

If $\mathfrak{g}$ is either $\mathfrak{gl}(m|n)$ or $\mathfrak{osp}(2|2n)$, then $\mathfrak{g}$ is of Type I and there is a compatible $\mathbb{Z}$-grading $\mathfrak{g} = \mathfrak{g}_{-1} \oplus \mathfrak{g}_0 \oplus \mathfrak{g}_1$. Given a simple $\mathfrak{g}_0$-supermodule S, one makes it into a $\mathfrak{g}_0 \oplus \mathfrak{g}_1$-supermodule by having $\mathfrak{g}_1$ act trivially, and one then defines the associated *Kac supermodule*

$$K(S) = U(\mathfrak{g}) \otimes_{U(\mathfrak{g}_0 \oplus \mathfrak{g}_1)} S.$$

In the cases $\mathfrak{g} = \mathfrak{gl}(m|n)$ and $\mathfrak{g} = \mathfrak{osp}(2|2n)$, the previous theorem also holds true if L is replaced with $K(S)$. Similarly, the type P Lie superalgebra has so-called *thin* Kac supermodules, and the above formulas hold for these as well [**5**].

Unfortunately, the above results are obtained by determining explicit formulas for the complexity and z-complexity and, separately, the dimensions of the relevant varieties. The proof is then completed by nodding sagely and observing that the two numbers in each equality are indeed equal. This leads to the following important open question:

QUESTION 3.2.4. Do the formulas for $c_{\mathcal{F}}(L)$ and $z_{\mathcal{F}}(L)$ given in Theorem 3.2.3 hold for arbitrary classical Lie superalgebras and arbitrary supermodules in $\mathcal{F}$?

Answering this question will no doubt require new insights. In particular, the Duflo–Serganova associated variety is not (yet) known to have a cohomological description in characteristic zero. The fact that it appears alongside the support variety in these complexity formulas remains unexplained. For Type I simple Lie superalgebras, a hint that it has a role to play can be found in the fact that the associated variety vanishes if and only if the supermodule in question is projective [**31**].

It is worth remarking that an interesting theory of relative support varieties can also be developed for certain non-classical Lie superalgebras. If $\mathfrak{g} = \bigoplus_{k\in\mathbb{Z}} \mathfrak{g}_k$ has a suitable $\mathbb{Z}$-grading, then one can instead consider relative cohomology for the pair $(\mathfrak{g}, \mathfrak{g}_0)$. This setup applies to the simple Lie superalgebras of type WSH in the Kac classification. For example, in [**1**] the theory is developed for the Lie superalgebras $W(n)$ and, once again, the support variety of a simple supermodule is shown to capture its atypicality.

3.3. The Balmer spectrum. Let $\mathcal{K} := \operatorname{Stab}(\mathcal{F})$ be the stable supermodule category for $\mathcal{F}$ obtained by factoring out the homomorphisms that factor through a projective supermodule. Since the projective and injective supermodules in $\mathcal{F}$ coincide, $\mathcal{K}$ is a triangulated category. The tensor structure inherited from $\mathcal{F}$ makes $\mathcal{K}$ a symmetric tensor triangulated category. To such a category, Balmer [**2**] introduced the notion of the spectrum of $\mathcal{K}$, denoted $\operatorname{Spec}(\mathcal{K})$.

Briefly, a *(thick) tensor ideal* of $\mathcal{K}$ is a full, replete, triangulated subcategory $\mathcal{I}$ of $\mathcal{K}$ that satisfies the following conditions:

- If X and Y are objects of $\mathcal{K}$ and $X \oplus Y$ is an object of $\mathcal{I}$, then X and Y are objects of $\mathcal{I}$;
- If X is an object of $\mathcal{K}$ and Y is an object of $\mathcal{I}$, then $X \otimes Y$ is an object of $\mathcal{I}$.

A tensor ideal $\mathcal{I}$ is *prime* if it further satisfies the condition that whenever $X \otimes Y$ is an object of $\mathcal{I}$, then either X or Y is an object of $\mathcal{I}$.

The Balmer spectrum $\operatorname{Spec}(\mathcal{K})$ is the collection of all proper prime tensor ideals in $\mathcal{K}$, considered as a topological space via the Zariski topology. The *support* of an object X in $\mathcal{K}$ is defined by

$$\operatorname{supp}(X) = \{\mathcal{P} \in \operatorname{Spec}(\mathcal{K}) : X \text{ is not an object of } \mathcal{P}\}.$$

The Balmer spectrum and support are known to have many of the desirable properties of a support variety theory. They satisfy a certain universal property for such support variety theories, and they can be used to classify the tensor ideals of $\mathcal{K}$. For classical Lie superalgebras we have seen four interrelated support theories: the support varieties of $\mathfrak{g}$, $\mathfrak{f}$, $\mathfrak{e}$, and the associated variety of Duflo and Serganova. It is an obvious question to ask how these are related to Balmer's 'universal' support theory, and to hopefully give a more concrete description of Balmer's theory for $\mathrm{Stab}(\mathcal{F})$.

As we explain, the detecting subalgebra $\mathfrak{f}$ plays a key role. First, recall that N is the normalizer of the action of $G_{\bar{0}}$ on $\mathfrak{f}_{\bar{1}}$. There is an action of N on $S^\bullet(\mathfrak{f}_{\bar{1}}) = \mathrm{H}^\bullet(\mathfrak{f}, \mathfrak{f}_{\bar{0}}; \mathbb{C})$. Let $X = \mathrm{Proj}(S^\bullet(\mathfrak{f}_{\bar{1}}))$ be the set of homogeneous (with respect to the cohomological $\mathbb{Z}$-grading) prime ideals of $S^\bullet(\mathfrak{f}_{\bar{1}})$, and let $X_N = N\text{-}\mathrm{Proj}(S^\bullet(\mathfrak{f}_{\bar{1}})) \subseteq X$ be the set of homogeneous (with respect to the $\mathbb{Z}$-grading) N-prime ideals of $S^\bullet(\mathfrak{f}_{\bar{1}})$. By an N-prime ideal we mean an N-stable ideal of $S^\bullet(\mathfrak{f}_{\bar{1}}^*)$ that is prime among the set of all N-stable ideals; see for example [**41**]. There is a canonical map $\pi : X \to X_N$ given by $P \mapsto \bigcap_{n \in N} n \cdot P$.

The key feature of X_N is that its closed sets are in bijection, via π, with the N-stable closed sets of X. In particular, given a $\mathfrak{g}$-supermodule M in $\mathcal{F}$, the projectivization of the support variety $\mathcal{V}_{(\mathfrak{f},\mathfrak{f}_{\bar{0}})}(M)$ is an N-stable closed set in X, and its image under π is a closed set in X_N. Recall that in a topological space a subset is called *specialization closed* if it is a union of (arbitrarily many) closed subsets. Then the main result of [**9**] is the following:

THEOREM 3.3.1 ([**9**, Theorems 5.2.1 and 5.2.2]). *Let $\mathfrak{g}$ be $\mathfrak{gl}(m|n)$ or $\mathfrak{osp}(2|2n)$. Then there is a homeomorphism*

$$X_N \to \mathrm{Stab}(\mathcal{F}(\mathfrak{g}, \mathfrak{g}_{\bar{0}}))$$

that identifies the cohomological support variety (with respect to the detecting subalgebra $\mathfrak{f}$) with the Balmer support.

Moreover, the thick tensor ideals of $\mathrm{Stab}(\mathcal{F}(\mathfrak{g}, \mathfrak{g}_{\bar{0}}))$ *are in bijection with the specialization closed subsets of X_N.*

QUESTION 3.3.2. Can one obtain a similar description of the Balmer spectrum for classical Lie superalgebras in other types?

If one has a support variety theory, then there are results that can be used to identify it with Balmer's support theory and, in turn, to identify the Balmer spectrum; see [**9**, Theorem 3.5.1] or [**11**, Theorem 1.5]. Using these results requires that the support variety theory in question satisfy a list of properties, including the 'realizability' condition that whenever W is a closed subset of the ambient space, then there is an object M in $\mathcal{K}$ for which the support theory's value on M is W. In the context of the previous theorem, this says, roughly, that given a closed N-stable subset of $\mathrm{Proj}(S^\bullet(\mathfrak{f}_{\bar{1}}))$, we must find a finite-dimensional $\mathfrak{g}$-supermodule that has the given closed subset as its $\mathfrak{f}$-support variety. The standard realization arguments are not helpful here. Instead, the necessary closed sets are constructed 'by hand' using Kac supermodules. Since Kac supermodules only exist for Type I Lie superalgebras, new methods would be required for the other types, see Question 3.1.5.

In other contexts (e.g., modular representations of finite groups) it is known that support varieties can be used to classify and relate other families of subcategories. For example, if $\mathcal{C}(\mathfrak{g}, \mathfrak{g}_{\bar{0}})$ is the category of all $\mathfrak{g}$-supermodules which are finitely semisimple as $\mathfrak{g}_{\bar{0}}$-modules, then $\mathrm{Stab}(\mathcal{F}(\mathfrak{g}, \mathfrak{g}_{\bar{0}}))$ is the compact part of $\mathrm{Stab}(\mathcal{C}(\mathfrak{g}, \mathfrak{g}_{\bar{0}}))$ and in this setting the Telescope Conjecture asks if the smashing localizing tensor ideals of $\mathrm{Stab}(\mathcal{C}(\mathfrak{g}, \mathfrak{g}_{\bar{0}}))$ are generated by the compact objects they contain. Or one can ask which (co)localizing subcategories of $\mathrm{Stab}(\mathcal{C}(\mathfrak{g}, \mathfrak{g}_{\bar{0}}))$

and related categories are classified by (co)supports. These types of classification results—beyond the thick tensor ideals of $\mathrm{Stab}(\mathcal{F}(\mathfrak{g},\mathfrak{g}_{\overline{0}}))$ discussed above—remain an open problem. It is natural to ask which of these results have anologues for Lie superalgebras.

QUESTION 3.3.3. Can one develop analogues for $\mathcal{F}(\mathfrak{g},\mathfrak{g}_{\overline{0}})$ and $\mathcal{C}(\mathfrak{g},\mathfrak{g}_{\overline{0}})$ of results known to hold for other support theories?

4. Support varieties for finite supergroup schemes in characteristic zero

4.1. Finite supergroup schemes in characteristic zero. The data of a finite supergroup scheme is equivalent to the data of a finite-dimensional cocommutative Hopf superalgebra. Over an algebraically closed field of characteristic zero, these have a particularly simple form:

THEOREM 4.1.1 ([**15**, Corollary 3.1.2]). *Let k be an algebraically closed field of characteristic zero, and let A be a finite-dimensional cocommutative Hopf superalgebra over k. Then there exists a finite group G, a finite-dimensional odd superspace V, and a representation of G on V such that A is isomorphic as a Hopf superalgebra to the smash product algebra $\Lambda(V)\#kG$. Here kG denotes the group algebra of G over k, considered as a purely even superalgebra.*

We denote the finite supergroup scheme corresponding to the algebra $\Lambda(V)\#kG$ by $V\rtimes G$, and we write V/G for the set of G-orbits in V. Over a field of characteristic zero, the group algebra kG is semisimple. This leads to the following calculation of the cohomological spectrum of $V \rtimes G$:

THEOREM 4.1.2 ([**15**, Theorem 3.2.2]). *Let G be a finite group, and let V be a finite-dimensional purely odd kG-module. Then there exist isomorphisms of varieties*

$$|V \rtimes G| \cong \mathrm{MaxSpec}(S(V^*)^G) \cong V/G.$$

Write $[v]$ for the G-orbit of an element $v \in V$, and write $\langle v\rangle$ for the k-subalgebra of $\Lambda(V)$ generated by v. By convention, $\langle 0\rangle = k$. The category of $V \rtimes G$-supermodules is equivalent to the category of supermodules over the superalgebra $\Lambda(V)\#kG$. Now one gets the following rank variety description for the support varieties of finite-dimensional $V \rtimes G$-supermodules:

THEOREM 4.1.3 ([**15**, Theorem 3.2.3]). *Let G be a finite group, and let V be a finite-dimensional purely odd kG-module. Let M be a finite-dimensional $V \rtimes G$-supermodule. Then*

$$|V \rtimes G|_M \cong \left\{[v] \in V/G : M|_{\langle v\rangle} \text{ is not free}\right\} \cup \{0\}.$$

Part 2. Positive characteristic

In Part 2, let k be a field of characteristic $p > 0$. Usually $p \neq 2$, although some results can be extended to the case $p = 2$; see [**19**]. For convenience we will assume that k is algebraically closed, although some results may hold without this assumption. Given a k-vector space V and an integer $r \geq 1$, let $V^{(r)} = V \otimes_{\varphi_r} k$ be the r-th Frobenius twist of V, i.e., the k-vector space obtained from V via base change along the (r-th) Frobenius endomorphism $\varphi_r : k \to k$, $\lambda \mapsto \lambda^{p^r}$. If $W = V^{(r)}$, then we may write $V = W^{(-r)}$. More generally, if X is an affine k-scheme (resp. affine algebraic variety) with coordinate algebra $k[X]$, then we may write $X^{(r)}$ for the scheme (resp. variety) with coordinate algebra $k[X^{(r)}] = k[X]^{(r)}$, and if $Y = X^{(r)}$, then we may write $X = Y^{(-r)}$.

5. Support varieties via ordinary Lie superalgebra cohomology

In this section we describe the support variety theory that arises for an arbitrary finite-dimensional Lie superalgebra $\mathfrak{g}$ over a field of (odd) positive characteristic. One gets a rank variety theory that nicely parallels the classical results for finite-dimensional restricted Lie algebras (discussed later in Section 6.1), and that exhibits surprising connections to the non-cohomological associated variety discussed earlier in (3.2.1).

5.1. Lie superalgebra cohomology in positive characteristic. Let $\mathfrak{g}$ be a finite-dimensional Lie superalgebra over the field k. As discussed in Section 2.2, the Lie superalgebra cohomology ring $\mathrm{H}^\bullet(\mathfrak{g}, k)$ can be computed as the cohomology of the superexterior algebra $\boldsymbol{\Lambda}(\mathfrak{g}^*)$ with respect to the Koszul differential ∂. Since ∂ makes $\boldsymbol{\Lambda}(\mathfrak{g}^*)$ into a differential graded superalgebra, and since $\boldsymbol{\Lambda}(\mathfrak{g}^*)$ is graded-commutative in the sense of (1.2.1), it follows for all $f \in \mathfrak{g}_{\overline{1}}^* \subseteq \boldsymbol{\Lambda}^1(\mathfrak{g}^*)$ that $\partial(f^p) = 0$. Then there exists a map of graded superalgebras

$$\varphi : S(\mathfrak{g}_{\overline{1}}^*[p])^{(1)} \to \mathrm{H}^\bullet(\mathfrak{g}, k) \tag{5.1.1}$$

induced by the p-power map on $S(\mathfrak{g}_{\overline{1}}^*)$ and the inclusion $S(\mathfrak{g}_{\overline{1}}^*) \subset \boldsymbol{\Lambda}(\mathfrak{g}^*)$. If M and N are finite-dimensional $\mathfrak{g}$-supermodules, then $\mathrm{Ext}_{\mathfrak{g}}^\bullet(M, N)$ is a finite module over the image of φ. In particular, $\mathrm{H}^\bullet(\mathfrak{g}, k)$ is a finitely-generated k-algebra, and $\mathrm{Ext}_{\mathfrak{g}}^\bullet(M, N)$ is a finite $\mathrm{H}^\bullet(\mathfrak{g}, k)$-module.

Given a $\mathfrak{g}$-supermodule M, let $J_{\mathfrak{g}}(M)$ be the kernel of the composite map

$$S(\mathfrak{g}_{\overline{1}}^*)^{(1)} \xrightarrow{\varphi} \mathrm{H}^\bullet(\mathfrak{g}, k) \xrightarrow{-\otimes M} \mathrm{Ext}_{\mathfrak{g}}^\bullet(M, M),$$

and set

$$\mathcal{X}_{\mathfrak{g}}(M) = \mathrm{MaxSpec}\left(S(\mathfrak{g}_{\overline{1}}^*)^{(1)}/J_{\mathfrak{g}}(M)\right).$$

Then φ induces for each $\mathfrak{g}$-supermodule M a homeomorphism of varieties

$$\varphi^* : |U(\mathfrak{g})|_M \simeq \mathcal{X}_{\mathfrak{g}}(M). \tag{5.1.2}$$

In the case $M = k$, this provides the following explicit description of the cohomological spectrum for $U(\mathfrak{g})$ in terms of the *odd nullcone* of $\mathfrak{g}$.

THEOREM 5.1.1 ([**15**, Theorem 4.2.4]). *Let $\mathfrak{g}$ be a finite-dimensional Lie superalgebra over k. Then*

$$\mathcal{X}_{\mathfrak{g}}(k)^{(-1)} = \mathcal{N}_{\mathrm{odd}}(\mathfrak{g}) := \{x \in \mathfrak{g}_{\overline{1}} : [x, x] = 0\}.$$

5.2. Support varieties as rank varieties. One of the main results of [**18**] was the following 'rank variety' description for $\mathcal{X}_{\mathfrak{g}}(M)$ when M is finite-dimensional:

THEOREM 5.2.1 ([**18**, Corollary 3.3.2]). *Let $\mathfrak{g}$ be a finite-dimensional Lie superalgebra over k, and let M be a finite-dimensional $\mathfrak{g}$-supermodule. Then*

$$\mathcal{X}_{\mathfrak{g}}(M)^{(-1)} = \left\{x \in \mathcal{N}_{\mathrm{odd}}(\mathfrak{g}) : M|_{\langle x \rangle} \text{ is not free}\right\} \cup \{0\}. \tag{5.2.1}$$

Here $M|_{\langle x \rangle}$ denotes the restriction of M to the k-subalgebra of $U(\mathfrak{g})$ generated by x; for $x = 0$ this subalgebra is just the field k (and hence $M|_{\langle x \rangle}$ is free), while for $0 \neq x \in \mathcal{N}_{\mathrm{odd}}(\mathfrak{g})$ this subalgebra is of the form $k[x]/(x^2)$. The rank variety on the right-hand side of (5.2.1) is equivalent in definition to the 'associated variety' of a

$\mathfrak{g}$-supermodule originally defined—without reference to cohomology—by Duflo and Serganova for Lie superalgebras in characteristic zero [**20**], and mentioned earlier in Section 3.2.

Let us say a few words about the strategy used in [**18**] to prove (5.2.1). Let $\mathcal{X}'_{\mathfrak{g}}(M)$ be the closed subset of $\mathcal{X}_{\mathfrak{g}}(k) = \mathcal{N}_{\text{odd}}(\mathfrak{g})$ such that $\mathcal{X}'_{\mathfrak{g}}(M)^{(-1)}$ is the right-hand side of (5.2.1). First, using naturality it is relatively easy to show that $\mathcal{X}'_{\mathfrak{g}}(M) \subseteq \mathcal{X}_{\mathfrak{g}}(M)$. Equality for arbitrary $\mathfrak{g}$ and M can then be reduced to showing for all $m \geq 1$ that equality holds for the general linear Lie superalgebra $\mathfrak{gl}(m|m)$ and its natural representation $M = k^{m|m}$. To prove equality for $\mathfrak{g} = \mathfrak{gl}(m|m)$, we consider the Clifford filtration on $\mathfrak{g}$, that is, the Lie superalgebra filtration $F^0\mathfrak{g} \subseteq F^1\mathfrak{g} \subseteq F^2\mathfrak{g}$ defined by $F^0\mathfrak{g} = 0$, $F^1\mathfrak{g} = \mathfrak{g}_{\overline{1}}$, and $F^2\mathfrak{g} = \mathfrak{g}$. The associated graded Lie superalgebra $\widetilde{\mathfrak{g}}$ is concentrated in $\mathbb{Z}$-degrees 1 and 2, with $\widetilde{\mathfrak{g}}_1 \cong \mathfrak{g}_{\overline{1}}$ and $\widetilde{\mathfrak{g}}_2 \cong \mathfrak{g}_{\overline{0}}$ as superspaces. The Lie bracket on $\widetilde{\mathfrak{g}}_1$ identifies with the original Lie bracket on $\mathfrak{g}_{\overline{1}}$ (and hence retains information about the odd nullcone of $\mathfrak{g}$), while $\widetilde{\mathfrak{g}}_2$ is now central in $\widetilde{\mathfrak{g}}$. Given a choice of generators for the $\mathfrak{g}$-supermodule M, one can also define an associated graded $\widetilde{\mathfrak{g}}$-supermodule $\widetilde{M}$. Now the Clifford filtration gives rise to a spectral sequence relating $\mathrm{H}^\bullet(\widetilde{\mathfrak{g}}, k)$ and $\mathrm{H}^\bullet(\mathfrak{g}, k)$, and by studying this spectral sequence one can relate the support of M as a $\mathfrak{g}$-supermodule and the support of $\widetilde{M}$ as a $\widetilde{\mathfrak{g}}$-supermodule. Finally, the support of $\widetilde{M}$ as a $\widetilde{\mathfrak{g}}$-supermodule is calculable, and thus can be used to help put an upper bound on the size of $\mathcal{X}_{\mathfrak{g}}(M)$. The support of the $\widetilde{\mathfrak{g}}$-supermodule $\widetilde{M}$ is easier to calculate partly because $\widetilde{\mathfrak{g}}$ has a simpler Lie superalgebra structure than $\widetilde{\mathfrak{g}}$, and partly because the $\widetilde{\mathfrak{g}}$-action on $\widetilde{M}$ factors through the action of a p-nilpotent restricted Lie superalgebra (equivalently, through the action of an infinitesimal unipotent supergroup scheme), for which we have a clearer picture of the support theory; see Remark 7.6.3.

An immediate consequence of the 'rank variety' description (5.2.1) and the representation theory of the Hopf algebra $k[x]/(x^2)$ is that support varieties satisfy the Tensor Product Property:

PROPOSITION 5.2.2. *Let $\mathfrak{g}$ be a finite-dimensional Lie superalgebra over a field k of characteristic $p \geq 3$, and let M and N be finite-dimensional $\mathfrak{g}$-supermodules. Then*

$$\mathcal{X}_{\mathfrak{g}}(M \otimes N) = \mathcal{X}_{\mathfrak{g}}(M) \cap \mathcal{X}_{\mathfrak{g}}(N).$$

Consequently, $|U(\mathfrak{g})|_{M\otimes N} = |U(\mathfrak{g})|_M \cap |U(\mathfrak{g})|_N$.

As mentioned in Section 2.1, if A is a *finite-dimensional* (and hence self-injective) Hopf superalgebra whose cohomology satisfies the finiteness properties (FG1) and (FG2), then for each finite-dimensional A-supermodule M one gets $\dim(|A|_M) = \mathrm{cx}_A(M)$. In particular, one gets that $|A|_M = \{0\}$ if and only if M is projective. The cohomology of $U(\mathfrak{g})$ does satisfy (FG1) and (FG2), but $U(\mathfrak{g})$ is not finite-dimensional unless $\mathfrak{g}$ is a purely odd (abelian) Lie superalgebra. While $U(\mathfrak{g})$ is not self-injective in general, it is Gorenstein (i.e., it has finite injective dimension as a module over itself), and a finitely-generated $U(\mathfrak{g})$-supermodule M has finite projective dimension if and only if it has finite injective dimension; see [**18**, §2.4]. For a finite-dimensional $U(\mathfrak{g})$-supermodule M, one can show that $|U(\mathfrak{g})|_M = \{0\}$ if and only if M is of finite projective dimension. Reformulated in terms of rank varieties, this yields the following statement:

THEOREM 5.2.3 ([**18**, Theorem 3.4.2, Corollary 3.4.3]). *Let $\mathfrak{g}$ be a finite-dimensional Lie superalgebra over an algebraically closed field k of characteristic $p \geq 3$,* *and let M be a finite-dimensional $\mathfrak{g}$-supermodule. Then*

$$\{x \in \mathcal{N}_{\text{odd}}(\mathfrak{g}) : M|_{\langle x \rangle} \text{ is not free}\} = \emptyset \quad \text{if and only if} \quad \operatorname{projdim}_{U(\mathfrak{g})}(M) < \infty.$$

In particular, if $\mathcal{N}_{\text{odd}}(\mathfrak{g}) = \{0\}$, then $U(\mathfrak{g})$ has finite global dimension (and conversely).

For the last statement of the theorem, one observes that $\operatorname{projdim}_{U(\mathfrak{g})}(M) \leq \operatorname{projdim}_{U(\mathfrak{g})}(k)$ for any $U(\mathfrak{g})$-supermodule M, because a projective resolution for M can be obtained by applying the functor $- \otimes M$ to a projective resolution for the trivial module k. For the converse of the last statement, one observes that $\operatorname{gldim}(U(\mathfrak{g}')) \leq \operatorname{gldim}(U(\mathfrak{g}))$ for any Lie sub-superalgebra $\mathfrak{g}'$ of $\mathfrak{g}$, as a consequence of the fact that $U(\mathfrak{g})$ is free over $U(\mathfrak{g}')$. So if $U(\mathfrak{g})$ has finite global dimension, it cannot have any subalgebras of the form $k[x]/(x^2)$ for $x \in \mathfrak{g}_{\overline{1}}$.

REMARK 5.2.4. The last statement of Theorem 5.2.3 can also be derived from a theorem of Bøgvad [**10**] concerning the classification of solvable graded Lie algebras of finite global dimension. Luchezar Avramov kindly pointed out to us that (some of our assertions in [**18**] notwithstanding) Bøgvad makes no assumption on the characteristic of the ground field, and this generality together with some straightforward reduction arguments implies that Bøgvad's theorem extends to arbitrary finite-dimensional Lie superalgebras. For a recent treatment of Bøgvad's theorem in characteristic zero, see [**42**, Theorem 17.1.2].

PROBLEM 5.2.5. Let $\mathfrak{g}$ be a finite-dimensional Lie superalgebra over k, and let M be a finite-dimensional $\mathfrak{g}$-supermodule. Provide a general representation-theoretic interpretation for the geometric dimension of the support variety

$$|U(\mathfrak{g})|_M \simeq \mathcal{X}_{\mathfrak{g}}(M).$$

6. Finite group schemes: recollections from the non-super theory

To aid the reader in comparing the super theory with its classical (non-super) counterpart, in this section we summarize the classical results for restricted Lie algebras and infinitesimal group schemes (as developed by Friedlander and Parshall and by Suslin, Friedlander, and Bendel), before discussing in Section 7 our work to date related to restricted Lie superalgebras and infinitesimal supergroup schemes. Section 6.1 summarizes the general support variety theory for finite-dimensional restricted Lie algebras, and Section 6.2 describes the cohomological finite-generation result of Friedlander and Suslin for infinitesimal group schemes. In Section 6.3 we describe the theory of support varieties for infinitesimal group schemes as developed by Suslin, Friedlander, and Bendel using infinitesimal one-parameter subgroups, and in Section 6.4 we describe the generalization to arbitrary finite group schemes developed by Friedlander and Pevtsova using π-points.

6.1. Restricted Lie algebras. Let $\mathfrak{g}$ be a finite-dimensional restricted Lie algebra over k. The restricted enveloping algebra of $\mathfrak{g}$, denoted $V(\mathfrak{g})$, is a finite-dimensional cocommutative Hopf algebra over k. The dual Hopf algebra $V(\mathfrak{g})^* = \operatorname{Hom}_k(V(\mathfrak{g}), k)$ is then a finite-dimensional commutative Hopf algebra over k with the property that $f^p = 0$ for each element f of the augmentation ideal I_ε of $V(\mathfrak{g})^*$. Thus $V(\mathfrak{g})^*$ is the coordinate algebra $k[G]$ of a height-1 infinitesimal group scheme G, i.e., a finite k-group scheme that is equal to its own first Frobenius kernel $G_{(1)}$. The category of (left) $V(\mathfrak{g})$-modules is then equivalent to the category of rational (left) G-modules (i.e., the category of right $k[G]$-comodules), and cohomology for $V(\mathfrak{g})$ may be identified with (rational) cohomology for G. For more details, see [**35**, I.2.7, I.2.8, I.8, I.9].

Let M be a $V(\mathfrak{g})$-module (equivalently, a rational G-module). The powers of the augmentation ideal $I_\varepsilon \subset k[G]$ give rise to a filtration on the Hochschild complex $C^\bullet(G, M)$ that computes the cohomology group $\mathrm{H}^\bullet(G, M) = \mathrm{H}^\bullet(V(\mathfrak{g}), M)$. The filtration gives rise in turn to a spectral sequence, which for $p > 2$ can be written in the form

$$E_2^{i,j}(M) = S^{i/2}(\mathfrak{g}^*)^{(1)} \otimes \mathrm{H}^j(\mathfrak{g}, M) \Rightarrow \mathrm{H}^{i+j}(V(\mathfrak{g}), M) \tag{6.1.1}$$

Here $\mathrm{H}^\bullet(\mathfrak{g}, M)$ is the (ordinary) Lie algebra cohomology group of M, and the superscript $i/2$ means that $E_2^{i,j}(M) = 0$ for odd i. For details on the construction and indexing of this spectral sequence, see [**35**, I.9.20]. In the case $M = k$, the horizontal edge map of (6.1.1) defines a finite map of graded k-algebras

$$\Phi^* : S(\mathfrak{g}^*[2])^{(1)} \to \mathrm{H}^\bullet(V(\mathfrak{g}), k). \tag{6.1.2}$$

Here $S(\mathfrak{g}^*[2])$ means that we consider the symmetric algebra $S(\mathfrak{g}^*)$ as generated in cohomological degree 2. Passing to maximal ideal spectra, one gets a finite morphism of varieties

$$\Phi : |V(\mathfrak{g})| := \operatorname{MaxSpec}\left(\mathrm{H}^\bullet(V(\mathfrak{g}), k)\right) \to \operatorname{MaxSpec}\left(S(\mathfrak{g}^*[2])^{(1)}\right) = \mathfrak{g}^{(1)}$$

Jantzen [**34**] showed, as a consequence of explicit calculations in the case $\mathfrak{g} = \mathfrak{gl}_n$, that the image of the morphism Φ is the restricted nullcone of $\mathfrak{g}$:

$$\Phi(|V(\mathfrak{g})|)^{(-1)} = \mathcal{N}_1(\mathfrak{g}) := \{x \in \mathfrak{g} : x^{[p]} = 0\}. \tag{6.1.3}$$

Then for each finite-dimensional $V(\mathfrak{g})$-module M, Friedlander and Parshall [**25**] computed the image of the support variety $|V(\mathfrak{g})|_M$ under the morphism Φ:

$$\Phi(|V(\mathfrak{g})|_M)^{(-1)} = \left\{x \in \mathcal{N}_1(\mathfrak{g}) : M|_{\langle x\rangle} \text{ is not free}\right\} \cup \{0\}. \tag{6.1.4}$$

Here $M|_{\langle x\rangle}$ denotes the restriction of M to the k-subalgebra of $V(\mathfrak{g})$ generated by x. If $x = 0$ this subalgebra is just the field k, while for $x \neq 0$ the subalgebra has the form $k[x]/(x^p)$.

6.2. Cohomology of infinitesimal group schemes. For this section, fix an integer $r \geq 1$.

Using the theory of strict polynomial functors, Friedlander and Suslin [**28**] established the existence of certain 'universal extension classes' for the rational cohomology of GL_n:

$$e_i^{(r-i)} \in \mathrm{H}^{2p^{i-1}}(GL_n, \mathfrak{gl}_n^{(r)}), \quad 1 \leq i \leq r. \tag{6.2.1}$$

Here $\mathfrak{gl}_n$ denotes the adjoint representation of GL_n, and the superscript (r) indicates both that the vector space structure of $\mathfrak{gl}_n$ has been twisted by the r-th Frobenius morphism of the field, and also that the rational GL_n-module structure has been twisted via pre-composition with the r-th Frobenius morphism $F^r : GL_n \to (GL_n)^{(r)}$ of the scheme GL_n. The restriction of $e_i^{(r-i)}$ to the r-th Frobenius kernel $GL_{n(r)}$ of GL_n determines an element of

$$\mathrm{H}^{2p^{i-1}}(GL_{n(r)}, \mathfrak{gl}_n^{(r)}) \cong \mathrm{H}^{2p^{i-1}}(GL_{n(r)}, k) \otimes \mathfrak{gl}_n^{(r)} \cong \operatorname{Hom}_k(\mathfrak{gl}_n^{*(r)}, \mathrm{H}^{2p^{i-1}}(GL_{n(r)}, k)),$$

i.e., it determines a linear map $\mathfrak{gl}_n^{*(r)} \to \mathrm{H}^{2p^{i-1}}(GL_{n(r)}, k)$, which then extends multiplicatively to a homomorphism of graded k-algebras

$$S(\mathfrak{gl}_n^{*(r)}[2p^{i-1}]) \to \mathrm{H}^\bullet(GL_{n(r)}, k).$$

Taking the product of these homomorphisms, one gets a map of graded k-algebras

$$\phi_{GL_{n(r)}} : \bigotimes_{i=1}^{r} S(\mathfrak{gl}_n^{*(r)}[2p^{i-1}]) \to \mathrm{H}^\bullet(GL_{n(r)}, k). \tag{6.2.2}$$

Now let G be an infinitesimal k-group scheme of height $\leq r$. Thus G is a finite k-group scheme with the property that $f^{p^r} = 0$ for each element f of the augmentation ideal of $k[G]$. Equivalently, G is equal to its own r-th Frobenius kernel $G_{(r)}$. Set $\mathfrak{g} = \mathrm{Lie}(G)$, and fix for some n a choice of closed embedding $\iota : G \hookrightarrow GL_n$, i.e., an embedding of G as a closed subgroup scheme of GL_n. Then the image of G is contained in $GL_{n(r)}$, and ι differentiates to an injective homomorphism $d\iota : \mathfrak{g} \hookrightarrow \mathfrak{gl}_n$ of restricted Lie algebras. Composing $\phi_{GL_{n(r)}}$ with the restriction map in cohomology, one gets a homomorphism of graded k-algebras

$$\phi_G : \bigotimes_{i=1}^{r} S(\mathfrak{gl}_n^{*(r)}[2p^{i-1}]) \to \mathrm{H}^\bullet(G, k).$$

One of the main results of **[28]** then states that ϕ_G is a finite algebra map, and that $\mathrm{H}^\bullet(G, M)$ is finite over the image of ϕ_G whenever M is a finite-dimensional rational G-module. In particular, $\mathrm{H}^\bullet(G, k)$ is a finitely-generated k-algebra, and $\mathrm{H}^\bullet(G, M)$ is a finite $\mathrm{H}^\bullet(G, k)$-module. In the special case $r = 1$, the map ϕ_G reduces to the map (6.1.2) considered by Friedlander and Parshall.

6.3. Support varieties for infinitesimal group schemes. Given an affine k-group scheme G with coordinate algebra $k[G]$, and given a commutative k-algebra A, let $G_A = G \otimes_k A$ be the affine A-group scheme with coordinate algebra $A[G_A] := k[G] \otimes_k A$. If M is a rational G-module, then $M_A := M \otimes_k A$ is a rational G_A-module.

Let $\mathbb{G}_a$ be the additive group scheme, and $\mathbb{G}_{a(r)}$ its r-th Frobenius kernel. Write $\mathsf{CAlg}(k)$ for the category of commutative (unital, associative) k-algebras. The functor

$$V_r(G) : \mathsf{CAlg}(k) \to \mathsf{Set}$$

is defined by

$$V_r(G)(A) = \mathrm{Hom}_{\mathsf{Grp}/A}(\mathbb{G}_{a(r)} \otimes_k A, G \otimes_k A), \tag{6.3.1}$$

the set of A-group scheme homomorphisms $\nu : \mathbb{G}_{a(r)} \otimes_k A \to G \otimes_k A$. If G is an affine algebraic k-group scheme (i.e., if $k[G]$ is a finitely-generated k-algebra), then by [**48**, Theorem 1.5], $V_r(G)$ admits the structure of an affine k-scheme of finite type, called the scheme of one-parameter subgroups of height $\leq r$ in G. In the special case $r = 1$, $V_1(G)$ identifies with $\mathcal{N}_1(\mathfrak{g})$, the restricted nullcone of $\mathfrak{g} = \mathrm{Lie}(G)$. More generally, for $G = GL_{n(r)}$ one gets a natural identification

$$\begin{aligned} &V_r(GL_{n(r)})(A) = \\ &\qquad \{(\alpha_0, \cdots, \alpha_{r-1}) \in \mathfrak{gl}_n(A) : \alpha_i^p = 0 = [\alpha_j, \alpha_\ell] \text{ for all } 0 \leq i, j, \ell < r\}. \end{aligned} \tag{6.3.2}$$

The 'group algebra' of a finite k-group scheme, denoted kG, is the finite-dimensional Hopf algebra that is dual to the coordinate algebra of G: $kG = k[G]^*$. In the notation of [**35**, I.8], $kG = M(G)$, and if G is infinitesimal, then $kG = \mathrm{Dist}(G)$. For a finite k-group scheme, the category of rational (left) G-modules is equivalent to the category of (left) kG-modules [**35**, I.8.6], and rational cohomology for G identifies with cohomology for the Hopf algebra kG.

For $G = \mathbb{G}_{a(r)}$, one has $k[\mathbb{G}_{a(r)}] = k[T]/(T^{p^r})$, so

$$k\mathbb{G}_{a(r)} \cong k[u_0, \dots, u_{r-1}]/(u_0^p, \dots, u_{r-1}^p). \tag{6.3.3}$$

Here u_i is the functional such that $u_i(f)$ is equal to the coefficient of T^{p^i}, for each $f \in k[T]/(T^{p^r})$. Given an integer $0 \leq j < p^r$, let $j = \sum_{\ell=0}^{r-1} j_\ell p^\ell$ be its base-p decomposition (so $0 \leq j_\ell < p$), and set

$$\gamma_j = \frac{u_0^{j_0} \cdot u_1^{j_1} \cdots u_{r-1}^{j_{r-1}}}{j_0! \cdot j_1! \cdots j_{r-1}!} \in k\mathbb{G}_{a(r)}.$$

Then the γ_j form a k-basis for $k\mathbb{G}_{a(r)}$. The coproduct Δ and antipode S on $k\mathbb{G}_{a(r)}$ are given by

$$\Delta(\gamma_\ell) = \sum_{i+j=\ell} \gamma_i \otimes \gamma_j \quad \text{and} \quad S(\gamma_j) = (-1)^j \gamma_j. \tag{6.3.4}$$

Given a point $\mathfrak{s} \in V_r(G)$ (i.e., a prime ideal $\mathfrak{s} \in \operatorname{Spec} k[V_r(G)]$), let $k(\mathfrak{s})$ be the residue field of $V_r(G)$ at $\mathfrak{s}$, and let $\phi_{\mathfrak{s}} \in \operatorname{Hom}_{\mathsf{CAlg}(k)}(k[V_r(G)], k(\mathfrak{s}))$ be the canonical k-algebra homomorphism. Then $\phi_{\mathfrak{s}}$ defines a $k(\mathfrak{s})$-point of $V_r(G)$, and hence determines a homomorphism of $k(\mathfrak{s})$-group schemes $\nu_{\mathfrak{s}} : \mathbb{G}_{a(r)} \otimes_k k(\mathfrak{s}) \to G \otimes_k k(\mathfrak{s})$. Now given a rational G-module M, let $\nu_{\mathfrak{s}}^*(M \otimes_k k(\mathfrak{s}))$ be the rational $\mathbb{G}_{a(r)} \otimes_k k(\mathfrak{s})$-module (equivalently, the $k(\mathfrak{s})\mathbb{G}_{a(r)}$-module) obtained via pullback along $\nu_{\mathfrak{s}}$. Then the subset $V_r(G)_M$ of $V_r(G)$ is defined by

(6.3.5)

$$\begin{aligned} V_r(G)_M = \{\mathfrak{s} \in V_r(G) : {}& \nu_{\mathfrak{s}}^*(M \otimes_k k(\mathfrak{s})) \text{ is not projective over the subalgebra} \\ & k(\mathfrak{s})[u_{r-1}]/(u_{r-1}^p) \subset k(\mathfrak{s})[u_0, \dots, u_{r-1}]/(u_0^p, \dots, u_{r-1}^p) = k(\mathfrak{s})\mathbb{G}_{a(r)}\}. \end{aligned}$$

By [**49**, Proposition 6.1], if G is an infinitesimal k-group scheme and if M is a finite-dimensional rational G-module, then $V_r(G)_M$ is a Zariski closed conical subset of $V_r(G)$. If $r = 1$, then the set of k-points of $V_r(G)_M$ identifies with the 'rank variety' defined by the right-hand side of (6.1.4).

The identity map on $k[V_r(G)]$ defines a $k[V_r(G)]$-point of $V_r(G)$, and hence determines a universal homomorphism $u : \mathbb{G}_{a(r)} \otimes_k k[V_r(G)] \to G \otimes_k k[V_r(G)]$ of group schemes over $k[V_r(G)]$. In [**48**], Suslin, Friedlander, and Bendel (SFB) used this map to define, for each affine (algebraic) k-group scheme G, a natural k-algebra homomorphism

$$\psi_r : H(G, k) \to k[V_r(G)] \tag{6.3.6}$$

Here $H(G, k) = \mathrm{H}^{\mathrm{ev}}(G, k)$ if p is odd, and $H(G, k) = \mathrm{H}^\bullet(G, k)$ if $p = 2$. Using the algebra relations among the extension classes (6.2.1), they also showed that $\phi_{GL_{n(r)}}$ factors through a map

$$\overline{\phi} : k[V_r(GL_{n(r)})] \to H(GL_{n(r)}, k). \tag{6.3.7}$$

Let Ψ and Φ denote the morphisms of affine schemes induced by (6.3.6) and (6.3.7), respectively. Through a detailed analysis of how the universal extension classes (6.2.1) restrict to the Frobenius kernels of $\mathbb{G}_a$, SFB determined that the composite morphism

$$\Theta : V_r(GL_{n(r)}) \xrightarrow{\Psi} \operatorname{Spec} H(GL_{n(r)}, k) \xrightarrow{\Phi} V_r(GL_{n(r)})$$

equals the r-th Frobenius twist morphism for the scheme $V_r(GL_{n(r)}) = V_r(GL_n)$ [**48**, Theorem 5.2]. In particular, the map $\overline{\phi} : k[V_r(GL_{n(r)})] \to H(GL_{n(r)}, k)$ has nilpotent kernel, and by naturality it follows that the map $\psi_r : H(G,k) \to k[V_r(G)]$ is surjective onto p^r-th powers for any infinitesimal k-group scheme G of height $\leq r$.

Next, SFB proved the following detection theorem for infinitesimal groups:

THEOREM 6.3.1 ([**49**, Theorem 4.3]). *Let G be an infinitesimal group scheme of height $\leq r$ over k and let Λ be an associative unital rational G-algebra. Then the following conditions on a cohomology class $z \in \mathrm{H}^n(G, \Lambda)$ are equivalent:*

(1) *z is nilpotent.*
(2) *For every field extension K/k and every K-group scheme homomorphism $\nu : \mathbb{G}_{a(r)} \otimes_k K \to G \otimes_k K$, the cohomology class $\nu^*(z_K) \in \mathrm{H}^n(\mathbb{G}_{a(r)} \otimes_k K, \Lambda_K)$ is nilpotent.*
(3) *For every point $\mathfrak{s} \in V_r(G)$, the class $\nu_{\mathfrak{s}}^*(z_{k(\mathfrak{s})}) \in \mathrm{H}^n(\mathbb{G}_{a(r)} \otimes_k k(\mathfrak{s}), \Lambda_{k(\mathfrak{s})})$ is nilpotent.*

The detection theorem was proved first for unipotent infinitesimal groups.[1] The general case was then handled with the help of a spectral sequence relating the cohomology of a Frobenius kernel $G_{(r)}$ of a reductive group G to the cohomology of the Frobenius kernel $B_{(r)}$ of a Borel subgroup B of G. Similar ideas were also used in the proof of the equality (6.1.4).

Using the detection theorem, SFB deduced (for G infinitesimal of height $\leq r$) that the kernel of the k-algebra map $\psi_r : H(G,k) \to k[V_r(G)]$ is also nilpotent, and hence ψ_r induces a finite universal homeomorphism of schemes

$$\Psi : V_r(G) \simeq |G| := \operatorname{Spec} H(G,k).$$

Furthermore, they deduced for each finite-dimensional rational G-module M that Ψ restricts to a homeomorphism

$$\Psi : V_r(G)_M \simeq |G|_M := \operatorname{Spec}\big(H(G,k)/I_G(M)\big).$$

The rank variety description for $|G|_M$ afforded by $V_r(G)_M$ implies the Tensor Product Property for finite-dimensional rational G-modules M and N:

$$|G|_{M\otimes N} = |G|_M \cap |G|_N .$$

6.4. Finite group schemes. Let G be a finite k-group scheme. A *π-point* of G is a flat map of K-algebras $\alpha : K[t]/(t^p) \to KG_K$, for some field extension K of k, which factors through the group algebra KE of an abelian unipotent subgroup scheme E of G_K. Two π-points $\alpha : K[t]/(t^p) \to KG_K$ and $\beta : L[t]/(t^p) \to LG_L$ are equivalent, denoted $\alpha \sim \beta$, if for any finite-dimensional kG-module M, the pullback $\alpha^*(M_K)$ is projective over $K[t]/(t^p)$ if and only if the pullback $\beta^*(M_L)$ is projective over $L[t]/(t^p)$. Denote the equivalence class of a π-point α by $[\alpha]$, and denote the set of equivalence classes of π-points in G by $\Pi(G)$.

Given a finite-dimensional rational G-module M, set

$$\Pi(G)_M = \{[\alpha : K[t]/(t^p) \to KG_K] \in \Pi(G) : \alpha^*(M_K) \text{ is not projective}\}.$$

[1]Recall that an affine group scheme G is unipotent if each nonzero rational G-module has a nontrivial fixed vector. If G is a finite group scheme, then G is unipotent if and only if the augmentation ideal of the group ring $kG = k[G]^*$ is nilpotent. For more on unipotent group schemes, see for example [**50**, Chapter 8].

Then the class of subsets $\{\Pi(G)_M : M \text{ finite-dimensional } kG\text{-module}\}$ is the class of closed subsets of a Noetherian topology on $\Pi(G)$.

Given a π-point $\alpha : K[t]/(t^p) \to KG$, let $\mathrm{H}^\bullet(\alpha)$ be the composite morphism

$$\mathrm{H}^\bullet(G,k) = \mathrm{Ext}^\bullet_G(k,k) \xrightarrow{K\otimes_k -} \mathrm{Ext}^\bullet_{G_K}(K,K) \xrightarrow{\alpha^*} \mathrm{Ext}^\bullet_{K[t]/(t^p)}(K,K).$$

Then $\sqrt{\ker(\mathrm{H}^\bullet(\alpha))}$ is a homogeneous prime ideal in $\mathrm{H}^\bullet(G,k)$, and hence defines a point $\mathfrak{p}(\alpha)$ in the projective spectrum $\mathrm{Proj}\,\mathrm{H}^\bullet(G,k)$. Conversely, each point $\mathfrak{p} \in \mathrm{Proj}\,\mathrm{H}^\bullet(G,k)$ can be realized as $\mathfrak{p} = \mathfrak{p}(\alpha_{\mathfrak{p}})$ for some π-point $\alpha_{\mathfrak{p}} : K[t]/(t^p) \to KG_K$ of G. In this way, one gets a homeomorphism of topological spaces

$$\Psi_G : \Pi(G) \simeq \mathrm{Proj}\,\mathrm{H}^\bullet(G,k), \quad [\alpha] \mapsto \mathfrak{p}(\alpha),$$

which satisfies

$$\Psi_G(\Pi(G)_M) = \mathrm{Proj}(|G|_M) = \mathrm{Proj}\left(\mathrm{H}^\bullet(G,k)/I_G(M)\right)$$

for each finite-dimensional kG-module M [**27**, Theorem 3.6].

For an arbitrary (i.e., not necessarily finite-dimensional) kG-module M, the π-support of M, denoted $\pi\text{-}\mathrm{supp}_G(M)$, is the subset of $\mathrm{Proj}\,\mathrm{H}^\bullet(G,k)$ defined by

$$\pi\text{-}\mathrm{supp}_G(M) = \left\{\mathfrak{p} \in \mathrm{Proj}\,\mathrm{H}^\bullet(G,k) : \alpha_{\mathfrak{p}}^*(M_K) \text{ is not projective}\right\}.$$

If M is finite-dimensional, then $\pi\text{-}\mathrm{supp}_G(M) = \Psi_G(\Pi(G)_M)$. For arbitrary kG-modules, one can show (see [**4, 27**]) that π-support detects projectivity,

$$\pi\text{-}\mathrm{supp}_G(M) = \emptyset \quad \text{if and only if} \quad M \text{ is projective},$$

and π-support satisfies the Tensor Product Property:

$$\pi\text{-}\mathrm{supp}_G(M \otimes N) = \pi\text{-}\mathrm{supp}_G(M) \cap \pi\text{-}\mathrm{supp}_G(N).$$

Let $\iota : k[t]/(t^p) \to k\mathbb{G}_{a(r)}$ be the k-algebra map defined by $\iota(t) = u_{r-1}$. If $\nu : \mathbb{G}_{a(r)} \otimes_k K \to G \otimes_k K$ is a homomorphism defined over a field extension K/k, then the induced K-algebra map $\nu \circ \iota : K[t]/(t^p) \to K\mathbb{G}_{a(r)} \to KG_K$ is a π-point of G. In this way, one gets for each infinitesimal k-group scheme G of height $\leq r$ a bijection $\mathrm{Proj}\,V_r(G) \simeq \Pi(G)$, which restricts for each finite-dimensional kG-module M to a bijection $\mathrm{Proj}\,V_r(G)_M \simeq \Pi(G)_M$; see [**26**, Proposition 3.8].

7. Infinitesimal supergroup schemes

In this section we summarize what we currently know about support varieties for infinitesimal supergroup schemes. In the unipotent case the theory is very satisfactory: By a theorem of Benson, Iyengar, Krause, and Pevtsova (BIKP), one gets that nilpotence of cohomology classes and projectivity of modules can be detected via restriction to 'elementary' sub-supergroup schemes (Theorem 7.5.2). From this we derive a description for support varieties analogous to the descriptions in the non-super case using varieties of one-parameter subgroups (Theorem 7.6.2). But in the general (not necessarily unipotent) case, the final shape of the theory remains unclear.

We start in Section 7.1 describing why the cohomology ring $\mathrm{H}^\bullet(G,k)$ of a finite supergroup scheme G is a finitely-generated algebra, and hence why it provides an appropriate setting for support varieties. In Section 7.2 we describe a family of finite supergroup schemes that includes the infinitesimal elementary supergroup schemes of BIKP, and which are candidate replacements for $\mathbb{G}_{a(r)}$ vis-à-vis its role in the non-super theory described in Section 6.3. Sections 7.3 and 7.4 discuss super analogues of the varieties of one-parameter subgroups and the commuting varieties that appeared in (6.3.1) and (6.3.2). Then Sections 7.5 and 7.6 discuss how these pieces fit together to describe the cohomological spectrum and support varieties for infinitesimal unipotent supergroup schemes (and conjecturally, for arbitrary infinitesimal supergroup schemes).

In this section, let $r \geq 1$ be a fixed integer.

7.1. Cohomology of infinitesimal supergroups. The general linear Lie superalgebra $\mathfrak{gl}(m|n)$ identifies with the space $\mathrm{Hom}_k(k^{m|n}, k^{m|n})$ of k-linear endomorphisms of the superspace $k^{m|n}$. Then $\mathfrak{gl}(m|n)_{\overline{0}} = \mathfrak{gl}_m \oplus \mathfrak{gl}_n$ and $\mathfrak{gl}(m|n)_{\overline{1}} = \mathfrak{g}_{+1} \oplus \mathfrak{g}_{-1}$, where $\mathfrak{g}_{+1} = \mathrm{Hom}_k(k^{0|n}, k^{m|0})$ and $\mathfrak{g}_{-1} = \mathrm{Hom}_k(k^{m|0}, k^{0|n})$. The adjoint action of the general linear supergroup $GL_{m|n}$ restricts to a rational action of $GL_m \times GL_n$ on each of $\mathfrak{gl}_m$, $\mathfrak{gl}_n$, $\mathfrak{g}_{+1}$, and $\mathfrak{g}_{-1}$. Given a (super)commutative k-superalgebra A and a matrix $X \in GL_{m|n}(A)$, the r-th Frobenius morphism $F^r : GL_{m|n} \to (GL_{m|n})^{(r)}$ acts on X by raising the coefficients to the p^r-th power. In particular, this annihilates the odd blocks in X, implying that the image of F^r is actually contained in the underlying purely even subgroup scheme $(GL_m \times GL_n)^{(r)}$ of $(GL_{m|n})^{(r)}$. Then via the r-th Frobenius morphism, the Frobenius twists $\mathfrak{gl}_m^{(r)}$, $\mathfrak{gl}_n^{(r)}$, $\mathfrak{g}_{+1}^{(r)}$, and $\mathfrak{g}_{-1}^{(r)}$ each become rational $GL_{m|n}$-supermodules.

In [**13**], the first author used the theory of strict polynomial superfunctors to establish the existence of certain universal extension classes for the general linear supergroup:

(7.1.1)
$$\left.\begin{aligned} \boldsymbol{e}_i^{(r-i)} &\in \mathrm{H}^{2p^{i-1}}(GL_{m|n}, \mathfrak{gl}_m^{(r)}) \\ (\boldsymbol{e}_i^{(r-i)})^{\Pi} &\in \mathrm{H}^{2p^{i-1}}(GL_{m|n}, \mathfrak{gl}_n^{(r)}) \end{aligned}\right\} \text{ for } 1 \leq i \leq r, \text{ and } \left\{\begin{aligned} \boldsymbol{c}_r &\in \mathrm{H}^{p^r}(GL_{m|n}, \mathfrak{g}_{+1}^{(r)}), \\ \boldsymbol{c}_r^{\Pi} &\in \mathrm{H}^{p^r}(GL_{m|n}, \mathfrak{g}_{-1}^{(r)}). \end{aligned}\right.$$

In a manner similar to (6.2.2), the restriction of these classes to the r-th Frobenius kernel of $GL_{m|n}$ can be used to construct a homomorphism of graded superalgebras

(7.1.2)
$$\phi_{GL_{m|n(r)}} : \left(\bigotimes_{i=1}^r S(\mathfrak{gl}(m|n)_{\overline{0}}^*[2p^{i-1}])^{(r)}\right) \otimes S(\mathfrak{gl}(m|n)_{\overline{1}}^*[p^r])^{(r)} \to \mathrm{H}^\bullet(GL_{m|n(r)}, k).$$

Given a closed sub-supergroup scheme G of $GL_{m|n(r)}$, one can compose with the restriction map in cohomology $\mathrm{H}^\bullet(GL_{m|n(r)}, k) \to \mathrm{H}^\bullet(G, k)$ to get a homomorphism of graded superalgebras

$$\phi_G : \left(\bigotimes_{i=1}^r S(\mathfrak{gl}(m|n)_{\overline{0}}^*[2p^{i-1}])^{(r)}\right) \otimes S(\mathfrak{gl}(m|n)_{\overline{1}}^*[p^r])^{(r)} \to \mathrm{H}^\bullet(G, k).$$

One of the main consequences of the work in [**13**] was that, if G is an infinitesimal k-supergroup scheme of height $\leq r$, embedded into some $GL_{m|n(r)}$, then ϕ_G is a finite map and $\mathrm{H}^\bullet(G, M)$ is finite over the image of ϕ_G whenever M is a finite-dimensional rational G-supermodule. In particular, $\mathrm{H}^\bullet(G, k)$ is a finitely-generated k-superalgebra, and $\mathrm{H}^\bullet(G, M)$ is a finite $\mathrm{H}^\bullet(G, k)$-module.

In the case $r = 1$, the map ϕ_G can be interpreted via the edge maps of a spectral sequence. Indeed, let G be a height-1 infinitesimal k-supergroup scheme, and let $\mathfrak{g}$ be its restricted Lie superalgebra. Arguing in exactly the same manner as for (6.1.1), one gets a spectral sequence

(7.1.3)
$$E_2^{i,j} = S^{i/2}(\mathfrak{g}_{\overline{0}}^*)^{(1)} \otimes \mathrm{H}^j(\mathfrak{g}, k) \Rightarrow \mathrm{H}^{i+j}(G, k).$$

Fixing a choice of embedding $G \hookrightarrow GL_{m|n(1)}$, one gets a map of restricted Lie superalgebras $\mathfrak{g} \hookrightarrow \mathfrak{gl}(m|n)$. Then up to a nonzero scalar, the restricted map

$$\phi_G\big|_{S(\mathfrak{gl}(m|n)_{\overline{0}}^*)} : S(\mathfrak{gl}(m|n)_{\overline{0}}^*)^{(1)} \to \mathrm{H}^\bullet(G,k)$$

is the composition of the quotient map $S(\mathfrak{gl}(m|n)_{\overline{0}}^*)^{(1)} \twoheadrightarrow S(\mathfrak{g}_{\overline{0}}^*)^{(1)}$ and the horizontal edge map of (7.1.3). Similarly, the composition of the restriction

$$\phi_G\big|_{S(\mathfrak{gl}(m|n)_{\overline{1}}^*)} : S(\mathfrak{gl}(m|n)_{\overline{1}}^*)^{(1)} \to \mathrm{H}^\bullet(G,k)$$

with the horizontal edge map $\mathrm{H}^\bullet(G,k) \to \mathrm{H}^\bullet(\mathfrak{g},k)$ of (7.1.3) is equal, up to a nonzero scalar, to the composition of the quotient map $S(\mathfrak{gl}(m|n)_{\overline{1}}^*)^{(1)} \twoheadrightarrow S(\mathfrak{g}_{\overline{1}}^*)^{(1)}$ and the homomorphism $\varphi : S(\mathfrak{g}_{\overline{1}}^*)^{(1)} \to \mathrm{H}^\bullet(\mathfrak{g},k)$ of (5.1.1).

Unlike the classical situation of (6.1.1), in general the E_2-page of (7.1.3) is not finite over the evident subalgebra of permanent cycles generated by the terms on the horizontal axis. This shows how, starting immediately with the base case of $r = 1$, the problem of establishing the basic finite-generation property (FG1) is more subtle for infinitesimal supergroups than for ordinary infinitesimal group schemes.

7.2. Multiparameter supergroups. Define $\mathbb{M}_r$ to be the affine k-supergroup scheme whose coordinate algebra $k[\mathbb{M}_r]$ is generated by the odd element τ and the even elements θ and σ_i for $i \in \mathbb{N}$, such that $\tau^2 = 0$, $\sigma_0 = 1$, $\theta^{p^{r-1}} = \sigma_1$, and $\sigma_i\sigma_j = \binom{i+j}{i}\sigma_{i+j}$, i.e.,

$$k[\mathbb{M}_r] = k[\tau,\theta,\sigma_1,\sigma_2,\sigma_3,\dots]/\left(\tau^2, \theta^{p^{r-1}} - \sigma_1, \sigma_i\sigma_j - \tbinom{i+j}{i}\sigma_{i+j} : i,j \in \mathbb{N}\right).$$

Then the set of monomials $\left\{\theta^i\sigma_j, \tau\theta^i\sigma_j : 0 \le i < p^{r-1}, j \in \mathbb{N}\right\}$ is a homogeneous basis for $k[\mathbb{M}_r]$. The coproduct Δ and the antipode S on $k[\mathbb{M}_r]$ are defined on generators by the formulas

$$\begin{aligned}
\Delta(\tau) &= \tau\otimes 1 + 1\otimes\tau, & S(\tau) &= -\tau,\\
\Delta(\theta) &= \theta\otimes 1 + 1\otimes\theta, & S(\theta) &= -\theta,\\
\Delta(\sigma_i) &= \textstyle\sum_{u+v=i}\sigma_u\otimes\sigma_v + \sum_{u+v+p=i}\sigma_u\tau\otimes\sigma_v\tau, & S(\sigma_i) &= (-1)^i\sigma_i.
\end{aligned}$$

The $\mathbb{Z}_2$-grading on $k[\mathbb{M}_r]$ lifts to a $\mathbb{Z}$-grading such that $\deg(\tau) = p^r$, $\deg(\theta) = 2$, and $\deg(\sigma_i) = 2ip^{r-1}$. This makes $k[\mathbb{M}_r]$ into a $\mathbb{Z}$-graded Hopf algebra of finite type (i.e., each $\mathbb{Z}$-graded component of $k[\mathbb{M}_r]$ is of finite dimension over k). The supergroup $\mathbb{M}_r$ arose out of our first attempts to find an appropriate super analogue of the one-parameter subgroups discussed in Section 6.3. We can think of it as a kind of inverse limit of the supergroups $\mathbb{M}_{r,s}$ and $\mathbb{M}_{r,s,\eta}$ defined below.

Let $k\mathbb{M}_r = k[\mathbb{M}_r]^*$ be the 'group algebra' of $\mathbb{M}_r$, i.e., the k-algebra that is dual to the coalgebra structure on $k[\mathbb{M}_r]$. For $0 \le i \le r-1$, let $u_i \in k\mathbb{M}_r$ be the (even) linear functional that is dual to the basis vector θ^{p^i} (so in particular, u_{r-1} is dual to $\theta^{p^{r-1}} = \sigma_1$), and let $v \in k\mathbb{M}_r$ be the (odd) linear functional that is dual to the basis vector τ. Then

$$k\mathbb{M}_r = \frac{k[u_0,\dots,u_{r-2},v][[u_{r-1}]]}{(u_0^p,\dots,u_{r-2}^p,u_{r-1}^p+v^2)}.$$

The graded dual of the $\mathbb{Z}$-graded vector space $k[\mathbb{M}_r]$ is the subalgebra

$$\mathbb{P}_r = \frac{k[u_0, \dots, u_{r-1}, v]}{(u_0^p, \dots, u_{r-2}^p, u_{r-1}^p + v^2)}.$$

The algebra $\mathbb{P}_r$ can be thought of as the 'polynomial part' of $k\mathbb{M}_r$. Since $k[\mathbb{M}_r]$ is a $\mathbb{Z}$-graded Hopf algebra of finite type, its graded dual is also a $\mathbb{Z}$-graded Hopf algebra (and hence a Hopf superalgebra, by reducing the $\mathbb{Z}$-grading modulo 2). The coproduct Δ and antipode S for $\mathbb{P}_r$ are defined on the generator v by $\Delta(v) = v \otimes 1 + 1 \otimes v$ and $S(v) = -v$, and are defined on $u_0, \dots, u_{r-1}$ by the same formulas as in (6.3.4), using the monomials $\gamma_0, \gamma_1, \dots, \gamma_{p^r-1}$, considered now as elements of $\mathbb{P}_r$. In particular, $\Delta(u_{r-1}^p) = u_{r-1}^p \otimes 1 + 1 \otimes u_{r-1}^p$.

Let $f = \sum_{i=1}^t c_i T^{p^i}$ be an inseparable p-polynomial (i.e., a p-polynomial without a linear term), and let $\eta \in k$. Since u_0 and u_{r-1} are both primitive in $\mathbb{P}_r$, the sum $f(u_{r-1}) + \eta \cdot u_0$ is also primitive in $\mathbb{P}_r$. Assuming that $f \neq 0$ if $r \geq 2$, and that either $f \neq 0$ or $\eta \neq 0$ if $r = 1$, the quotient

$$k\mathbb{M}_{r,f,\eta} := \mathbb{P}_r / (f(u_{r-1}) + \eta \cdot u_0)$$

is then a finite-dimensional Hopf superalgebra. Let $\mathbb{M}_{r,f,\eta}$ be the finite k-supergroup scheme whose group algebra is $k\mathbb{M}_{r,f,\eta}$. For $s \geq 1$, set $\mathbb{M}_{r,s,\eta} = \mathbb{M}_{r,T^{p^s},\eta}$, and set $\mathbb{M}_{r,s} = \mathbb{M}_{r,s,0}$. Then

$$k\mathbb{M}_{r,s,\eta} = \frac{k[u_0, \dots, u_{r-1}, v]}{(u_0^p, \dots, u_{r-2}^p, u_{r-1}^p + v^2, u_{r-1}^{p^s} + \eta \cdot u_0)}, \quad \text{and}$$

$$k\mathbb{M}_{r,s} = \frac{k[u_0, \dots, u_{r-1}, v]}{(u_0^p, \dots, u_{r-2}^p, u_{r-1}^p + v^2, u_{r-1}^{p^s})}.$$

Observe also that $\mathbb{P}_{r-1} \cong \mathbb{P}_r/(u_0)$, $k\mathbb{G}_{a(r)} \cong \mathbb{P}_r/(v)$, and

$$k\mathbb{G}_a^- = k[v]/(v^2) \cong \mathbb{P}_r/(u_0, \dots, u_{r-1}).$$

Here $\mathbb{G}_a^-$ is the odd additive supergroup scheme. Its coordinate superalgebra is $k[\mathbb{G}_a^-] = k[t]/(t^2)$ with $\bar{t} = \bar{1}$. As a functor $\mathbb{G}_a^- : \mathsf{CSAlg}(k) \to \mathsf{Grp}$, it is defined by $\mathbb{G}_a^-(A) = (A_{\bar{1}}, +)$. Here $\mathsf{CSAlg}(k)$ denotes the category of (super)commutative k-superalgebras.

We say that a finite k-supergroup scheme G is a *multiparameter supergroup of height* $\leq r$ if its group algebra $kG = k[G]^*$ is isomorphic to a (finite-dimensional) Hopf superalgebra quotient of $\mathbb{P}_r$. By [**17**, Proposition 2.2.1], each multiparameter supergroup of height $\leq r$ is isomorphic to (at least) one of the following:

(MP1) $\mathbb{G}_{a(s)}$ for some $0 \leq s \leq r$,

(MP2) $\mathbb{G}_{a(s)} \times \mathbb{G}_a^-$ for some $0 \leq s \leq r$ (note that $\mathbb{G}_{a(s)} \times \mathbb{G}_a^- = \mathbb{M}_{s,1}$), or

(MP3) $\mathbb{M}_{s,f,\eta}$ for some $1 \leq s \leq r$, some inseparable p-polynomial $0 \neq f \in k[T]$, and some $\eta \in k$.

Among these supergroups, the *unipotent*[2] multiparameter supergroups have the forms:

(UMP1) $\mathbb{G}_{a(r)}$ for some $r \geq 0$,

[2]As for ordinary affine group schemes, and affine supergroup scheme G is unipotent if each nonzero rational G-supermodule has a nontrivial fixed vector. If G is a finite supergroup scheme, then G is unipotent if and only if the augmentation ideal of the group ring $kG = k[G]^*$ is nilpotent. By [**52**], an affine algebraic supergroup G is unipotent if and only if its largest purely even subgroup scheme G_{ev} is unipotent (as an ordinary affine group scheme).

(UMP2) $\mathbb{G}_{a(r)} \times \mathbb{G}_a^-$ for some $r \geq 0$,
(UMP3) $\mathbb{M}_{r,s}$ for some $r, s \geq 1$, or
(UMP4) $\mathbb{M}_{r,s,\eta}$ for some integers $r \geq 2$, $s \geq 1$, and some scalar $0 \neq \eta \in k$.

By definition, $\mathbb{G}_{a(0)}$ is the trivial (constant) group scheme.

7.3. The functor of multiparameter supergroups. Given affine k-supergroup schemes G and H, let $\operatorname{Hom}(G, H) : \mathsf{CAlg}(k) \to \mathsf{Set}$ be the functor defined by

$$\operatorname{Hom}(G, H)(A) = \operatorname{Hom}_{\mathsf{sGrp}/A}(G \otimes_k A, H \otimes_k A),$$

the set of A-supergroup scheme homomorphisms $\rho : G \otimes_k A \to H \otimes_k A$. Similarly, given Hopf k-superalgebras R and S, let $\operatorname{Hom}(R, S) : \mathsf{CAlg}(k) \to \mathsf{Set}$ be the functor defined by

$$\operatorname{Hom}(R, S)(A) = \operatorname{Hom}_{\mathsf{sHopf}/A}(R \otimes_k A, S \otimes_k A),$$

the set of Hopf A-superalgebra homomorphisms $\rho : R \otimes_k A \to S \otimes_k A$. If G and H are finite k-supergroup schemes, then $\operatorname{Hom}(G, H) \cong \operatorname{Hom}(kG, kH)$.

DEFINITION 7.3.1. Given a finite k-supergroup scheme G, define

$$V_r(G) : \mathsf{CAlg}(k) \to \mathsf{Set} \quad \text{by} \quad V_r(G) = \operatorname{Hom}(\mathbb{P}_r, kG).$$

We call $V_r(G)$ the *functor of multiparameter supergroups of height $\leq r$ in G.*

We consider the objects in $\mathsf{CAlg}(k)$ as superalgebras concentrated in even superdegree. If G is an ordinary (purely even) finite k-group scheme, then any $\rho \in V_r(G)(A)$ will automatically factor through the canonical quotient map $\mathbb{P}_r \otimes_k A \twoheadrightarrow \mathbb{P}_r/(v) \otimes_k A \cong k\mathbb{G}_{a(r)} \otimes_k A$, and hence will define a map of A-group schemes $\rho : \mathbb{G}_{a(r)} \otimes_k A \to G \otimes_k A$ (and conversely). Thus when G is purely even, our new use of the notation $V_r(G)$ agrees with that already established in Section 6.3 (although our new usage requires G to be finite, whereas before G was merely assumed to be affine algebraic, to get a scheme structure on $V_r(G)$).

By [**17**, Lemma 4.1.1], the functor $V_r(G)$ admits the structure of an affine k-scheme of finite type, and the assignment $G \mapsto V_r(G)$ is then a covariant functor from finite k-supergroup schemes to affine k-schemes of finite type, which takes closed embeddings (of supergroup schemes) to closed embeddings.[3] If E is a multiparameter supergroup of height $\leq r$ (i.e., if kE is a Hopf superalgebra quotient of $\mathbb{P}_r$), then the functor $\operatorname{Hom}(kE, kG) \cong \operatorname{Hom}(E, G)$ also admits an affine k-scheme structure, and identifies via the quotient map $\mathbb{P}_r \twoheadrightarrow kE$ with a closed subscheme of $V_r(G)$.[4] Identifying $\operatorname{Hom}(E, G)$ with its image in $V_r(G)$ via this embedding, one then has

$$V_r(G)(k) = \bigcup_{(E \leq G)/\cong} \operatorname{Hom}(E, G)(k), \tag{7.3.1}$$

[3]In [**17**], we defined a k-*super*functor, denoted $\boldsymbol{V}_r(G)$, which admitted the structure of an affine k-*super*scheme of finite type, and whose purely underlying even subscheme is $V_r(G)$. The superfunctor and superscheme structures can be ignored as far as the applications to cohomological support varieties are concerned, so we dispense with them in the current article.

[4]More generally, if G is an affine algebraic k-group scheme, i.e., if $k[G]$ is finitely-generated as a k-algebra but not necessarily finite-dimensional, and if E is a multiparameter supergroup, then $\operatorname{Hom}(E, G)$ admits an affine k-scheme structure by [**14**, Lemma 3.3.6].

where the union is over all isomorphism classes of multiparameter k-supergroups of height $\leq r$ that occur as closed subsupergroups of G. Similarly one has

$$V_r(G)(k) = \bigcup_{E \leq G} V_r(E)(k), \tag{7.3.2}$$

where the union is over all multiparameter subsupergroups of height $\leq r$ in G [**17**, Lemma 4.1.4].

REMARK 7.3.2. For all $1 \leq s \leq s'$, there are canonical quotient maps $\mathbb{M}_r \twoheadrightarrow \mathbb{M}_{r,s'} \twoheadrightarrow \mathbb{M}_{r,s}$. If G is an algebraic k-supergroup scheme and if $\rho \in \mathrm{Hom}(\mathbb{M}_r, G)(A)$, then ρ must factor through the quotient $\mathbb{M}_r \otimes_k A \twoheadrightarrow \mathbb{M}_{r,s} \otimes_k A$ for some $s \geq 1$ [**15**, Remark 3.1.3(4)]. In this way one gets

$$\mathrm{Hom}(\mathbb{M}_r, G) = \bigcup_{s \geq 1} \mathrm{Hom}(\mathbb{M}_{r,s}, G). \tag{7.3.3}$$

If G is a finite unipotent k-supergroup scheme, this union is finite; see Section 7.5.2. In the papers [**15, 16**] we initially considered the functor $\mathrm{Hom}(\mathbb{M}_r, G)$ as a possible superization of the classical functor of one-parameter subgroups (6.3.1), before deciding while writing [**17**] that $\mathrm{Hom}(\mathbb{P}_r, kG)$ seems more likely to provide a better generalization for arbitrary finite supergroups, at least as far as applications to support varieties are concerned.

QUESTION 7.3.3. Is there an affine (though perhaps not algebraic) k-supergroup scheme $\mathbb{G}$ such that for all finite k-supergroup schemes G,

$$\mathrm{Hom}(\mathbb{G}, G) \cong \mathrm{Hom}(\mathbb{P}_r, kG)\ ?$$

Given a multiparameter k-supergroup E and a finite k-supergroup scheme G, set

$$V_E(G) = \mathrm{Hom}(E, G).$$

Then $V_E(G)$ is a closed subscheme of $V_r(G)$. In particular, set

$$V_{r,f,\eta}(G) = \mathrm{Hom}(\mathbb{M}_{r,f,\eta}, G) \quad \text{and} \quad V_{r,s}(G) = \mathrm{Hom}(\mathbb{M}_{r,s}, G).$$

7.4. Commuting varieties. If G is an affine algebraic k-supergroup scheme with restricted Lie superalgebra $\mathfrak{g}$, the Lie superalgebra of $G_A = G \otimes_k A$ is then given by $\mathfrak{g}_A = \mathfrak{g} \otimes_k A$. Under this identification, the p-operation on $(\mathfrak{g}_A)_{\overline{0}} = \mathfrak{g}_{\overline{0}} \otimes_k A$ is given by $(x \otimes a)^{[p]} = x^{[p]} \otimes a^p$.

DEFINITION 7.4.1. Given an affine k-supergroup scheme G with restricted Lie superalgebra $\mathfrak{g}$, let $C_r(G) : \mathsf{CAlg}(k) \to \mathsf{Set}$ be the functor defined by

$$\begin{aligned} C_r(G)(A) = \Big\{ (\alpha_0, \ldots, \alpha_{r-1}, \beta) \in [(\mathfrak{g}_A)_{\overline{0}}]^{\times r} \times (\mathfrak{g}_A)_{\overline{1}} :& \\ [\alpha_i, \alpha_j] = [\alpha_i, \beta] = 0 \text{ for all } i \text{ and } j,& \\ \alpha_i^{[p]} = 0 \text{ for all } 0 \leq i \leq r-2, \text{ and } \alpha_{r-1}^{[p]} + \tfrac{1}{2}[\beta, \beta] = 0 \Big\}.& \end{aligned}$$

REMARK 7.4.2. In [**14, 16**], we considered a k-superfunctor $\boldsymbol{V}_r(GL_{m|n})$ whose restriction to $\mathsf{CAlg}(k)$ we denoted there by $V_r(GL_{m|n})$, but which is denoted in this paper by $C_r(GL_{m|n})$. We advise the reader to exercise caution when comparing the notation in [**14, 16**] with the sometimes inconsistent notation in [**17**] and in the present paper.

Since the p-operation on $(\mathfrak{gl}(m|n)_A)_{\overline{0}} \cong \mathfrak{gl}_m(A) \oplus \mathfrak{gl}_n(A)$ is given simply by raising matrices to the p-th power, it readily follows that $C_r(GL_{m|n})$ is represented by an affine scheme of finite type over k, which we also denote $C_r(GL_{m|n})$. More generally, if G is an affine algebraic k-supergroup scheme, one can use a choice of closed embedding $G \hookrightarrow GL_{m|n}$ to show that $C_r(G)$ admits a scheme structure, and that $G \mapsto C_r(G)$ is then a covariant functor from affine algebraic k-supergroup schemes to affine k-schemes of finite type, which takes closed embeddings (of supergroup schemes) to closed embeddings.

Given an inseparable p-polynomial $0 \neq f \in k[T]$ and a scalar $\eta \in k$, one has

$$V_{r,f,\eta}(GL_{m|n(r)})(A) \cong \left\{(\alpha_0, \dots, \alpha_{r-1}, \beta) \in C_r(GL_{m|n(r)})(A) : f(\alpha_{r-1}) + \eta \cdot \alpha_0 = 0\right\}$$

by [**15**, Proposition 3.3.5]. In particular, given an integer $s \geq 1$, one has

$$V_{r,s}(GL_{m|n(r)})(A) \cong \left\{(\alpha_0, \dots, \alpha_{r-1}, \beta) \in C_r(GL_{m|n(r)})(A) : \alpha_{r-1}^{p^s} = 0\right\}.$$

For the ordinary (purely even) k-group scheme $GL_{n(r)}$, one has

$$V_r(GL_{n(r)}) = \operatorname{Hom}(\mathbb{P}_r, kGL_{n(r)}) \cong \operatorname{Hom}(\mathbb{G}_{a(r)}, GL_{n(r)}) \cong C_r(GL_{n(r)})$$

by (6.3.2). More generally, if G is a finite purely even k-group scheme that admits an embedding $G \hookrightarrow GL_n$ of *exponential type* [**48**, p. 697], then $V_r(G) \cong C_r(G)$ by [**48**, Lemma 1.7].

REMARK 7.4.3. Let G be a height-1 infinitesimal k-supergroup scheme, and let $\mathfrak{g}$ be its (restricted) Lie superalgebra. Then $kG = V(\mathfrak{g})$. In this situation, a Hopf superalgebra homomorphism $\sigma : \mathbb{P}_1 = k[u,v]/(u^p + v^2) \to kG = V(\mathfrak{g})$ is determined by the data of the even primitive element $\sigma(u) \in \mathfrak{g}_{\overline{0}}$ and the odd primitive element $\sigma(v) \in \mathfrak{g}_{\overline{1}}$, which must satisfy $\sigma(u)^p + \sigma(v)^2 = 0$. Then

$$V_1(G)(k) = \operatorname{Hom}_{\mathsf{sHopf}/k}(\mathbb{P}_1, V(\mathfrak{g})) \cong \left\{(\alpha, \beta) \in \mathfrak{g}_{\overline{0}} \times \mathfrak{g}_{\overline{1}} : \alpha^{[p]} + \tfrac{1}{2}[\beta, \beta] = 0\right\}.$$

In particular, $V_1(G)(k) \cong C_1(G)(k)$ for G of height 1.

7.5. The cohomological spectrum.

7.5.1. *The general case.* Let G be a finite k-supergroup scheme. The identity map on $k[V_r(G)]$ defines a $k[V_r(G)]$-point of $V_r(G)$, and hence determines a universal homomorphism $u_G : \mathbb{P}_r \otimes_k k[V_r(G)] \to kG \otimes_k k[V_r(G)]$ of Hopf superalgebras over $k[V_r(G)]$. In [**17**, §4.2], we used this map to define a natural (with respect to G) k-algebra homomorphism

$$\psi_r : H(G,k) \to k[V_r(G)]. \tag{7.5.1}$$

The ring $k[V_r(G)]$ is naturally $\mathbb{Z}[\frac{p^r}{2}]$-graded, and the map ψ_r then multiplies $\mathbb{Z}$-degrees by $\frac{p^r}{2}$. Composing (7.5.1) with the quotient map $k[V_r(G)] \twoheadrightarrow k[V_E(G)]$ for a multiparameter k-supergroup E, one gets a k-algebra homomorphism

$$\psi_E : H(G,k) \to k[V_E(G)]. \tag{7.5.2}$$

Set $\psi_{r,f,\eta} = \psi_{\mathbb{M}_{r,f,\eta}}$, and set $\psi_{r,s} = \psi_{\mathbb{M}_{r,s}} = \psi_{r,T^{p^s},0}$.

The domain of the superalgebra homomorphism (7.1.2) identifies with the coordinate algebra of the affine space $(\bigoplus_{i=1}^r \mathfrak{gl}(m|n)_{\overline{0}}) \oplus \mathfrak{gl}(m|n)_{\overline{1}}$. In [**14**, Proposition 6.1.1], we used the algebra relations among the extension classes (7.1.1) to show that, analogously to (6.3.7), the map $\phi_{GL_{m|n(r)}}$ factors through a superalgebra homomorphism

$$\overline{\phi}_{GL_{m|n(r)}} : k[C_r(GL_{m|n(r)})] \to H(GL_{m|n(r)}, k). \tag{7.5.3}$$

Let Ψ_E and Φ denote the morphisms of affine schemes induced by (7.5.2) and (7.5.3), respectively. Through an analysis of how the universal extension classes (7.1.1) restrict to multiparameter supergroups, we showed in [**14**, Theorem 6.2.3] that the composite morphism

(7.5.4) $$\Theta_{r,f,\eta} : V_{r,f,\eta}(GL_{m|n(r)}) \xrightarrow{\Psi_{r,f,\eta}} \operatorname{Spec} H(GL_{m|n(r)}, k) \xrightarrow{\Phi} C_r(GL_{m|n(r)})$$

is equal to the composite of the closed embedding $V_{r,f,\eta}(GL_{m|n(r)}) \hookrightarrow V_r(GL_{m|n})$ and the r-th Frobenius twist morphism for the scheme $V_r(GL_{m|n(r)}) = V_r(GL_{m|n})$. In particular, the map $\psi_{r,f,\eta}$ is surjective onto p^r-th powers, and the kernel of $\overline{\phi}$ is nilpotent. Using naturality of $\psi_{r,f,\eta}$ with respect to G, we deduce for an arbitrary infinitesimal k-supergroup scheme G of height $\leq r$ that the map $\psi_{r,f,\eta} : H(G,k) \to k[V_{r,f,\eta}(G)]$ is surjective onto p^r-th powers.

REMARK 7.5.1. Allowing f to vary, we used the morphisms $\Theta_{r,f,0}$ to deduce in [**14**, Corollary 6.2.4] that the finite morphism of affine algebraic varieties

$$\Phi : \operatorname{MaxSpec} H(GL_{m|n(r)}, k) \to C_r(GL_{m|n(r)})(k) = C_r(GL_{m|n})(k)$$

induced by Φ is a surjection. Now let G be a closed subsupergroup of $GL_{m|n(r)}$. Composing (7.5.3) with the restriction map in cohomology, one gets a k-algebra map $\overline{\phi}_G : k[C_r(GL_{m|n(r)})] \to H(G,k)$. In the special case $r = 1$, we showed in [**15**, §5.4] that $\overline{\phi}_G$ factors through $k[C_1(G)]$, and that the induced finite morphism of varieties

$$\Phi_G : \operatorname{MaxSpec} H(G,k) \to$$
$$C_1(G)(k) = \left\{(\alpha,\beta) \in \mathfrak{g}_{\overline{0}} \times \mathfrak{g}_{\overline{1}} : [\alpha,\beta] = 0 \text{ and } \alpha^{[p]} + \tfrac{1}{2}[\beta,\beta] = 0\right\}$$

is a surjection. This provides an analogue for restricted Lie superalgebras of (6.1.3).

7.5.2. *The unipotent case.* Now suppose G is an infinitesimal *unipotent* k-supergroup scheme of height $\leq r$. Then by [**17**, Lemma 4.1.7], there exists an integer $s = s(G) \geq 1$ such that the canonical quotient maps $\mathbb{P}_r \twoheadrightarrow k\mathbb{M}_{r,s'} \twoheadrightarrow k\mathbb{M}_{r,s}$ induce identifications

(7.5.5)
$$\operatorname{Hom}(\mathbb{M}_{r,s}, G) = \operatorname{Hom}(\mathbb{M}_{r,s'}, G) = \operatorname{Hom}(\mathbb{P}_r, kG) = V_r(G) \quad \text{for all } s' \geq s = s(G).$$

So for $s = s(G)$ one has $V_r(G) = V_{r,s}(G)$, and the homomorphism ψ_r of (7.5.1) is equal to $\psi_{r,s}$, which as we stated above is surjective onto p^r-th powers.

Benson, Iyengar, Krause, and Pevtsova (BIKP) [**3**] define a finite supergroup scheme to be *elementary* if it is isomorphic, for some positive integers r, s, t, to a quotient of $\mathbb{M}_{r,s} \times (\mathbb{Z}/p)^t$. Here $\mathbb{Z}/p$ denotes the (purely even) constant group scheme corresponding to the cyclic group of order p. It turns out that a supergroup scheme is elementary if and only if it is a product of one of the unipotent multiparameter supergroups listed in (UMP1)–(UMP4) and the constant group scheme $(\mathbb{Z}/p)^t$ for some $t \geq 0$. BIKP then prove the following detection theorem:

THEOREM 7.5.2 ([**3**, Theorem 1.2, Theorem 11.2]). *Let G be a finite unipotent supergroup scheme over a field k of positive characteristic $p \geq 3$. Then the following hold:*

(1) *An element $x \in \mathrm{H}^\bullet(G,k)$ is nilpotent if and only if, for every extension field K/k and every elementary sub-supergroup scheme E of G_K, the restriction of $x_K \in \mathrm{H}^\bullet(G_K,K)$ to $\mathrm{H}^\bullet(E,K)$ is nilpotent.*

(2) *A kG-supermodule M is projective if and only if, for every extension field K/k and every elementary sub-supergroup scheme E of G_K, the restriction of M_K to E is projective.*

(3) *Let Λ be a unital G-algebra. Then an element $x \in \mathrm{H}^\bullet(G, \Lambda)$ is nilpotent if and only if, for every extension field K/k and every elementary sub-supergroup scheme E of G_K, the restriction of $x_K \in \mathrm{H}^\bullet(G_K, \Lambda_K)$ to $\mathrm{H}^\bullet(E, \Lambda_K)$ is nilpotent.*

Using part (1) of the theorem, one can then prove:

THEOREM 7.5.3 ([**17**, Theorem 5.1.3]). *Let G be an infinitesimal unipotent k-supergroup scheme of height $\leq r$. Then the kernel of the homomorphism $\psi_r : H(G,k) \to k[V_r(G)]$ is a locally nilpotent ideal, and its image contains all p^r-th powers. Consequently, the associated morphism of schemes Ψ_r defines a universal homeomorphism*

$$\Psi_r : V_r(G) \simeq |G| = \operatorname{Spec} H(G,k).$$

QUESTION 7.5.4. Let G be an arbitrary infinitesimal k-supergroup scheme of height $\leq r$. Does the map $\psi_r : H(G,k) \to k[V_r(G)]$ induce a homeomorphism $V_r(G) \simeq \operatorname{Spec} H(G,k)$?

A (or more likely, *the*) key step toward answering Question 7.5.4 would be to establish an analogue of Theorem 7.5.2 for non-unipotent G. As observed in [**14**, Example 1.4.2], the family of elementary supergroups is inadequate for detecting nilpotence of cohomology classes for the non-unipotent supergroup $\mathbb{M}_{1,T^p,-1}$, whose group algebra is $k\mathbb{M}_{1,T^p,-1} = k[u,v]/(u^p + v^2, u^p - u)$.

QUESTION 7.5.5. Does an analogue of Theorem 7.5.2 hold for non-unipotent G if the family of elementary supergroups is replaced by the family of supergroups of the form $\mathbb{M} \times (\mathbb{Z}/p)^t$, where $\mathbb{M}$ is one of the multiparameter supergroups defined in Section 7.2?

As a starting point for trying to answer Question 7.5.5, one might try to imitate in some way the proof of either Theorem 6.3.1 or its generalization to arbitrary finite group schemes given in [**47**]. Roughly, these arguments first establish a detection theorem for unipotent groups, and hence for Frobenius kernels $B_{(r)}$ of the Borel subgroups of GL_n. The arguments then exploit algebro-geometric relationships between GL_n and its Borel subgroups to construct a spectral sequence that allows one to bootstrap information from $B_{(r)}$ to all of $GL_{n(r)}$. There are non-trivial difficulties in trying to directly translate this strategy over to $GL_{m|n}$, however, because of such issues as the existence of non-conjugate Borel subgroups, and the lack in general of a super-analogue of the Kempf vanishing theorem. It could be necessary to replace the Borel subgroups in this picture with some other class of subgroups of $GL_{m|n}$ for whom the nilpotence of cohomology classes is detected by restriction to subgroups of the form $\mathbb{M} \times (\mathbb{Z}/p)^t$.

7.6. Support schemes for supermodules. Write $\mathbb{P}_1 = k[u,v]/(u^p + v^2)$, and let $\iota : \mathbb{P}_1 \to \mathbb{P}_r$ be the superalgebra map defined by $\iota(u) = u_{r-1}$ and $\iota(v) = v$. Let G be a finite k-supergroup scheme. Now given a kG-supermodule M and a point $\mathfrak{s} \in V_r(G)$ (i.e., a prime ideal in $\operatorname{Spec} k[V_r(G)]$), we consider $M \otimes_k k(\mathfrak{s})$ as a $\mathbb{P}_1 \otimes_k k(\mathfrak{s})$-supermodule by pulling back along the composite $k(\mathfrak{s})$-superalgebra homomorphism

$$\nu_{\mathfrak{s}} \circ (\iota \otimes 1) : \mathbb{P}_1 \otimes_k k(\mathfrak{s}) \hookrightarrow \mathbb{P}_r \otimes_k k(\mathfrak{s}) \to kG \otimes_k k(\mathfrak{s}).$$

Recall that the ring $k[V_r(G)]$ is naturally $\mathbb{Z}[\frac{p^r}{2}]$-graded. We say that a Zariski closed subset $X \subseteq V_r(G)$ is *conical* if it is defined by a $\mathbb{Z}[\frac{p^r}{2}]$-homogeneous ideal.

PROPOSITION 7.6.1 ([**17**, Proposition 4.3.1]). *Let G be a finite k-supergroup scheme, and let M be a kG-supermodule. Then*

$$V_r(G)_M := \{\mathfrak{s} \in V_r(G) : \operatorname{projdim}_{\mathbb{P}_1 \otimes_k k(\mathfrak{s})}(M \otimes_k k(\mathfrak{s})) = \infty\}$$

is a Zariski closed conical subset of $V_r(G)$.

Given a $\mathbb{P}_1$-supermodule M, the matrices

$$\varphi = \begin{pmatrix} u^{p-1} & v \\ v & -u \end{pmatrix} \quad \text{and} \quad \psi = \begin{pmatrix} u & v \\ v & -u^{p-1} \end{pmatrix}$$

naturally define operators $\varphi_M : M \oplus \Pi(M) \to M \oplus \Pi(M)$ and $\psi_M : M \oplus \Pi(M) \to M \oplus \Pi(M)$. Here $\Pi(M)$ denotes the parity shift of M. Then for M finite-dimensional, one gets

$$\text{(7.6.1)} \quad \operatorname{projdim}_{\mathbb{P}_1}(M) = \infty \quad \text{if and only if} \quad \operatorname{rank}(\varphi_M) = \operatorname{rank}(\psi_M) < \dim_k(M);$$

see [**4**, §8] or [**17**, Remark 4.3.4].

Again making crucial use of the BIKP detection theorem, we proved:

THEOREM 7.6.2 ([**17**, Theorem 5.4.1]). *Let G be an infinitesimal unipotent k-supergroup scheme of height $\leq r$, and let M be a finite-dimensional rational G-supermodule. Then the morphism of schemes $\Psi_r : V_r(G) \to |G|$ satisfies*

$$\Psi_r^{-1}(|G|_M) = V_r(G)_M.$$

Thus, Ψ_r restricts to a homeomorphism

$$\text{(7.6.2)} \qquad \Psi_r : V_r(G)_M \xrightarrow{\sim} |G|_M .$$

In particular, M is projective as a rational G-supermodule if and only if $V_r(G)_M = \{0\}$, i.e., if and only if $V_r(G)_M$ consists solely of the trivial homomorphism $\mathbb{P}_r \twoheadrightarrow k \hookrightarrow kG$.

REMARK 7.6.3. Combining Remark 7.4.3, the rank criterion (7.6.1), and the theorem, one gets a fairly explicit description of support varieties for height-1 infinitesimal unipotent supergroup schemes, or equivalently, for finite-dimensional p-nilpotent restricted Lie superalgebras.

The equalities (7.3.1) and (7.3.2) describe how $V_r(G)(k)$ is stratified by pieces coming from the multiparameter subsupergroup schemes of G. The next result, which is reminiscent of Quillen's stratification theorem for finite groups [**43**], translates this to $|G|_M$.

THEOREM 7.6.4 ([**17**, Theorem 6.1.5]). *Let G be an infinitesimal unipotent k-supergroup scheme of height $\leq r$, and let M be a finite-dimensional rational G-supermodule. Then the support variety $|G|_M$ (i.e., the set of k-points of the scheme of the same name) can be written as*

$$\text{(7.6.3)} \qquad |G|_M = \bigcup_{E \leq G} \operatorname{res}^*_{G,E}(|E|_M),$$

where the union is taken over all (closed) elementary k-subsupergroup schemes E of G, and $\operatorname{res}_{G,E} : H(G,k) \to H(E,k)$ is the restriction map induced by the embedding $E \hookrightarrow G$.

QUESTION 7.6.5. Are (7.6.2) and (7.6.3) true as written for non-unipotent G?

The general theory discussed in Section 2.1 implies for any finite k-supergroup scheme G and any finite-dimensional rational G-supermodule M that $|G|_M = \{0\}$ if and only if M is projective, i.e., the abstract support variety $|G|_M$ detects the projectivity of M. The real meat of Question 7.6.5 is whether or not there exists a more concrete rank variety type criterion for measuring projectivity, valid for non-unipotent G.

7.7. π-points for finite supergroup schemes. In [**4**], Benson, Iyengar, Krause, and Pevtsova (BIKP) recently introduced a notion of π-points for arbitrary finite supergroup schemes, generalizing the notion of π-points for finite group schemes recalled in Section 6.4.

Set $A_k = k[u,v]/(u^p - v^2)$. Modulo the rescaling $u \mapsto -u$, this is the superalgebra $\mathbb{P}_1$ defined in Section 7.2. Given a field extension K/k, set $A_K = A \otimes_k K$. Given a finite k-supergroup scheme G, BIKP define a π-point of G to be a K-algebra map of finite flat (equivalently, projective) dimension $\alpha : A_K \to KG_K$, for some field extension K/k, such that α factors through the group algebra KE of an elementary sub-supergroup scheme E of G_K. Two π-points $\alpha : A_K \to KG_K$ and $\beta : A_L \to LG_L$ are defined to be equivalent if and only if, for all finite-dimensional kG-supermodules M, the module $\alpha^*(M_K)$ is of finite flat dimension if and only if $\beta^*(M_L)$ is of finite flat dimension.

In [**4**, §9], BIKP show for an arbitrary finite unipotent k-supergroup scheme G, there is a bijection Φ_G between $\operatorname{Proj} \mathrm{H}^\bullet(G,k)$ and the set $\Pi(G)$ of equivalence classes of π-points in G.[5] They also define the π-support of a module, and show (for unipotent G) that π-support detects projectivity of modules, and identifies with support defined via cohomology.

QUESTION 7.7.1. In light of the elementary supergroup schemes being inadequate for detecting projectivity for non-unipotent finite supergroup schemes: What modifications should be made, if any, to the definition of π-points for an arbitrary (non-unipotent) finite supergroup scheme, to enable an identification between π-support and cohomological support?

Acknowledgments

The authors thank Daniel Nakano for helpful conversations. The authors also thank the referees for their comments. In particular, we thank the referee for bringing to our attention the example discussed in Remark 3.1.3.

References

[1] I. Bagci, J. R. Kujawa, and D. K. Nakano, *Cohomology and support varieties for Lie superalgebras of type W(n)*, Int. Math. Res. Not. IMRN, posted on 2008, Art. ID rnn115, 42, DOI 10.1093/imrn/rnn115. MR2448087

[2] P. Balmer, *The spectrum of prime ideals in tensor triangulated categories*, J. Reine Angew. Math. **588** (2005), 149–168, DOI 10.1515/crll.2005.2005.588.149. MR2196732

[5] There are some inaccuracies at the beginning of [**4**, §9] in the discussion of the results from our paper [**14**], though these can be corrected by strategically replacing instances of the functor $\mathrm{Hom}_{\mathsf{sGrp}}(\mathbb{M}_r, G)$ with either $C_r(G)$ or $\mathrm{Hom}_{\mathsf{sHopf}}(\mathbb{P}_r, kG)$. The discussion is essentially correct as-is once the authors specialize to finite connected unipotent supergroups, which is their main case of interest. See Remark 7.4.2, Remark 7.3.2, and (7.5.5) for details.

[3] D. Benson, S. B. Iyengar, H. Krause, and J. Pevtsova, *Detecting nilpotence and projectivity over finite unipotent supergroup schemes*, Selecta Math. (N.S.) **27** (2021), no. 2, Paper No. 25, 59, DOI 10.1007/s00029-021-00632-7. MR4247925

[4] D. Benson, S. B. Iyengar, H. Krause, and J. Pevtsova, *Stratification and duality for unipotent finite supergroup schemes*, Equivariant topology and derived algebra, London Math. Soc. Lecture Note Ser., vol. 474, Cambridge Univ. Press, Cambridge, 2022, pp. 241–275. MR4327102

[5] B. D. Boe and J. R. Kujawa, *Complexity and support varieties for type P Lie superalgebras*, Math. Res. Lett. **29** (2022), no. 1, 59–100. MR4477681

[6] B. D. Boe, J. R. Kujawa, and D. K. Nakano, *Cohomology and support varieties for Lie superalgebras. II*, Proc. Lond. Math. Soc. (3) **98** (2009), no. 1, 19–44, DOI 10.1112/plms/pdn019. MR2472160

[7] B. D. Boe, J. R. Kujawa, and D. K. Nakano, *Cohomology and support varieties for Lie superalgebras*, Trans. Amer. Math. Soc. **362** (2010), no. 12, 6551–6590, DOI 10.1090/S0002-9947-2010-05096-2. MR2678986

[8] B. D. Boe, J. R. Kujawa, and D. K. Nakano, *Complexity for modules over the classical Lie superalgebra* $\mathfrak{gl}(m|n)$, Compos. Math. **148** (2012), no. 5, 1561–1592, DOI 10.1112/S0010437X12000231. MR2982440

[9] B. D. Boe, J. R. Kujawa, and D. K. Nakano, *Tensor triangular geometry for classical Lie superalgebras*, Adv. Math. **314** (2017), 228–277, DOI 10.1016/j.aim.2017.04.022. MR3658717

[10] R. Bøgvad, *Some elementary results on the cohomology of graded Lie algebras*, Algebraic homotopy and local algebra (Luminy, 1982), Astérisque, vol. 113, Soc. Math. France, Paris, 1984, pp. 156–166. MR749051

[11] I. Dell'Ambrogio, *Tensor triangular geometry and KK-theory*, J. Homotopy Relat. Struct. **5** (2010), no. 1, 319–358. MR2812924

[12] C. M. Drupieski, *Cohomological finite generation for restricted Lie superalgebras and finite supergroup schemes*, Represent. Theory **17** (2013), 469–507, DOI 10.1090/S1088-4165-2013-00440-5. MR3096330

[13] C. M. Drupieski, *Cohomological finite-generation for finite supergroup schemes*, Adv. Math. **288** (2016), 1360–1432, DOI 10.1016/j.aim.2015.11.017. MR3436410

[14] C. M. Drupieski and J. R. Kujawa, *Graded analogues of one-parameter subgroups and applications to the cohomology of* $GL_{m|n(r)}$, Adv. Math. **348** (2019), 277–352, DOI 10.1016/j.aim.2019.03.014. MR3925933

[15] C. M. Drupieski and J. R. Kujawa, *On support varieties for Lie superalgebras and finite supergroup schemes*, J. Algebra **525** (2019), 64–110, DOI 10.1016/j.jalgebra.2018.12.033. MR3909818

[16] C. M. Drupieski and J. R. Kujawa, *On the cohomological spectrum and support varieties for infinitesimal unipotent supergroup schemes*, Advances in algebra, Springer Proc. Math. Stat., vol. 277, Springer, Cham, 2019, pp. 121–167, DOI 10.1007/978-3-030-11521-0_7. MR3922643

[17] C. M. Drupieski and J. R. Kujawa, *Support schemes for infinitesimal unipotent supergroups*, Adv. Math. **384** (2021), Paper No. 107754, 61 pp., DOI 10.1016/j.aim.2021.107754. MR4246972

[18] C. M. Drupieski and J. R. Kujawa, *Support varieties and modules of finite projective dimension for modular Lie superalgebras*, Algebra Number Theory **15** (2021), no. 5, 1157–1180, DOI 10.2140/ant.2021.15.1157. With an appendix by Luchezar L. Avramov and Srikanth B. Iyengar. MR4283100

[19] Christopher M. Drupieski and Jonathan R. Kujawa, *Support varieties for Lie superalgebras in characteristic 2*, preprint, 2022, `arXiv:2205.08672`.

[20] Michel Duflo and Vera Serganova, *On associated variety for Lie superalgebras*, unpublished, 2005. `arXiv:0507198`

[21] H. El Turkey, *Complexity of modules over classical Lie superalgebras*, J. Algebra **445** (2016), 365–393, DOI 10.1016/j.jalgebra.2015.08.022. MR3418063

[22] H. El Turkey, *Complexity of simple modules over the Lie superalgebra* $\mathfrak{osp}(k|2)$, J. Pure Appl. Algebra **222** (2018), no. 1, 181–190, DOI 10.1016/j.jpaa.2017.03.010. MR3681001

[23] I. Entova-Aizenbud and V. Serganova, *Kac-Wakimoto conjecture for the periplectic Lie superalgebra*, J. Algebra Appl. **20** (2021), no. 1, Paper No. 2140015, 13 pp., DOI 10.1142/S0219498821400156. MR4209969

[24] I. Entova-Aizenbud and V. Serganova, *Duflo-Serganova functor and superdimension formula for the periplectic Lie superalgebra*, Algebra Number Theory **16** (2022), no. 3, 697–729, DOI 10.2140/ant.2022.16.697. MR4449396

[25] E. M. Friedlander and B. J. Parshall, *Support varieties for restricted Lie algebras*, Invent. Math. **86** (1986), no. 3, 553–562, DOI 10.1007/BF01389268. MR860682

[26] E. M. Friedlander and J. Pevtsova, *Representation-theoretic support spaces for finite group schemes*, Amer. J. Math. **127** (2005), no. 2, 379–420. MR2130619

[27] E. M. Friedlander and J. Pevtsova, *Π-supports for modules for finite group schemes*, Duke Math. J. **139** (2007), no. 2, 317–368, DOI 10.1215/S0012-7094-07-13923-1. MR2352134

[28] E. M. Friedlander and A. Suslin, *Cohomology of finite group schemes over a field*, Invent. Math. **127** (1997), no. 2, 209–270, DOI 10.1007/s002220050119. MR1427618

[29] D. B. Fuks, *Cohomology of infinite-dimensional Lie algebras*, Contemporary Soviet Mathematics, Consultants Bureau, New York, 1986. Translated from the Russian by A. B. Sosinskiĭ. MR874337

[30] N. Geer, J. Kujawa, and B. Patureau-Mirand, *Generalized trace and modified dimension functions on ribbon categories*, Selecta Math. (N.S.) **17** (2011), no. 2, 453–504, DOI 10.1007/s00029-010-0046-7. MR2803849

[31] Maria Gorelik, Crystal Hoyt, Vera Serganova, and Alexander Sherman, *The Duflo–Serganova functor, vingt ans après*, J. Indian Inst. Sci. **102** (2022), no. 3, 961–1000.

[32] D. Grantcharov, N. Grantcharov, D. K. Nakano, and J. Wu, *On BBW parabolics for simple classical Lie superalgebras*, Adv. Math. **381** (2021), Paper No. 107647, 44, DOI 10.1016/j.aim.2021.107647. MR4219889

[33] G. Hochschild, *Relative homological algebra*, Trans. Amer. Math. Soc. **82** (1956), 246–269, DOI 10.2307/1992988. MR80654

[34] J. C. Jantzen, *Kohomologie von p-Lie-Algebren und nilpotente Elemente* (German), Abh. Math. Sem. Univ. Hamburg **56** (1986), 191–219, DOI 10.1007/BF02941516. MR882415

[35] J. C. Jantzen, *Representations of algebraic groups*, 2nd ed., Mathematical Surveys and Monographs, vol. 107, American Mathematical Society, Providence, RI, 2003. MR2015057

[36] V. G. Kac, *Lie superalgebras*, Advances in Math. **26** (1977), no. 1, 8–96, DOI 10.1016/0001-8708(77)90017-2. MR486011

[37] V. G. Kac and M. Wakimoto, *Integrable highest weight modules over affine superalgebras and number theory*, Lie theory and geometry, Progr. Math., vol. 123, Birkhäuser Boston, Boston, MA, 1994, pp. 415–456, DOI 10.1007/978-1-4612-0261-5_15. MR1327543

[38] J. Kujawa, *The generalized Kac-Wakimoto conjecture and support varieties for the Lie superalgebra* $\mathfrak{osp}(m|2n)$, Recent developments in Lie algebras, groups and representation theory, Proc. Sympos. Pure Math., vol. 86, Amer. Math. Soc., Providence, RI, 2012, pp. 201–215, DOI 10.1090/pspum/086/1419. MR2977005

[39] S. Kumar, *Kac-Moody groups, their flag varieties and representation theory*, Progress in Mathematics, vol. 204, Birkhäuser Boston, Inc., Boston, MA, 2002, DOI 10.1007/978-1-4612-0105-2. MR1923198

[40] G. I. Lehrer, D. K. Nakano, and R. Zhang, *Detecting cohomology for Lie superalgebras*, Adv. Math. **228** (2011), no. 4, 2098–2115, DOI 10.1016/j.aim.2011.06.033. MR2836115

[41] M. Lorenz, *Algebraic group actions on noncommutative spectra*, Transform. Groups **14** (2009), no. 3, 649–675, DOI 10.1007/s00031-009-9059-8. MR2534802

[42] I. M. Musson, *Lie superalgebras and enveloping algebras*, Graduate Studies in Mathematics, vol. 131, American Mathematical Society, Providence, RI, 2012, DOI 10.1090/gsm/131. MR2906817

[43] D. Quillen, *The spectrum of an equivariant cohomology ring. I, II*, Ann. of Math. (2) **94** (1971), 549–572; ibid. (2) 94 (1971), 573–602, DOI 10.2307/1970770. MR298694

[44] V. Serganova, *On the superdimension of an irreducible representation of a basic classical Lie superalgebra*, Supersymmetry in mathematics and physics, Lecture Notes in Math., vol. 2027, Springer, Heidelberg, 2011, pp. 253–273, DOI 10.1007/978-3-642-21744-9_12. MR2906346

[45] Vera Serganova and Alexander Sherman, *Splitting quasireductive supergroups and volumes of supergrassmannians*, 2022.

[46] Vera Serganova and Dmitry Vaintrob, *Localization for cs manifolds and volume of homogeneous superspaces*, preprint, 2022.

[47] A. Suslin, *Detection theorem for finite group schemes*, J. Pure Appl. Algebra **206** (2006), no. 1-2, 189–221, DOI 10.1016/j.jpaa.2005.03.017. MR2220088

[48] A. Suslin, E. M. Friedlander, and C. P. Bendel, *Infinitesimal 1-parameter subgroups and cohomology*, J. Amer. Math. Soc. **10** (1997), no. 3, 693–728, DOI 10.1090/S0894-0347-97-00240-3. MR1443546

[49] A. Suslin, E. M. Friedlander, and C. P. Bendel, *Support varieties for infinitesimal group schemes*, J. Amer. Math. Soc. **10** (1997), no. 3, 729–759, DOI 10.1090/S0894-0347-97-00239-7. MR1443547

[50] W. C. Waterhouse, *Introduction to affine group schemes*, Graduate Texts in Mathematics, vol. 66, Springer-Verlag, New York-Berlin, 1979. MR547117

[51] S. J. Witherspoon, *Hochschild cohomology for algebras*, Graduate Studies in Mathematics, vol. 204, American Mathematical Society, Providence, RI, 2019, DOI 10.1090/gsm/204. MR3971234

[52] A. N. Zubkov and P. A. Ulyashev, *Solvable and unipotent supergroups* (Russian, with English and Russian summaries), Algebra Logika **53** (2014), no. 3, 323–339, 419, 422, DOI 10.1007/s10469-014-9284-7; English transl., Algebra Logic **53** (2014), no. 3, 206–216. MR3288440

Department of Mathematical Sciences, DePaul University, Chicago, Illinois 60614
Email address: `c.drupieski@depaul.edu`

Department of Mathematics, University of Oklahoma, Norman, Oklahoma 73019
Current address: Department of Mathematics, Oregon State University, Corvallis, Oregon 97331
Email address: `kujawaj@oregonstate.edu`

Contemporary Mathematics
Volume **804**, 2024
https://doi.org/10.1090/conm/804/16113

Wreath Macdonald polynomials, a survey

Daniel Orr and Mark Shimozono

ABSTRACT. Wreath Macdonald polynomials arise from the geometry of Γ-fixed loci of Hilbert schemes of points in the plane, where Γ is a finite cyclic group of order $r \geq 1$. For $r = 1$, they recover the classical (modified) Macdonald symmetric functions through Haiman's geometric realization of these functions. The existence, integrality, and positivity of wreath Macdonald polynomials for $r > 1$ was conjectured by Haiman and first proved in work of Bezrukavnikov and Finkelberg by means of an equivalence of derived categories. Despite the power of this approach, a lack of explicit tools providing direct access to wreath Macdonald polynomials—in the spirit of Macdonald's original works—has limited progress in the subject.

A recent result of Wen provides a remarkable set of such tools, packaged in the representation theory of quantum toroidal algebras. In this article, we survey Wen's result along with the basic theory of wreath Macdonald polynomials, including its geometric foundations and the role of bigraded reflection functors in the construction of wreath analogs of the ∇ operator. We also formulate new conjectures on the values of important constants arising in the theory of wreath Macdonald P-polynomials. A variety of examples are used to illustrate these objects and constructions throughout the paper.

1. Introduction

1.1. Classical Macdonald theory. The Macdonald polynomials P_μ of [**Mac**, VI] were originally conceived as a distinguished family of symmetric orthogonal polynomials in several (or infinitely many) variables and depending on two parameters, q and t. They were designed to interpolate, through specialization of the parameters, between the major families of multivariable symmetric orthogonal polynomials arising in representation theories associated with the general linear groups GL_n and symmetric groups $\mathfrak{S}_n$—namely, the Schur, Hall-Littlewood, and Jack polynomials. The existence of the P_μ was originally derived from the digaonalization of explicit q-difference operators [**Mac**] or, equivalently, the symmetric function vertex operators of [**GH**].

Macdonald's famous integrality and positivity conjectures [**Mac**, VI.8] drove the study of Macdonald polynomials into deeper territory and inspired many tremendous efforts toward their proof. Among these, Haiman's proof [**H1**] of the Macdonald positivity conjecture established a remarkable connection between Macdonald polynomials and the geometry of Hilbert schemes of points in the plane. Central to Haiman's approach is an equivalence of derived categories of equivariant coherent sheaves

$$D^b(\mathrm{Coh}^{\mathbb{T}}(\mathrm{Hilb}_n)) \cong D^b(\mathrm{Coh}^{\mathbb{T}\times\mathfrak{S}_n}(\mathbb{C}^{2n})), \tag{1.1}$$

where $\mathrm{Hilb}_n = \mathrm{Hilb}_n(\mathbb{C}^2)$ is the Hilbert scheme of n points in the plane $\mathbb{C}^2$ and $\mathbb{T} = (\mathbb{C}^\times)^2$ is a two-dimensional torus acting naturally on Hilb_n and on $\mathbb{C}^{2n} = (\mathbb{C}^2)^n$. At the level of equivariant K-theory, the equivalence (1.1) leads to an isomorphism of vector spaces

$$\Phi : \bigoplus_{n\geq 0} K^{\mathbb{T}}(\mathrm{Hilb}_n)_{\mathrm{loc}} \xrightarrow{\sim} \Lambda, \tag{1.2}$$

where Λ denotes the $\mathbb{K} = \mathbb{Q}(q,t)$-algebra of symmetric functions in infinitely many variables and loc stands for localization with respect to the $\mathbb{T}$-action. (This localization is achieved by extending of scalars to the field of fractions $\mathbb{K}$ of the representation ring $K^{\mathbb{T}}(\mathrm{pt}) \cong R(\mathbb{T}) \cong \mathbb{Z}[q^{\pm 1}, t^{\pm 1}]$ of the torus $\mathbb{T}$.)

For each $n \geq 0$ and each partition μ of size n, we have a $\mathbb{T}$-fixed point $I_\mu \in \mathrm{Hilb}_n$ given by a monomial ideal, and the isomorphism Φ sends the corresponding class $[I_\mu]$ in $K^{\mathbb{T}}(\mathrm{Hilb}_n)_{\mathrm{loc}}$ to the modified Macdonald symmetric function $\tilde{H}_\mu$ (which is a cousin of P_μ). Moreover, the equivalence (1.1) realizes $\tilde{H}_\mu$ as the Frobenius character of a bigraded $\mathfrak{S}_n$-module, namely the fiber $P_n|_{I_\mu}$ of a rank $n!$ vector bundle P_n on Hilb_n. This vector bundle P_n, which induces (1.1) and has many special properties, is known as the Procesi bundle.

It was later realized that the isomorphism Φ carries further representation-theoretic significance. In particular, both sides of (1.2) admit natural actions of the quantum toroidal $\mathfrak{gl}_1$ algebra U_1^{tor}, also known as the elliptic Hall algebra [**BS**]. The space on the left side of (1.2) affords a well-known geometrically-defined action of U_1^{tor} [**SV2**, **FT**], which is constructed along the lines of earlier work of Nakajima and Grojnowski. The action of U_1^{tor} on the right-hand side Λ is expressed via vertex operators going back to [**GH**, **GT**] and arises as limit of polynomial representations of type GL spherical double affine Hecke algebras [**SV1**]. A result of [**SV2**, **FT**] then asserts that the isomorphism Φ arising from Haiman's theorem interwines these two actions of U_1^{tor}, up to an automorphism of U_1^{tor} known as the Miki automorphism [**M2**] (or, equivalently, the limit of Cherednik's Fourier transform [**C**]). This perspective on Φ provided by the methods of [**SV1**, **SV2**, **FT**] has proven to be tremendously fruitful in Macdonald theory; see, for instance, the works [**GN**] and [**BHMPS**], both of which also employ the powerful shuffle realization of U_1^{tor} [**SV2**, **FT**, **Neg1**].

The particular developments in Macdonald theory which we have described above will serve as a model for our treatment of wreath Macdonald polynomials. To summarize, these developments proceeded as follows:

(1) *Eigenoperators.* Macdonald polynomials P_μ (and their variants $\tilde{H}_\mu$) are constructed as orthogonal polynomials by means of explicit eigenoperators.
(2) *Geometry.* Haiman's theorem on the Hilbert scheme realization of Macdonald polynomials implies the Macdonald positivity conjecture.
(3) *Representation theory.* The isomorphism (1.2) arising from Haiman's theorem is understood as an intertwining map between two representations of the quantum toroidal $\mathfrak{gl}_1$ algebra.

1.2. Wreath Macdonald theory. Our aim in this article is to describe developments in the theory of *wreath Macdonald polynomials*. These are a generalization of Macdonald polynomials arising from the geometry of the Γ-fixed loci $\mathrm{Hilb}_n^\Gamma = \mathrm{Hilb}_n(\mathbb{C}^2)^\Gamma$ for a finite subgroup $\Gamma \subset SL_2(\mathbb{C})$. Although the original framework for wreath Macdonald theory proposed by Haiman [**H2**] allows for an arbitrary finite subgroup Γ, we shall assume throughout that $\Gamma \cong \mathbb{Z}/r\mathbb{Z}$ is a finite cyclic group, as essentially all developments in the subject to date are limited to this case.

While we shall use the developments in classical Macdonald theory described above as a model, it is interesting that wreath Macdonald theory has proceeded in a different order, beginning with geometry. In broad strokes, its development has proceeded to date as follows:

(1) *Geometry.* Following Haiman's proposal of the geometric foundations for wreath Macdonald theory [**H2**], Bezrukavnikov and Finkelberg [**BF**] prove that a wreath analog of the equivalence (1.1) leads to a geometric realization of wreath Macdonald polynomials via the analog of the map (1.2). This establishes the existence, integrality, and positivity of wreath Macdonald polynomials, all conjectured by Haiman [**H2**]. This was later established and extended by Losev [**L1, L2**].

(2) *Representation theory.* Wen [**W1**] proves that the wreath analog of the isomorphism (1.2) is, up to nontrivial scalars, an intertwining map between two representations of the quantum toroidal $\mathfrak{gl}_r$ algebra U_r^{tor}. This involves a twist by the remarkable *Miki automorphism* of U_r^{tor}. Wen's result, which builds upon on earlier work of Negut [**Neg2**] and Tsymbaliuk [**T**], constitutes an highly nontrivial extension of the result of [**SV2, FT**] described above.

(3) *Eigenoperators.* By virtue of [**W1**], the horizontal Heisenberg subalgebra of U_r^{tor} acts diagonally on wreath Macdonald polynomials. This leads to explicit vertex operators having the wreath Macdonald polynomials as their joint eigenbasis.

In somewhat more detail, for each element β in the root lattice Q of $\mathfrak{sl}_r$, one has a basis of wreath Macdonald polynomials $\{\tilde{H}_{\mu^\bullet}^{t_{-\beta^\vee}}\}_{\mu^\bullet}$ for the r-fold tensor power $\Lambda^{\otimes r}$ of symmetric functions. These are indexed arbitrary by r-tuples of partitions $\mu^\bullet = (\mu^{(0)}, \ldots, \mu^{(r-1)})$, with $\tilde{H}_{\mu^\bullet}^{t_{-\beta^\vee}}$ homogeneous of total degree $|\mu^\bullet| = \sum_i |\mu^{(i)}|$. The root lattice element β corresponds to an r-core partition γ, and once this is fixed we can think of the multipartition $\mu^\bullet$ as the r-quotient of an arbitrary (single) partition of size $N = |\gamma| + r|\mu^\bullet|$ (see §3.1.2).

The wreath analog of (1.1) is an equivalence

$$D^b(\mathrm{Coh}^{\mathbb{T}}(\mathfrak{M}_{\beta,n})) \cong D^b(\mathrm{Coh}^{\mathbb{T}\times\Gamma_n}(\mathbb{C}^{2n})) \tag{1.3}$$

studied in [**BF**] and arising from more general result of Bezrukavnikov and Kaledin [**BK**]. Here $\Gamma_n = \mathfrak{S}_n \wr \Gamma$ is the wreath product group, $\beta \in Q$ is an element of root lattice of $\mathfrak{sl}_r$, and $\mathfrak{M}_{\beta,n}$ is a Nakajima quiver variety for the cyclic quiver with r vertices and one-dimensional framing space. As we recall in detail below, $\mathfrak{M}_{\beta,n}$ is isomorphic to an irreducible component of Hilb_N^Γ for N as above.

The analog of (1.2) is then an isomorphism

$$\Phi_{\beta,n} : K^{\mathbb{T}}(\mathfrak{M}_{\beta,n})_{\mathrm{loc}} \xrightarrow{\sim} (\Lambda^{\otimes r})_n, \tag{1.4}$$

where $(\Lambda^{\otimes r})_n$ is the component of $\Lambda^{\otimes r}$ consisting of multisymmetric functions of total degree n. Bezrukavnikov and Finkelberg [**BF**] established that the isomorphism $\Phi_{\beta,n}$ sends the fixed-point basis $\{[I_{\mu^\bullet}^{t_{-\beta^\vee}}]\}$ of $K^{\mathbb{T}}(\mathfrak{M}_{\beta,n})_{\mathrm{loc}}$, where $\mu^\bullet$ runs over r-multipartitions of total size n, to the sought-after wreath Macdonald basis $\{\tilde{H}_{\mu^\bullet}^{t_{-\beta^\vee}}\}$ of $(\Lambda^{\otimes r})_n$ conjectured to exist in [**H2**]. In this way, the wreath Macdonald polynomial $\tilde{H}_{\mu^\bullet}^{t_{-\beta^\vee}}$ is realized as the bigraded Γ_n-Frobenius character of the fiber of a rank $r^n n!$ vector bundle (the wreath Procesi bundle) over the corresponding $\mathbb{T}$-fixed point, establishing integrality and positivity of wreath Macdonald-Kostka coefficients.

Taking the direct sum of all maps $\Phi_{\beta,n}$ from (1.4), one obtains an isomorphism

$$\Phi : \bigoplus_{\substack{\beta\in Q \\ n\geq 0}} K^{\mathbb{T}}(\mathfrak{M}_{\beta,n})_{\mathrm{loc}} \xrightarrow{\sim} \bigoplus_{\beta\in Q} \Lambda^{\otimes r} \tag{1.5}$$

of $\mathbb{K}$-vector spaces. This is where the representation theory of the quantum toroidal algebra U_r^{tor} enters the picture. In particular, (1.5) is the setting for Wen's work [**W1**], which involves a comparison between the geometrically-defined action of U_r^{tor} on the left-hand side [**VV2, Nak2**] and Saito's vertex representation [**S**] on the right-hand side, twisted by the Miki automorphism. The main result of [**W1**] can be interpreted as the assertion that the map (1.5) arising from the equivalence (1.3) agrees with Tsymbaliuk's isomorphism [**T**] between two realizations of the Fock representation of U_r^{tor}, up to a rescaling of the fixed-point basis.

Under the U_r^{tor} action of [**VV2, Nak2**], it is easy to identify eigenoperators for the fixed-point basis on the left-hand side of (1.5). A key point of Wen's result is that it allows one to transfer these to the right-hand side, thus obtaining eigenoperators for wreath Macdonald polynomials $\{\tilde{H}_{\mu^\bullet}^{t_{-\beta^\vee}}\}$ in the vertex representation [**S**]. What makes this transfer highly nontrivial and difficult to discover directly is that it must go through the Miki automorphism of U_r^{tor}.

More generally, for any element $w = ut_{-\beta^\vee} \in \mathfrak{S}_r \ltimes Q$ in the affine Weyl group of type A_{r-1}, one has a wreath Macdonald basis $\{\tilde{H}_{\mu^\bullet}^{w}\}$ of $\Lambda^{\otimes r}$ indexed by multipartitions $\mu^\bullet$ of arbitrary size. There are corresponding equivalences (1.3) and isomorphisms $\Phi_n^w : K^{\mathbb{T}}(\mathfrak{M}_{\beta,n})_{\mathrm{loc}} \xrightarrow{\sim} (\Lambda^{\otimes r})_n$ realizing each $\tilde{H}_{\mu^\bullet}^{w}$ as the bigraded Γ_n-Frobenius character of a fiber of a different wreath Procesi bundle (depending on the finite permutation u). These more general Procesi bundles were introduced and classifed by Losev [**L1, L2**], who gave an independent proof of the main result of [**BF**]. At present, we do not know if it is possible to characterize the more general wreath Macdonald polynomials $\tilde{H}_{\mu^\bullet}^{w}$ for $w = ut_{-\beta^\vee}$ with $u \neq 1$ using quantum toroidal algebras.

In forthcoming work [**OS, OSW**], we use Wen's result to extract explicit q-difference eigenoperators for the finite variable specializations of wreath Macdonald polynomials, providing a characterization of these polynomials true to the spirit of Macdonald's original approach to the subject.

1.3. Overview. We provide a survey of the developments described above, culminating in a statement of Wen's result and a completely concrete description of the vertex eigenoperators it provides for the wreath Macdonald polynomials $\tilde{H}_{\mu^\bullet}^{t_{-\beta^\vee}}$ indexed by translations. Along the way, and with the intention of making the subject more widely accessible, we provide an explicit and detailed exposition of fundamental properties and symmetries of wreath Macdonald polynomials $\tilde{H}^w_{\mu^\bullet}$ for arbitrary affine Weyl group elements w. We explain the role of bigraded reflection functors in the theory (as explained to us by Haiman [**H3**]), and in particular how these lead to wreath analogs of the ∇ operator from classical Macdonald theory. We illustrate these constructions with a variety of examples. As an original contribution, we also formulate new conjectures on important constants arising in the theory of wreath Macdonald P-polynomials (Conjectures 3.36 and 3.38).

To set the stage, we begin with a review of some major results from classical Macdonald theory which have influenced these developments.

2. Classical Macdonald theory

2.1. Partitions. Let $\mathbb{Y}$ be Young's lattice of partitions. The diagram of a partition $\mu = (\mu_1, \mu_2, \dots) \in \mathbb{Y}$ is the subset of $\mathbb{Z}^2_{\geq 0}$ given by

$$D(\mu) = \{(a,b) \mid 0 \leq a < \mu_{b+1}\}.$$

Its elements are called cells. We often write μ to mean $D(\mu)$. For any $\mu \in \mathbb{Y}$, let $\mu^t \in \mathbb{Y}$ be the transposed partition, defined by $(a,b) \in \mu^t$ if and only if $(b,a) \in \mu$. Let $|\mu| = \sum_i \mu_i$. Let $\trianglerighteq$ be the dominance partial order on $\mathbb{Y}$ [**Mac**, §I.1]. Let $\mathbb{Y}_N \subset \mathbb{Y}$ be the set of partions with at most N parts.

For any $\mu \in \mathbb{Y}$, define the following quantities:

$$B_\mu = B_\mu(q,t) = \sum_{(a,b)\in\mu} q^a t^b \tag{2.1}$$

$$A_\mu = A_\mu(q,t) = 1 - (1-q)(1-t)B_\mu. \tag{2.2}$$

One has

$$A_\mu = \sum_{(a,b)\in A(\mu)} q^a t^b - qt \sum_{(a,b)\in R(\mu)} q^a t^b,$$

where $A(\mu)$ is the set of μ-addable cells, the cells $s \in \mathbb{Z}^2_{\geq 0} \setminus D(\mu)$ such that $D(\mu) \cup \{s\}$ is the diagram of a partition, and $R(\mu)$ is the set of μ-removable cells, those cells $s \in D(\mu)$ such that $D(\mu) \setminus \{s\}$ is the diagram of a partition.

EXAMPLE 2.1. For the partition $\mu = (3,2)$

$$\begin{aligned} D(\mu) &= \{(0,0),(1,0),(2,0),(0,1),(1,1)\}, \\ A(\mu) &= \{(3,0),(2,1),(0,2)\}, \\ R(\mu) &= \{(2,0),(1,1)\}. \end{aligned}$$

The addable (resp. removable) cells of μ are depicted with $+$ (resp. $-$) entries.

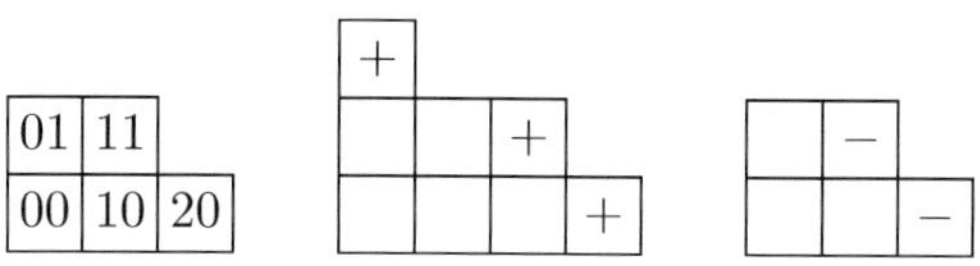

We have

$$\begin{aligned} B_\mu(q,t) &= 1 + q + q^2 + t + qt \\ A_\mu(q,t) &= q^3 + q^2 t + t^2 - qt(q^2 + qt). \end{aligned}$$

The arm $a_\mu(s)$ (resp. leg $l_\mu(s)$) of a cell $s = (a, b) \in \mu$ is by definition the number of cells in μ that are strictly to the right (resp. above) s in its row (resp. column). The hook of s in μ is given by $h_\mu(s) = 1 + a_\mu(s) + l_\mu(s)$.

EXAMPLE 2.2. For $\mu = (6, 4, 2, 1)$ and $s = (1, 1)$ we have $a_\mu(s) = 2$, $l_\mu(s) = 1$, and $h_\mu(s) = 4$.

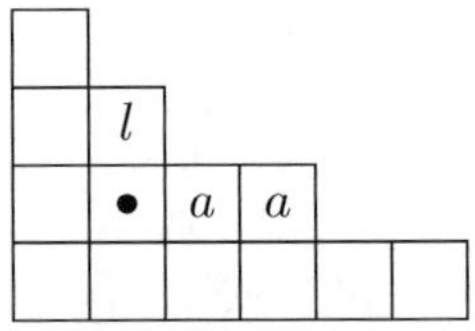

2.2. Symmetric functions. For any integer $N \geq 0$, let $\mathfrak{S}_N$ be the symmetric group and let $\Lambda_{N,\mathbb{Z}} = \mathbb{Z}[x_1, \ldots, x_N]^{\mathfrak{S}_N}$ be the ring of symmetric polynomials in the variables $x_1, \ldots, x_N$. This is a graded ring: $\Lambda_{N,\mathbb{Z}} = \bigoplus_{n=0}^{\infty} \Lambda^n_{N,\mathbb{Z}}$ where $\Lambda^n_{N,\mathbb{Z}} \subset \Lambda_{N,\mathbb{Z}}$ consists of symmetric polynomials which are homogeneous of total degree n.

Let $\Lambda_{\mathbb{Z}} = \bigoplus_{n=0}^{\infty} \Lambda^n_{\mathbb{Z}}$ be the ring of symmetric functions [**Mac**, §I.2], which is by definition the projective limit of the $\Lambda_{N,\mathbb{Z}}$ in the category of graded rings. In concrete terms, one may regard elements of $\Lambda_{\mathbb{Z}}$ as formal power series in infinitely many variables $x_1, x_2, \ldots$ which are symmetric and of bounded degree. For $f \in \Lambda_{\mathbb{Z}}$, we use the plethystic notation $f[X_N]$, where $X_N = x_1 + \ldots + x_N$, to denote the image of f in $\Lambda_{N,\mathbb{Z}}$.

Let $\{m_\lambda\}_{\lambda \in \mathbb{Y}}$ and $\{s_\lambda\}_{\lambda \in \mathbb{Y}}$ be the bases of monomial and Schur symmetric functions, respectively. We continue to denote the images of m_λ and s_λ in $\Lambda_{N,\mathbb{Z}}$ by the same letters; for $\lambda \in \mathbb{Y}_N$ we obtain bases of $\Lambda_{N,\mathbb{Z}}$.

For any commutative ring R, let $\Lambda_R = \Lambda_{\mathbb{Z}} \otimes_{\mathbb{Z}} R$, $\Lambda^n_R = \Lambda^n_{\mathbb{Z}} \otimes_{\mathbb{Z}} R$, and similarly define $\Lambda_{N,R}$ and $\Lambda^n_{N,R}$. We have the power sum basis $\{p_\lambda\}_{\lambda \in \mathbb{Y}}$ for $\Lambda_{\mathbb{Q}}$.

Since we will typically work over the field $\mathbb{K} = \mathbb{Q}(q, t)$ of rational functions in indeterminates q and t, we reserve the notations Λ, Λ^n, Λ_N, and Λ^n_N when working over this base field, i.e., $\Lambda = \Lambda_{\mathbb{K}}$, etc.

For any $A = A(q, t) \in \mathbb{K}^\times$ and $B = B(q, t) \in \mathbb{K}$, let $\psi_{A,B}$ be the $\mathbb{K}$-algebra automorphism of Λ defined by

$$\psi_{A,B}(p_n) = A(q^n, t^n) p_n + B(q^n, t^n) \qquad \text{for } n \in \mathbb{Z}_{>0}.$$

We also define the plethystic notation $f[AX+B]$ by means of the equality $\psi_{A,B}(f) = f[AX + B]$ for any $f \in \Lambda$, where $X = x_1 + x_2 + \cdots$.

The Hall inner product $\langle \cdot, \cdot \rangle$ on $\Lambda_{\mathbb{Z}}$ (or, by extension of scalars, on any Λ_R) is defined by $\langle s_\lambda, s_\mu \rangle = \delta_{\lambda\mu}$. Let $\mathrm{End}(\Lambda)$ denote the $\mathbb{K}$-algebra of $\mathbb{K}$-linear endomorphisms of Λ. For $T \in \mathrm{End}(\Lambda)$, let $T^\perp \in \mathrm{End}(\Lambda)$ denote the adjoint of T with respect to the Hall inner product. We often regard $g \in \Lambda$ as the multiplication operator $f \mapsto gf$ on Λ, and then $g^\perp$ denotes the adjoint of this operator. For instance, we have

$$p_n^\perp = n \frac{\partial}{\partial p_n} \qquad \text{for } n \in \mathbb{Z}_{>0}.$$

Finally, denoting by $R(\mathfrak{S}_n)$ the representation ring of the symmetric group, let $\mathrm{Frob} : R(\mathfrak{S}_n) \to \Lambda^n_{\mathbb{Z}}$ be the Frobenius character map [**Mac**, §I.7], which is a linear isomorphism sending the irreducible character χ^λ of $\mathfrak{S}_n$ for $|\lambda| = n$ to the Schur function s_λ. For a bigraded $\mathfrak{S}_n$-module $V = \bigoplus_{a,b \in \mathbb{Z}} V_{a,b}$ with each

$V_{a,b}$ finite-dimensional, we denote by $\mathrm{Frob}_{q,t}(V)$ the bigraded Frobenius series $\sum_{a,b} \mathrm{Frob}(V_{a,b}) q^a t^b$.

2.3. Macdonald P-functions. The Macdonald symmetric functions $\{P_\mu\}_{\mu\in\mathbb{Y}}$ form an orthogonal basis of Λ with respect to Macdonald's (q,t)-deformation of the Hall inner product [**Mac**, §VI]. For any $\mu \in \mathbb{Y}$, P_μ is homogeneous of degree $|\mu|$ and satisfies the unitriangularity (with $\lhd$ meaning $\unlhd$ and not equal):

$$P_\mu \in m_\mu + \bigoplus_{\lambda \lhd \mu} \mathbb{K} m_\lambda.$$

The images $P_\mu[X_N]$ for $\mu \in \mathbb{Y}_N$ form a basis for Λ_N. They can be characterized as the symmetric polynomial eigenfunctions of the Macdonald operator

$$M_N = \sum_{k=1}^{N} \prod_{\substack{\ell=1 \\ \ell \neq k}}^{N} \frac{t x_k - x_\ell}{x_k - x_\ell} T_{q,x_k}, \tag{2.3}$$

where $T_{q,x_k} f(x_1, \ldots, x_N) = f(x_1, \ldots, q x_k, \ldots, x_N)$, with eigenvalues given by:

$$M_N P_\mu[X_N] = \left(\sum_{k=1}^{N} q^{\mu_k} t^{N-k} \right) P_\mu[X_N]. \tag{2.4}$$

EXAMPLE 2.3. Let $N = 2$. The monomial symmetric polynomials

$$m_{(0,0)} = 1, \quad m_{(1,0)} = x_1 + x_2, \quad m_{(1,1)} = x_1 x_2, \quad m_{(2,0)} = x_1^2 + x_2^2$$

form a basis for the subspace $\Lambda_2^0 \oplus \Lambda_2^1 \oplus \Lambda_2^2 \subset \Lambda_2$. With respect to this basis, the restriction of the Macdonald operator M_2 has the following matrix:

$$\begin{pmatrix} 1+t & 0 & 0 & 0 \\ 0 & 1+qt & 0 & 0 \\ 0 & 0 & q(1+t) & (1-q^2)(1-t) \\ 0 & 0 & 0 & 1+q^2 t \end{pmatrix}.$$

One finds that

$$\begin{aligned} P_{(0,0)} &= 1 = m_{(0,0)} \\ P_{(1,0)} &= x_1 + x_2 = m_{(1,0)} \\ P_{(1,1)} &= x_1 x_2 = m_{(1,1)} \\ P_{(2,0)} &= x_1^2 + x_2^2 + \frac{(1+q)(1-t)}{1-qt} x_1 x_2 = m_{(2,0)} + \frac{(1+q)(1-t)}{1-qt} m_{(1,1)}. \end{aligned}$$

2.4. Vertex operators. Let Ω be the plethystic exponential.[1] Thus, in the completion $\widehat{\Lambda} = \prod_{n=0}^{\infty} \Lambda^n$ of symmetric functions, one has the elements

$$\Omega[X] = \prod_{i=1}^{\infty} \frac{1}{1-x_i} = \sum_{k=0}^{\infty} h_k = \exp\left(\sum_{k=1}^{\infty} \frac{p_k}{k} \right)$$

$$\Omega[-X] = \prod_{i=1}^{\infty} (1-x_i) = \sum_{k=0}^{\infty} (-1)^k e_k = \exp\left(-\sum_{k=1}^{\infty} \frac{p_k}{k} \right)$$

[1] For an introduction to Ω in the context of symmetric functions, see [**LR**, §5.3].

where $e_k = s_{(1^k)}$ and $h_k = s_{(k)}$ are the elementary and complete homogeneous symmetric functions, respectively. Generally, for plethystic alphabets X and Y, one has the identity $\Omega[X+Y] = \Omega[X]\Omega[Y]$.

By a symmetric function *vertex operator*, we mean a formal operator series of the form

$$V(z) = \sum_{k=-\infty}^{\infty} V_k z^k = \Omega[zAX]\Omega[z^{-1}BX]^{\perp} \in \mathrm{End}(\Lambda)[[z, z^{-1}]]$$

for some $A, B \in \mathbb{K}$. Here we regard the series

$$\Omega[zAX] = \sum_{k=0}^{\infty} h_k[AX]z^k = \exp\left(\sum_{k=1}^{\infty} \frac{p_k[AX]}{k} z^k\right)$$

as an element of $\mathrm{End}(\Lambda)[[z]]$ by viewing each symmetric function $h_k[AX]$ (or $p_k[AX]$) as a multiplication operator. For the other "half" of $V(z)$, we have the operator series

$$\Omega[z^{-1}BX]^{\perp} = \sum_{k=0}^{\infty} h_k[BX]^{\perp} z^{-k} = \exp\left(\sum_{k=1}^{\infty} \frac{p_k[BX]^{\perp}}{k} z^{-k}\right) \in \mathrm{End}(\Lambda)[[z^{-1}]].$$

Formally multiplying these two operator series, one obtains the following formula for the coefficients of $V(z)$:

$$V_k = \sum_{\substack{\ell, m \geq 0 \\ \ell - m = k}} h_\ell[AX] h_m[BX]^{\perp}.$$

Since $g^{\perp}\Lambda^n = 0$ for any $g \in \Lambda^m$ such that $m > n$, the expression for V_k above reduces to a finite sum when acting on any fixed Λ^n. In particular, each V_k for $k \in \mathbb{Z}$ is a well-defined endomorphism of Λ (of degree k).

By acting on the reproducing kernel $\Omega[XY]$ for the Hall pairing, one can also show that the annihilating half of a vertex operator acts on a symmetric function $f \in \Lambda$ by the plethystic formula:

$$\Omega[z^{-1}BX]^{\perp} \cdot f = f[X + z^{-1}B].$$

2.5. Macdonald vertex operator. In the context of Macdonald P-functions, the vertex operator

$$D(z) = \sum_{n=-\infty}^{\infty} D_k z^{-k} = \Omega[z(t-1)X]\Omega\Big[z^{-1}\frac{q-1}{t}X\Big]^{\perp} \tag{2.5}$$

is of particular importance. Here we have the two halves

$$\Omega[z(t-1)X] = \prod_{i=1}^{\infty} \frac{1 - zx_i}{1 - tzx_i} = \exp\left(\sum_{k=1}^{\infty} \frac{p_k}{k}(t^k - 1)z^k\right)$$

$$\Omega\Big[z^{-1}\frac{q-1}{t}X\Big]^{\perp} = \exp\left(\sum_{k=1}^{\infty} \frac{p_k^{\perp}}{k} \frac{q^k - 1}{t^k} z^{-k}\right).$$

THEOREM 2.4 ([**GH**, Theorem 3.2]). *The coefficient D_0 of z^0 in the vertex operator $D(z)$ acts diagonally on Macdonald symmetric functions*

$$D_0 P_\mu = A_\mu(q, t^{-1}) P_\mu, \tag{2.6}$$

with A_μ given by (2.2).

REMARK 2.5. The operator D_0 arises as the $N \to \infty$ limit of the operator $t^{-N}(1 + (1-t)M_N)$, and this is essential in the proof of Theorem 2.4. The renormalization of M_N is necessary for compatibility with the projective limit defining Λ (see [**Mac**, VI.4]). One sees this most transparently when deriving the eigenvalues in (2.6) from those in (2.4).

EXAMPLE 2.6. For the purpose of illustration, let us verify (2.6) for the Macdonald symmetric function $P_{(1,0)} = p_1$. First, we compute:

$$\Omega[z^{-1}\frac{q-1}{t}X]^{\perp} \cdot P_{(1,0)} = p_1 + z^{-1}\frac{q-1}{t}.$$

Since $\Omega[z(t-1)X] = 1 + z(t-1)p_1 + \cdots$, we then find:

$$\begin{aligned} D_0 P_{(1,0)} &= \left(1 + (t-1)\frac{q-1}{t}\right) p_1 \\ &= (1 - (1-q)(1-t^{-1}))p_1 \\ &= A_{(1,0)}(q, t^{-1})p_1. \end{aligned}$$

2.6. Modified Macdonald functions. The modified Macdonald symmetric functions $\{\tilde{H}_\mu\}_{\mu \in \mathbb{Y}}$ are defined by

$$\tilde{H}_\mu = t^{n(\mu)}(J_\mu[X/(1-t)]|_{t \mapsto t^{-1}}),$$

where $n(\mu) = \sum_i (i-1)\mu_i$ and $J_\mu = \prod_{s \in \lambda}(1 - q^{a_\mu(s)}t^{l_\mu(s)+1})P_\mu$.

The $\tilde{H}_\mu$ are uniquely characterized by the following three conditions [**H2**, §3.5]:

$$\tilde{H}_\mu[(1-q)X] \in \bigoplus_{\lambda \trianglerighteq \mu} \mathbb{K}s_\lambda \tag{2.7}$$

$$\tilde{H}_\mu[(1-t)X] \in \bigoplus_{\lambda \trianglerighteq \mu^t} \mathbb{K}s_\lambda \tag{2.8}$$

$$\langle s_n, \tilde{H}_\mu \rangle = 1. \tag{2.9}$$

REMARK 2.7. One can write (2.8) in the equivalent form

$$\tilde{H}_\mu[(1-t^{-1})X] \in \bigoplus_{\lambda \trianglelefteq \mu} \mathbb{K}s_\lambda, \tag{2.10}$$

since $\tilde{H}_\mu$ is homogeneous, $s_\lambda[-X] = (-1)^{|\lambda|}s_{\lambda^t}$, and $\lambda \trianglerighteq \mu^t$ if and only if $\lambda^t \trianglelefteq \mu$.

The $\tilde{H}_\mu$ are often preferred over P_μ due to their remarkable integrality and positivity properties. In particular, defining the Macdonald-Kostka coefficients $\tilde{K}_{\lambda\mu}(q,t)$ through the expansion

$$\tilde{H}_\mu = \sum_\lambda \tilde{K}_{\lambda\mu}(q,t)s_\lambda,$$

the Macdonald positivity conjecture [**Mac**, VI.8] (cf. Theorem 2.10 below) asserts that

$$\tilde{K}_{\lambda\mu}(q,t) \in \mathbb{Z}_{\geq 0}[q,t]. \tag{2.11}$$

EXAMPLE 2.8. One has

$$\begin{aligned} J_{(2)} &= (1-t)(1-qt)P_{(2)} \\ &= (1-t)(1-qt)m_{(2)} + (1+q)(1-t)^2 m_{(1,1)} \\ &= \frac{(1-q)(1-t^2)}{2}p_2 + \frac{(1+q)(1-t)^2}{2}p_1^2 \end{aligned}$$

and the corresponding Macdonald symmetric function

$$\begin{aligned} \tilde{H}_{(2)} &= t^0 \left(\frac{1-q}{2}p_2 + \frac{1+q}{2}p_1^2 \right)\bigg|_{t\mapsto t^{-1}} \\ &= \frac{1-q}{2}p_2 + \frac{1+q}{2}p_1^2 \\ &= s_{(2)} + qs_{(1,1)}. \end{aligned}$$

We observe (2.11) in this case. We also see that (2.7)–(2.9) hold, with the only nontrivial remaining computation being:

$$\begin{aligned} \tilde{H}_{(2)}[(1-q)X] &= \frac{(1-q)(1-q^2)}{2}p_2 + \frac{(1+q)(1-q)^2}{2}p_1^2 \\ &= \frac{(1-q)(1-q^2)}{2}(p_2 + p_1^2) \\ &= (1-q)(1-q^2)s_{(2)}. \end{aligned}$$

Conjugating the vertex operator of Theorem 2.4 by the transformation from P_μ to $\tilde{H}_\mu$ and making use of the identity $P_\mu|_{q\mapsto q^{-1},t\mapsto t^{-1}} = P_\mu$ [**Mac**, VI, (4.14)], we arrive at the following:

COROLLARY 2.9. *The coefficients* $\tilde{D}_0$ *and* $\tilde{D}_0^*$ *of* z^0 *in the vertex operators*

$$\tilde{D}(z) = \sum_{n=-\infty}^{\infty} \tilde{D}_k z^{-k} = \Omega[-zX]\Omega[z^{-1}(1-q)(1-t)X]^\perp \tag{2.12}$$

$$\tilde{D}^*(z) = \sum_{n=-\infty}^{\infty} \tilde{D}_k^* z^{-k} = \Omega[zX]\Omega[-z^{-1}(1-q^{-1})(1-t^{-1})X]^\perp \tag{2.13}$$

act diagonally on modified Macdonald functions:

$$\begin{aligned} \tilde{D}_0\tilde{H}_\mu &= A_\mu(q,t)\tilde{H}_\mu \\ \tilde{D}_0^*\tilde{H}_\mu &= A_\mu(q^{-1},t^{-1})\tilde{H}_\mu. \end{aligned}$$

2.7. Haiman's Positivity Theorem. Let $\mathrm{Hilb}_n = \mathrm{Hilb}_n(\mathbb{C}^2)$ be the Hilbert scheme of n points in the plane $\mathbb{C}^2$. This is a resolution of the symplectic quotient singularity $\mathbb{C}^{2n}/\mathfrak{S}_n$; in particular, Hilb_n is a smooth variety of dimension $2n$. Its points can be described by ideals $I \subset \mathbb{C}[x,y]$ such that $\dim \mathbb{C}[x,y]/I = n$. The natural action of $\mathbb{T} = (\mathbb{C}^\times)^2$ on $\mathbb{C}^2$ induces an action of $\mathbb{T}$ on Hilb_n with finitely many fixed points—namely, the monomial ideals $I_\mu = (y^{\mu_1}, xy^{\mu_2}, \dots)$ for $\mu \in \mathbb{Y}$ with $|\mu| = n$.

To connect with Macdonald theory, we identify $\mathbb{Z}[q^{\pm1}, t^{\pm1}]$ with the representation ring $R(\mathbb{T})$ and the field $\mathbb{K} = \mathbb{Q}(q,t)$ with $\mathrm{Frac}(R(\mathbb{T}))$. Here we follow the conventions of [**H2**], so that the standard coordinate functions $x(a_1,a_2) = a_1$ and $y(a_1,a_2) = a_2$ on $\mathbb{C}^2$ have $\mathbb{T}$-weights t and q, respectively.

Haiman's proof of the Macdonald positivity conjecture (2.11) centers on establishing the existence of a rank $n!$ vector bundle P_n on Hilb_n, the *Procesi bundle*, with the following special properties:

(1) P_n is $(\mathbb{T}\times\mathfrak{S}_n)$-equivariant, with respect to the trivial $\mathfrak{S}_n$-action on Hilb_n
(2) each fiber of P_n is isomorphic as an $\mathfrak{S}_n$-representation to the regular representation
(3) each fiber of P_n is an $\mathfrak{S}_n$-equivariant quotient of the polynomial ring $\mathbb{C}[x_1,\ldots,x_n,y_1,\ldots,y_n]$ with respect to the diagonal action of $\mathfrak{S}_n$ and with $\mathbb{T}$ acting so that each x_i (resp. each y_i) has weight t (resp. q)

In particular, any fiber $P_n|_{I_\mu}$ over a $\mathbb{T}$-fixed point in Hilb_n carries an action of $\mathbb{T}\times\mathfrak{S}_n$; equivalently, one may regard $P_n|_{I_\mu}$ as a *bigraded* $\mathfrak{S}_n$-module.

THEOREM 2.10 ([**H1**]). *There exists a vector bundle* P_n *on* Hilb_n *with the properties described above. This vector bundle induces an equivalence between derived categories of coherent sheaves*

$$D^b(\mathrm{Coh}^{\mathbb{T}}(\mathrm{Hilb}_n))\cong D^b(\mathrm{Coh}^{\mathbb{T}\times\mathfrak{S}_n}(\mathbb{C}^{2n})) \tag{2.14}$$

which, at the level of K-theory, realizes the modified Macdonald function $\tilde{H}_\mu$ as the bigraded Frobenius character of the fiber $P_n|_{I_\mu}$.

Let us describe how the equivalence (2.14) connects to symmetric functions in more detail. Passing to equivariant K-theory, i.e., applying the K_0-functor to (2.14), one obtains an isomorphism of $R(\mathbb{T})$-modules $K^{\mathbb{T}}(\mathrm{Hilb}_n)\cong K^{\mathbb{T}\times\mathfrak{S}_n}(\mathbb{C}^{2n})$. As explained in [**H2**, §5.4.3], there is an injection $K^{\mathbb{T}\times\mathfrak{S}_n}(\mathbb{C}^{2n})\hookrightarrow\Lambda^n$ of $R(\mathbb{T})$-modules given by the global sections map followed by the bigraded Frobenius series $\mathrm{Frob}_{q,t}$. The composite map $\Phi_n : K^{\mathbb{T}}(\mathrm{Hilb}_n)\hookrightarrow\Lambda^n$ is given explicitly by $\Phi_n([\mathcal{F}]) = \mathrm{Frob}_{q,t}(R\Gamma([\mathcal{F}]\otimes P_n))$ for any $\mathbb{T}$-equivariant coherent sheaf $\mathcal{F}$ on Hilb_n. Letting $[I_\mu]$ denote the class of the skyscraper sheaf at a $\mathbb{T}$-fixed point I_μ in Hilb_n, the second assertion in Theorem 2.10 is then stated precisely as follows:

$$\Phi_n([I_\mu]) = \mathrm{Frob}_{q,t}(P_n|_{I_\mu}) = \tilde{H}_\mu.$$

With $\tilde{H}_\mu$ realized as a bigraded Frobenius character, one immediately deduces that $\tilde{K}_{\lambda\mu}(q,t)$ is a *Laurent* polynomial in q and t with nonnegative integer coefficients for all $\lambda,\mu\in\mathbb{Y}$. Property (3) of the vector bundle P_n then implies the full assertion (2.11) of the Macdonald positivity conjecture.

EXAMPLE 2.11. Let us observe Macdonald positivity (2.11) in conjunction with property (2) of the Procesi bundle for $|\mu|=3$:

$$\begin{aligned}\mathrm{Frob}(\mathbb{C}\mathfrak{S}_3) &= s_{(3)}+2s_{(2,1)}+s_{(1,1,1)}\\ \tilde{H}_{(3)} &= s_{(3)}+(q+q^2)s_{(2,1)}+q^3s_{(1,1,1)}\\ \tilde{H}_{(2,1)} &= s_{(3)}+(q+t)s_{(2,1)}+qts_{(1,1,1)}\\ \tilde{H}_{(1,1,1)} &= s_{(3)}+(t+t^2)s_{(2,1)}+t^3s_{(1,1,1)}.\end{aligned}$$

2.8. Quantum toroidal $\mathfrak{gl}_1$. The maps Φ_n arising from Theorem 2.10 admit a further representation-theoretic interpretation involving an algebra U_1^{tor} known as the elliptic Hall algebra or quantum toroidal $\mathfrak{gl}_1$ algebra. This result—which beautifully unifies all of the material in this section—is due independently to Schiffmann and Vasserot [**SV2**] and Feigin and Tsymbaliuk [**FT**].

Let $\Phi : \bigoplus_{n\geq 0} K^{\mathbb{T}}(\mathrm{Hilb}_n)_{\mathrm{loc}} \to \Lambda$ be the direct sum of the maps $K^{\mathbb{T}}(\mathrm{Hilb}_n)_{\mathrm{loc}} \xrightarrow{\sim} \Lambda^n$ obtained from the Φ_n by extension of scalars to the localized equivariant K-groups $K^{\mathbb{T}}(\mathrm{Hilb}_n)_{\mathrm{loc}} = K^{\mathbb{T}}(\mathrm{Hilb}_n) \otimes_{R(T)} \mathbb{K}$. Then Φ is the isomorphism of $\mathbb{K}$-vector spaces given by $\Phi([I_\mu]) = \tilde{H}_\mu$ for all $\mu \in \mathbb{Y}$. The result of [**FT, SV2**] asserts the isomorphism Φ intertwines two actions of U_1^{tor}. Without giving a complete definition of U_1^{tor}, let us briefly describe the two actions:

- The action of U_1^{tor} on Λ is generated by all coefficients $\tilde{D}_k$ and $\tilde{D}_k^*$ of the vertex operators $\tilde{D}(z)$ and $\tilde{D}^*(z)$ from Corollary 2.9, together with all multiplication opertors f and all skewing operators $f^\perp$ for $f \in \Lambda$. In fact, the four operators $p_1, p_1^\perp, \tilde{D}_0$, and $\tilde{D}_0^*$ are sufficient to generate the entire action of U_1^{tor}.
- The action of U_1^{tor} on $\bigoplus_{n=0}^{\infty} K^{\mathbb{T}}(\mathrm{Hilb}_n)_{\mathrm{loc}}$ is defined geometrically, along the lines of earlier constructions of Nakajima and Grojnowski. This involves the nested Hilbert schemes $\mathrm{Hilb}_{n,n+1} \subset \mathrm{Hilb}_n \times \mathrm{Hilb}_{n+1}$ parametrizing pairs of ideals $(I, J) \in \mathrm{Hilb}_n \times \mathrm{Hilb}_{n+1}$ such that $J \subset I$. Each nested Hilbert scheme is equipped with natural projections $p_{n,n+1} : \mathrm{Hilb}_{n,n+1} \to \mathrm{Hilb}_n$ and $q_{n,n+1} : \mathrm{Hilb}_{n,n+1} \to \mathrm{Hilb}_{n+1}$, as well as a line bundle $L_{n,n+1}$ whose fiber over (I, J) is the quotient space I/J.

  Define convolution operators $e_k : K^{\mathbb{T}}(\mathrm{Hilb}_n) \to K^{\mathbb{T}}(\mathrm{Hilb}_{n+1})$ and $f_k : K^{\mathbb{T}}(\mathrm{Hilb}_n) \to K^{\mathbb{T}}(\mathrm{Hilb}_{n+1})$ for all $k \in \mathbb{Z}$ by

  $$e_k([\mathcal{F}]) = (q_{n,n+1})_*([L_{n,n+1}^{\otimes k}] \cdot p_{n,n+1}^*([\mathcal{F}]))$$
  $$f_k([\mathcal{F}]) = (p_{n,n+1})_*([L_{n,n+1}^{\otimes(k-1)}] \cdot q_{n,n+1}^*([\mathcal{F}])).$$

  Note that f_k is defined to be 0 on $K^{\mathbb{T}}(\mathrm{Hilb}_0)$. We continue denote the induced operators on $\bigoplus_{n=0}^{\infty} K^{\mathbb{T}}(\mathrm{Hilb}_n)_{\mathrm{loc}}$ by e_k and f_k.

  Finally, on each Hilbert scheme Hilb_n there is a tautological vector bundle T_n whose fibers are given by $T_n|_I = \mathbb{C}[x, y]/I$. For any symmetric function $f \in \Lambda$, define h_f and h_f^* to be the operators on $\bigoplus_{n=0}^{\infty} K^{\mathbb{T}}(\mathrm{Hilb}_n)_{\mathrm{loc}}$ which on each $K^{\mathbb{T}}(\mathrm{Hilb}_n)_{\mathrm{loc}}$ are given by K-theory multiplication with the virtual equivariant bundle $f[1 - (1 - q)(1 - t)T_n]$ and its dual, respectively. The $\mathbb{T}$-character of $1 - (1 - q)(1 - t)T_n$ over a fixed point I_μ is precisely $A_\mu(q, t)$ from (2.2), since that of T_n is given by $B_\mu(q, t)$. The $\mathbb{T}$-character of $f[1 - (1 - q)(1 - t)T_n]|_{I_\mu}$ is given by the plethysm $f[A_\mu(q, t)]$.

  The operators e_k, f_k, h_f, and h_f^* for $k \in \mathbb{Z}$ and $f \in \Lambda$ generate the action of U_1^{tor} on $\bigoplus_{n=0}^{\infty} K_{\mathbb{T}}(\mathrm{Hilb}_n)_{\mathrm{loc}}$.

The following result gives the connection these two U_1^{tor}-actions:

THEOREM 2.12 ([**FT**,**SV2**]). *The isomorphism Φ intertwines the two actions of U_1^{tor} described above as follows: the matrix coefficients of the operators $e_{-1}, f_1, h_{p_1}, h_{p_1}^*$ in the basis $\{[I_\mu]\}_{\mu\in\mathbb{Y}}$ are equal to those of $\frac{1}{(1-q)(1-t)}p_1, p_1^\perp, \tilde{D}_0, \tilde{D}_0^*$ in the basis $\{\tilde{H}_\mu\}_{\mu\in\mathbb{Y}}$, respectively.*

The nontrivial identification of generators arising in Theorem 2.12 is given by the *Miki automorphism* (or Fourier transform [**SV1**]) of U_1^{tor}. This automorphism exists for all quantum toroidal algebras U_r^{tor} of type $\mathfrak{gl}_r$ (see §5.1.5 for the $r \geq 3$ case) and it plays a key role in Wen's extension of Theorem 2.12 to the wreath setting (Theorem 5.18).

3. Wreath Macdonald Theory

The definition of wreath Macdonald polynomials has two main ingredients. The first is a partial order $\trianglerighteq_w$ on $\mathbb{Y}^I$ that depends on an affine Weyl group element $w \in \hat{\mathfrak{S}}_I$ (definitions and notation appear later in this section). This comes from the geometry of quiver varieties. The second is the construction of matrix plethysms $\mathcal{P}_M$, which are endomorphisms of the I-fold tensor product of symmetric functions $\Lambda^{\otimes I}$ determined by an $I \times I$ matrix M with coefficients in the base field $\mathbb{K} = \mathbb{Q}(q,t)$. The specific matrix plethysms come from tensoring with certain representations of the wreath product $\Gamma_n = \mathfrak{S}_n \wr \mathbb{Z}/r\mathbb{Z}$ via a wreath Frobenius map.

3.1. Combinatorics of the partial orders. Fix an integer $r \geq 1$ and let $I = \mathbb{Z}/r\mathbb{Z}$, which is regarded as the set of nodes of the affine Dynkin diagram of type $A_{r-1}^{(1)}$.

3.1.1. *Maya diagrams, 1-runner abaci, and charged partitions.* A Maya diagram or edge sequence is a function $b : \mathbb{Z} \to \{0,1\}$ such that $b(i) = 0$ for $i \ll 0$ and $b(i) = 1$ for $i \gg 0$. It can be viewed as a 1-runner abacus with integer positions which places a bead at i if $b(i) = 0$ and a hole at i if $b(i) = 1$. For $c \in \mathbb{Z}$ let $\varnothing_c$ be the Maya diagram defined by $\varnothing_c(i) = 0$ for $i < c$ and $\varnothing_c(i) = 1$ for $i \geq c$. The charge of b is the unique integer c such that

$$|\{k \in \mathbb{Z} \mid k < c \text{ and } b(k) = 1\}| = |\{k \in \mid k \geq c \text{ and } b(k) = 0\}|, \tag{3.1}$$

that is, the number of holes at positions before c, is equal to the number of beads at positions at least c. Equivalently, starting with b and repeatedly exchanging beads with holes to their left, one reaches an abacus with all beads to the left of all holes, that is, an abacus of the form $\varnothing_c$. Then $c = \operatorname{charge}(b)$. The Durfee square side length $d(b)$ of b is the common cardinality in (3.1).

EXAMPLE 3.1. For neatness in the diagram below we write $\overline{k}$ for $-k$. Let b be defined by

i	$\cdots$	$\overline{5}$	$\overline{4}$	$\overline{3}$	$\overline{2}$	$\overline{1}$	0	1	2	3	4	$\cdots$
$b(i)$	0	0	1	0	1	0	1	1	0	0	1	1

.

Its abacus is drawn below.

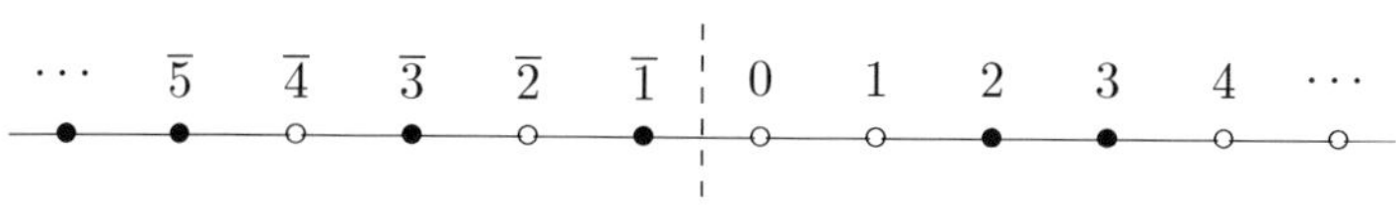

We have $\operatorname{charge}(b) = 0$; the dashed vertical line between $\overline{1}$ and 0 indicates this. We have $d(b) = 2$.

There is a partition defined by a Maya diagram b we denote by $\operatorname{shape}(b)$. It is defined as follows. We draw a polygonal path in the plane which follows a sequence of vectors (for increasing $i \in \mathbb{Z}$) in which the i-th vector is $(-1,0)$ if $b(i) = 0$ and $(0,1)$ if $b(i) = 1$, anchored such that the $(c-1)$-th vector enters the point $(d(b), d(b))$ and the c-th vector leaves it. The path is the top right border of the diagram of the partition which we call $\operatorname{shape}(b)$. Let $\mathbb{Y}$ be the set of partitions and $\mathcal{M}$ the set of Maya diagrams. There is a bijection

$$\begin{aligned} \mathcal{M} &\to \mathbb{Y} \times \mathbb{Z} \\ b &\mapsto (\operatorname{shape}(b), \operatorname{charge}(b)). \end{aligned}$$

We call the elements of $\mathbb{Y} \times \mathbb{Z}$ charged partitions.

EXAMPLE 3.2. The path associated to the Maya diagram b of the previous example is pictured below. We have shape$(b) = (4, 3, 2, 2)$.

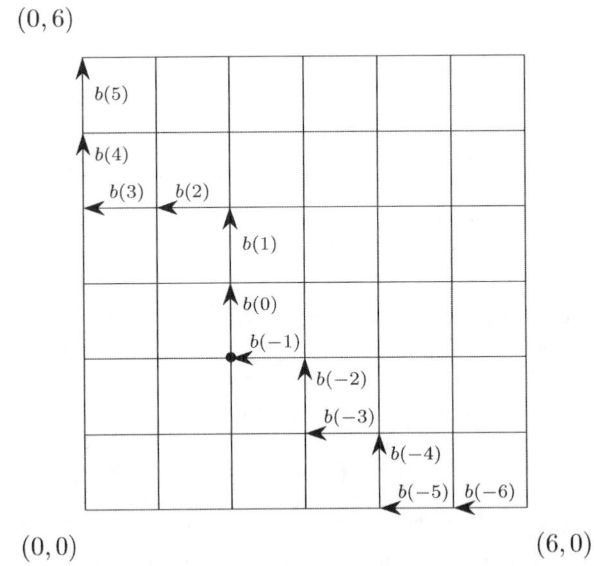

REMARK 3.3. Usually the bijection between Maya diagrams and charged partitions is defined in an opposite manner, by arrows pointing in the opposite directions and indexing them so as $i \in \mathbb{Z}$ increases, the arrows proceed from upper left to lower right. There are a number of notational consequences of this difference in indexing which we will need to point out from time to time.

3.1.2. *Cores and quotients.* Let $\mathcal{C} = \mathcal{C}_r \subset \mathbb{Y}$ be the set of r-cores, which by definition are the partitions λ with no cell $(a, b) \in \lambda$ of hook length $h_\lambda(a, b) = r$.

Let $\mathbb{Y}^I$ be the set of I-multipartitions, tuples $(\mu^{(0)}, \mu^{(1)}, \dots, \mu^{(r-1)})$ of partitions indexed by the affine Dynkin node set $I = \mathbb{Z}/r\mathbb{Z}$.

3.1.3. *Cores and quotients via (un)interleaving Maya diagrams.* Let b be a Maya diagram or equivalently 1-runner abacus. For $i \in I$ let $b^{(i)}$ be the 1-runner abacus obtained by selecting the subsequence of beads in positions in the coset of i mod $r\mathbb{Z}$, that is,

$$b^{(i)}(k) = b(kr + i) \qquad \text{for all } k \in \mathbb{Z} \text{ and } 0 \le i \le r - 1.$$

This gives a bijection

$$\begin{aligned} \mathcal{M} &\cong \mathcal{M}^I \\ b &\mapsto \mathrm{quot}_r(b) = (b^{(0)}, b^{(1)}, \dots, b^{(r-1)}), \end{aligned} \tag{3.2}$$

which we call the r-quotient for Maya diagrams. The inverse bijection is given by interleaving the r Maya diagrams $b^{(0)}, \dots, b^{(r-1)}$ using the same formula in (3.2). Pictorially we place $b^{(i)}$ as the i-th runner in an r-runner abacus and read the beads down the columns, proceeding from columns on the left to columns on the right.

EXAMPLE 3.4. For b in the previous example, the Maya diagram $b^{(i)}$ in $\mathrm{quot}_3(b)$ is pictured as the i-th runner in the 3-runner abacus pictured below. The vertical dashed lines picture the charge for each runner: charge$(b^{(0)}) = 1$, charge$(b^{(1)}) = -1$, and charge$(b^{(2)}) = 0$.

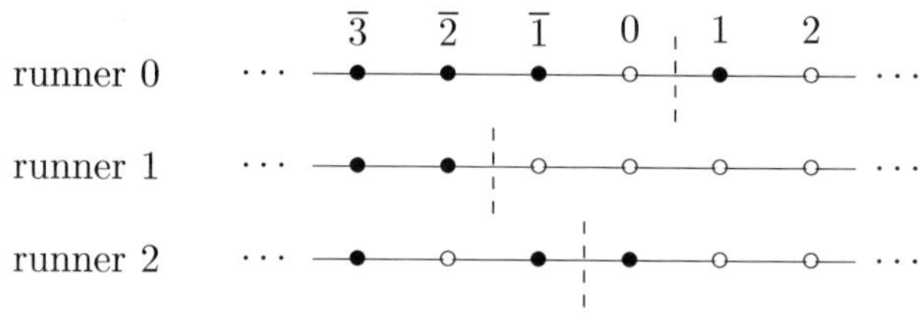

Under $\mathcal{M} \cong \mathbb{Y} \times \mathbb{Z}$ we have $b^{(0)} \mapsto (\square, 1)$, $b^{(1)} \mapsto (\varnothing, -1)$, and $b^{(2)} \mapsto (\square\square, 0)$; here $\varnothing$ denotes the empty partition.

We have the following commutative diagram in which all horizontal maps are bijections and vertical maps are obvious inclusions, where $(\varnothing^I)$ denotes the empty multipartition.

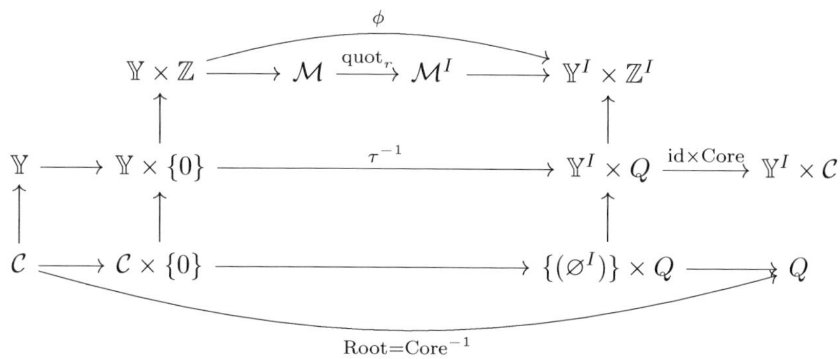

The top row of maps are given by bijections described above; we use the composite map $\mathcal{M}^I \cong (\mathbb{Y} \times \mathbb{Z})^I \cong \mathbb{Y}^I \times \mathbb{Z}^I$ using the I-fold product of the bijection $\mathcal{M} \to \mathbb{Y} \times \mathbb{Z}$ followed by reordering Cartesian factors. Let ϕ be the composite of the top row of maps. For $(\mu, c) \in \mathbb{Y} \times \mathbb{Z}$ let $\phi(\mu, c) = (\mu^\bullet, \beta)$ where $\mu^\bullet = (\mu^{(0)}, \dots, \mu^{(r-1)}) \in \mathbb{Y}^I$ and $\beta = (\beta_0, \dots, \beta_{r-1}) \in \mathbb{Z}^I$. It is straightforward to verify that

$$c = \sum_{i \in I} \beta_i.$$

Restricting ϕ to $\mathbb{Y} \times \{0\}$ we get a bijection $\mathbb{Y} \cong \mathbb{Y}^I \times Q$ where $Q \subset \mathbb{Z}^I$ is the root lattice of $\mathfrak{sl}_r$ realized by the zero-sum elements of $\mathbb{Z}^I$. The simple roots $\overline{\alpha}_i \in Q$ are realized by $\epsilon_{i-1} - \epsilon_i \in \mathbb{Z}^I$ for $1 \leq i \leq r-1$ where $\{\epsilon_i \mid i \in I\}$ is the standard basis of $\mathbb{Z}^I$. The inverse of this bijection we will denote by $\tau : \mathbb{Y}^I \times Q \to \mathbb{Y}$.

Let $(\varnothing^I) \in \mathbb{Y}^I$ denote the multipartition consisting of empty partitions. Then restricting τ to $\{(\varnothing^I)\} \times Q$ we get a bijection $Q \to \mathcal{C}$ called Core whose inverse bijection we will call Root $: \mathcal{C} \to Q$. Composing τ^{-1} with $\text{id}_{\mathbb{Y}^I} \times \text{Core}$ we obtain a bijection

$$\mathbb{Y} \to \mathbb{Y}^I \times \mathcal{C},$$

whose components are known as the r-quotient $\text{quot}_r : \mathbb{Y} \to \mathbb{Y}^I$ and the r-core $\text{core}_r : \mathbb{Y} \to \mathcal{C}$.

REMARK 3.5. The classical r-core map agrees with our map core_r, but the classical r-quotient map quot'_r (see [**Mac**, Example I.1.8] with m taken to be a multiple of r, or [**G**, §6.2]) produces the reverse sequence of partitions than ours:

$$\mathrm{quot}_r(\mu) = w_0\, \mathrm{quot}'_r(\mu).$$

where w_0 is the reversing permutation. This is due to our convention for the bijection $\mathcal{M} \cong \mathbb{Y} \times \mathbb{Z}$, namely, the direction of tracing of the edge of the partition from bottom right to upper left.

EXAMPLE 3.6. Continuing the above example, starting with $((4,3,2,2),0) \in \mathbb{Y} \times \mathbb{Z}$, we map to $\mathbb{Y}^I \times \mathbb{Z}^I$, getting $\mu^\bullet = (\square, \varnothing, \square\square)$ and $\beta = (1,-1,0)$.

For $\lambda \in \mathbb{Y}$ a λ-removable r-ribbon is a set of cells of the form $D(\lambda) \setminus D(\mu)$ for some $\mu \in \mathbb{Y}$ with $D(\mu) \subset D(\lambda)$, which is rookwise connected and has exactly one cell of each residue. Then λ is an r-core if and only if it has no removable r-ribbon.

One way to obtain $\mathrm{core}_r(\mu)$ is to repeatedly remove (removable) r-ribbons starting with μ until an r-core is reached. One obtains the same r-core independently of the order of removal of r-ribbons. This resulting r-core is $\mathrm{core}_r(\mu)$.

3.1.4. *Transposing partitions and the core-quotient bijection.* For all $\lambda \in \mathbb{Y}$ we have

$$\mathrm{core}_r(\lambda^t) = \mathrm{core}_r(\lambda)^t$$
$$\mathrm{quot}_r(\lambda^t) = \mathrm{quot}_r(\lambda)^*,$$

where $*$ denotes the reverse of the componentwise transpose of a multipartition.

3.1.5. *Residues.* We shall only require residues for boxes with respect to a partition shape(b) associated with a Maya diagram b of charge 0. In such a diagram the i-th step tracing the border of the partition shape(b), starts on the diagonal $y - x = i$. Therefore it is consistent to name this diagonal line as the i-th diagonal, that is, to define the content or diagonal index of (x,y) by $y-x$ and to define the residue of $(x,y) \in \mathbb{Z}^2$ as the residue of $y-x$ in $\mathbb{Z}/r\mathbb{Z}$. [2] For a partition $\mu \in \mathbb{Y}$ denote by $A_i(\mu) \subset A(\mu)$ and $R_i(\mu) \subset R(\mu)$ the subsets of μ-addable and μ-removable cells having residue i.

3.1.6. *Cores to root lattice, reprised.* The bijection $\mathrm{Root} : \mathcal{C} \to Q$ was defined previously in terms of the bijection ϕ. It can be computed directly as follows. Let $Q_{\mathrm{af}} = \bigoplus_{i\in I} \mathbb{Z}\alpha_i$ be the root lattice of the affine Kac-Moody algebra $\hat{\mathfrak{sl}}_r$ with simple roots $\{\alpha_i \mid i \in I\}$. Define the map $\kappa : \mathbb{Y} \to Q_{\mathrm{af}}^+ = \bigoplus_{i\in I} \mathbb{Z}_{\ge 0}\alpha_i$ by

$$\kappa(\mu) = \sum_{(x,y)\in\mu} \alpha_{y-x},$$

where the subscripts of the affine simple roots are taken modulo r as usual. Let $\mathrm{cl} : Q_{\mathrm{af}} \to Q$ be the restriction map. We have $\mathrm{cl}(\alpha_i) = \overline{\alpha}_i$ for $1 \le i \le r-1$ while $\mathrm{cl}(\alpha_0) = -\theta = -\sum_{i=1}^{r-1} \overline{\alpha}_i$ where $\theta \in Q$ is the highest root. The kernel of cl is $\mathbb{Z}\delta$ where $\delta = \sum_{i\in I} \alpha_i \in Q_{\mathrm{af}}$ is the null root. Define the map $\overline{\kappa} : \mathbb{Y} \to Q$ by

$$\overline{\kappa}(\mu) = -\mathrm{cl}(\kappa(\mu)). \tag{3.3}$$

Since each r-ribbon contributes δ in the computation of κ, it follows from [**G**, §6.5,7.5] that the restriction of $\overline{\kappa}$ to $\mathcal{C}$, which we shall by abuse of notation also call $\overline{\kappa}$, defines a bijection $\mathcal{C} \to Q$. The reason we use the negative sign in (3.3) is because with this definition, $\overline{\kappa}$ and Root define the same bijection $\mathcal{C} \to Q$.

[2] This is the negative of the standard convention.

EXAMPLE 3.7. The boxes of the partition $\mu = (4,3,2,2)$ and its 3-core $\gamma = (2)$ are labeled by their residues in $\mathbb{Z}/3\mathbb{Z}$.

0	2		
2	1		
1	0	2	
0	2	1	0

0	2

We have $\overline{\kappa}(\mu) = \overline{\kappa}(\gamma) = -\mathrm{cl}(\alpha_0 + \alpha_2) = \overline{\alpha}_1$.

3.1.7. *Affine Weyl action on partitions.* We now give another way to compute the bijection $Q \to \mathcal{C}$ inverse to $\overline{\kappa}$, based on affine crystal graph computations. Let $\hat{\mathfrak{S}}_I$ be the affine Weyl group for $\hat{\mathfrak{sl}}_r$, with Coxeter generators s_i for $i \in I$.

There is an action of the quantum affine $\mathfrak{sl}_r$ algebra on a certain Fock space $\mathcal{F}$,[3] inside which is the basic representation. The crystal graph of $\mathcal{F}$ may be realized by the set $\mathbb{Y}$ of all partitions. The empty partition $\varnothing$ corresponds to the vacuum vector. The orbit of $\varnothing$ under the affine crystal reflection operators s_i for $i \in I$, generates the extremal weight vectors in the subcrystal given by the basic representation; see [**MM**, §2] [**LLT**, §5]. Moreover the orbit $\hat{\mathfrak{S}}_I \cdot \varnothing$ is precisely the set $\mathcal{C}$ of r-core partitions.

The affine Weyl group $\hat{\mathfrak{S}}_I \cong Q^\vee \rtimes \mathfrak{S}_I$ acts on $\mathbb{Y}$: for all $i \in I$, s_i acts on $\lambda \in \mathbb{Y}$ by removing every removable cell of residue i and adding every addable cell of residue i. We have already seen this action: using the bijection between partitions and r-runner abaci described above, this action has the following pleasant description. The group $\mathfrak{S}_I$ of permutations of the set I, acts on r-runner abaci by permuting the runners, which are indexed by I. We adopt the convention that the simple reflection s_i for $1 \le i \le r-1$, acts on I by exchanging $i-1$ and i. We use $Q \cong Q^\vee$ for the translation lattice. The translation element $t_{\beta^\vee}$ for $\beta = (\beta_0, \beta_1, \dots, \beta_{r-1}) \in Q \subset \mathbb{Z}^I$ acts on the r-runner abaci by shifting the i-th runner by β_i positions for $i \in I$. Standard formulas for the affine Weyl group such as $s_0 = t_{\theta^\vee} s_\theta$ are consistent with our conventions.

We have $\mathcal{C} = \hat{\mathfrak{S}}_I \cdot \varnothing$ and the stabilizer of $\varnothing$ in $\hat{\mathfrak{S}}_I$ is $\mathfrak{S}_I$. Since there is a unique translation element in every coset of $\hat{\mathfrak{S}}_I/\mathfrak{S}_I$, there is a bijection $Q \to \mathcal{C}$ which agrees with the previously-defined map Core with

$$\mathrm{Core}(\beta) = t_{\beta^\vee} \cdot \varnothing.$$

This bijection coincides with $\mathrm{Root}^{-1} = \overline{\kappa}^{-1}$ [**G**, Lemma 7.5].

Let $w_0 \in \mathfrak{S}_I$ be the long element, sending i to $r-1-i$ for all $i \in I$. We have

$$\mathrm{Core}(\beta)^t = \mathrm{Core}(-w_0(\beta)) \qquad \text{for all } \beta \in Q.$$

Finally, let $w \mapsto w^*$ be the involutive group automorphism of $\hat{\mathfrak{S}}_I$ sending s_i to s_{-i} for all $i \in I$. For $u \in \mathfrak{S}_I$ and $\beta^\vee \in Q^\vee$ we have $u^* = w_0 u w_0$ and $t^*_{\beta^\vee} = t_{-w_0\beta^\vee}$ which is not equal to $w_0 t_{\beta^\vee} w_0 = t_{w_0\beta^\vee}$. We have

$$(w \cdot \lambda)^t = w^* \cdot \lambda^t \qquad \text{for all } w \in \hat{\mathfrak{S}}_I \text{ and } \lambda \in \mathbb{Y}.$$

[3]This corresponds to the *horizontal* quantum affine algebra action discussed in §5.2.1.

EXAMPLE 3.8. Let $r = 3$ and $\beta = \overline{\alpha}_1$. We have $t_{\overline{\alpha}_1^\vee} = s_2 s_0 s_2 s_1$. Computing in the crystal graph we obtain the 3-core

$$s_2 s_0 s_2 s_1 \cdot \varnothing = s_2 s_0 \cdot \varnothing = s_2 \cdot \boxed{0} = \boxed{0}\boxed{2}.$$

So $\mathrm{core}(\overline{\alpha}_1) = (2)$. Compare this with Example 3.7.

3.1.8. *Partial orders on multipartitions.* The previously-defined action of $\hat{\mathfrak{S}}_I$ on partitions or r-runner abaci is given explicitly on $\mathbb{Y}^I \times Q$ as follows: for $\beta^\vee \in Q^\vee$ and $u \in \mathfrak{S}_I$, define

$$t_{\beta^\vee} u \cdot (\mu^\bullet, \alpha) = (u(\mu^\bullet), \beta + u(\alpha)) \qquad \text{for all } (\mu^\bullet, \alpha) \in \mathbb{Y}^I \times Q. \tag{3.4}$$

The bijection τ of §3.1.3 is $\hat{\mathfrak{S}}_I$-equivariant using the $\hat{\mathfrak{S}}_I$-actions in (3.4) and §3.1.7.

For $w \in \hat{\mathfrak{S}}_I$ let $\tau_w : \mathbb{Y}^I \to \mathbb{Y}$ be defined by

$$\tau_w(\mu^\bullet) = \tau(w^{-1}(\mu^\bullet, 0)).$$

Define the partial order $\trianglerighteq_w$ on $\mathbb{Y}^I$ as in **[G]**[4]:

$$\lambda^\bullet \trianglerighteq_w \mu^\bullet \Leftrightarrow \tau_w(\lambda^\bullet) \trianglerighteq \tau_w(\mu^\bullet) \tag{3.5}$$

where $\trianglerighteq$ is the usual dominance order on $\mathbb{Y}$.

It is immediate from the definitions that (see **[G**, Lemma 7.11])

$$\lambda^\bullet \trianglerighteq_w \mu^\bullet \Leftrightarrow u(\lambda^\bullet) \trianglerighteq_{uw} u(\mu^\bullet) \qquad \text{for all } u \in \mathfrak{S}_I.$$

Since for $\lambda, \mu \in \mathbb{Y}$, $\lambda \trianglerighteq \mu$ if and only if $\lambda^t \trianglelefteq \mu^t$, it follows that

$$\begin{aligned} \lambda^\bullet \trianglerighteq_w \mu^\bullet &\Leftrightarrow (\lambda^\bullet)^* \trianglelefteq_{w^*} (\mu^\bullet)^* \\ &\Leftrightarrow \lambda^{\bullet t} \trianglelefteq_{w_0 w^*} \mu^{\bullet t}. \end{aligned}$$

To set some notation, let

$$w = u t_{-\beta^\vee} \in \hat{\mathfrak{S}}_I \qquad \text{with } u \in \mathfrak{S}_I \text{ and } \beta^\vee \in Q^\vee.$$

REMARK 3.9. We will need to apply $w^{-1} = t_{\beta^\vee} u^{-1}$ to the weight $\Lambda_0 - n\delta$ and want the core created by $w^{-1} \cdot \varnothing$ to be associated with β rather than $-\beta$. This explains our preference for the minus sign in the translation part of w.

Let

$$\mu := \tau_w(\mu^\bullet) = \tau(u^{-1}\mu^\bullet, \beta).$$

Equivalently we have

$$\begin{aligned} \mu^\bullet &= u \,\mathrm{quot}_r(\mu) \\ \beta &= \overline{\kappa}(\mathrm{core}_r(\mu)). \end{aligned}$$

EXAMPLE 3.10. Let $r = 3$, $n = 1$, $w = t_{-\overline{\alpha}_1^\vee}$. We consider the partial order $\trianglerighteq_w$ on the multipartitions of size 1. We have $u = \mathrm{id}$ and $\beta = \overline{\alpha}_1$. We compute $t_{\beta^\vee} \varnothing = t_{\overline{\alpha}_1} \varnothing = s_2 s_0 s_2 s_1 \varnothing = \square\square$. The partitions with 3-core $\square\square$ with quotients of size

[4]Gordon uses different conventions.

1 are $(5),(2,2,1),(2,1,1,1)$. We list them in reverse lex order, a total order which refines dominance order. The corresponding (u-permuted) quotients are listed.

μ			
$\operatorname{quot}_r(\mu)$	$(\cdot,\square,\cdot)$	$(\cdot,\cdot,\square)$	$(\square,\cdot,\cdot)$
$u\operatorname{quot}_r(\mu)$	$(\cdot,\square,\cdot)$	$(\cdot,\cdot,\square)$	$(\square,\cdot,\cdot)$

The table gives

$$(\cdot,\square,\cdot)\trianglerighteq_w(\cdot,\cdot,\square)\trianglerighteq_w(\square,\cdot,\cdot)\qquad\text{for } w=t_{-\overline{\alpha}_1^\vee}.$$

Now let $w=s_2s_1t_{-\overline{\alpha}_1^\vee}$. Then $u=s_2s_1$ and

μ	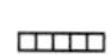		
$\operatorname{quot}_r(\mu)$	$(\cdot,\square,\cdot)$	$(\cdot,\cdot,\square)$	$(\square,\cdot,\cdot)$
$u\operatorname{quot}_r(\mu)$	$(\square,\cdot,\cdot)$	$(\cdot,\square,\cdot)$	$(\cdot,\cdot,\square)$

Therefore

(3.6) $$(\square,\cdot,\cdot)\trianglerighteq_w(\cdot,\square,\cdot)\trianglerighteq_w(\cdot,\cdot,\square)\qquad\text{for } w=s_2s_1t_{-\overline{\alpha}_1^\vee}.$$

3.2. Multisymmetric functions. Let $\Lambda^{\otimes I}$ be the I-fold tensor power of Λ over $\mathbb{K}=\mathbb{Q}(q,t)$. For $f\in\Lambda$ we write $f[X^{(i)}]$ to indicate an element in $\Lambda^{\otimes I}$ with 1 in tensor factors $j\neq i$ and f in factor i.

Then $\Lambda^{\otimes I}$ is a polynomial $\mathbb{K}$-algebra with generators $p_k[X^{(i)}]$ for $k\in\mathbb{Z}_{>0}$ and $i\in I$ where p_k is the power sum [**Mac**, §I.2].

We will sometimes write $f[X^\bullet]$ for an element $f\in\Lambda^{\otimes r}$ to remind the reader that f is symmetric separately in r sets of variables $X^{(0)},X^{(1)},\dots,X^{(r-1)}$.

The space $\Lambda^{\otimes I}$ acquires a grading by summing the degree of each factor in a tensor product of homogeneous elements. We write $(\Lambda^{\otimes I})_n$ for the subspace of $\Lambda^{\otimes I}$ consisting of elements of degree n.

3.2.1. *Tensor Schur basis.* For $\lambda^\bullet=(\lambda^{(0)},\lambda^{(1)},\dots,\lambda^{(r-1)})\in\mathbb{Y}^I$ define the tensor Schur function $s_{\lambda^\bullet}=\bigotimes_{i\in I}s_{\lambda^{(i)}}=\prod_{i\in I}s_{\lambda^{(i)}}[X^{(i)}]$. The tensor Hall pairing is the pairing on $\Lambda^{\otimes I}$ defined by

$$\langle s_{\lambda^\bullet},s_{\mu^\bullet}\rangle=\delta_{\lambda^\bullet,\mu^\bullet}\qquad\text{for all }\lambda^\bullet,\mu^\bullet\in\mathbb{Y}^I.$$

The Hall pairing on Λ has reproducing kernel

$$\Omega[XY]:=\prod_{i,j\geq1}(1-x_iy_j)^{-1}=\sum_{\lambda\in\mathbb{Y}}s_\lambda[X]s_\lambda[Y].$$

Let us regard $X^\bullet$ and $Y^\bullet$ as column vectors with i-th components $X^{(i)}$ and $Y^{(i)}$ for $i\in I$. Then the tensor Hall pairing has reproducing kernel

$$\sum_{\lambda^\bullet\in\mathbb{Y}^I}s_{\lambda^\bullet}[X^\bullet]s_{\lambda^\bullet}[Y^\bullet]=\Omega[Y^{\bullet t}X^\bullet]=\Omega[\sum_{i\in I}X^{(i)}Y^{(i)}].$$

3.2.2. *Vector and matrix plethysms.* Let $\mathrm{Mat}_{I\times I}(\mathbb{K})$ be the algebra of $I\times I$ matrices over $\mathbb{K}$. To any $M\in M_{I\times I}(\mathbb{K})$ with entries $M_{ij}(q,t)$ and $g\in\Lambda^{\otimes I}$ let $g[MX^\bullet]$ denote the image of g under the $\mathbb{K}$-algebra endomorphism of $\Lambda^{\otimes I}$ given by

$$p_k[X^{(i)}]\mapsto\sum_{j\in I}M_{ij}(q^k,t^k)p_k[X^{(j)}]\qquad\text{for all }k\in\mathbb{Z}_{>0},\ i\in I.$$

Let $\mathcal{P}_M \in \mathrm{End}(\Lambda^{\otimes I})$ be the $\mathbb{K}$-algebra homomorphism $g[X^\bullet] \mapsto g[M^t X^\bullet]$ where M^t is the transpose of M. The transpose makes the construction covariantly functorial:

$$\mathcal{P}_{MM'} = \mathcal{P}_M \circ \mathcal{P}_{M'} \qquad \text{for } M, M' \in \mathrm{Mat}_{I\times I}(\mathbb{K}).$$

EXAMPLE 3.11. Let $-\mathrm{id}$ be the negative of the $I \times I$ identity matrix. We have $\mathcal{P}_{-\mathrm{id}}(p_k[X^{(i)}]) = -p_k[X^{(i)}]$ for all $k > 0$ and $i \in I$. That is, $\mathcal{P}_{-\mathrm{id}}$ is the I-fold tensor product of the antipode map.

EXAMPLE 3.12. Suppose M is the matrix of a permutation u of the set I. Our convention is that this matrix sends the i-th standard basis column vector to the $u(i)$-th for all i. Then, writing $\mathcal{P}_u = \mathcal{P}_M$, we have

$$\mathcal{P}_u(p_k[X^{(i)}]) = p_k[X^{u(i)}] \qquad \text{for all } k \in \mathbb{Z}_{>0}, i \in I.$$

Examples of permutations of I that we will use are:

- neg (negate): $i \mapsto \mathrm{neg}(i) = i^* := r - i$;
- w_0 (reversal): $i \mapsto r - 1 - i$;
- χ (rotation): $i \mapsto i + 1$.

Often we just write $u \in \mathfrak{S}_I$ for the automorphism of $\Lambda^{\otimes I}$ rather than $\mathcal{P}_u$.

EXAMPLE 3.13. The following matrix plethysm is used in the definition of wreath Macdonald polynomials:

$$\mathcal{P}_{\mathrm{id}-q\chi^{-1}}(p_k[X^{(i)}]) = p_k[X^{(i)}] - q^k p_k[X^{(i-1)}] \qquad \text{for all } k \in \mathbb{Z}_{>0},\ i \in I.$$

3.3. Wreath Macdonald polynomials. We now define Haiman's wreath Macdonald polynomials [**H2**, **H3**]. For each $w \in \hat{\mathfrak{S}}_I$, there is a wreath Macdonald basis $\{\tilde{H}^w_{\mu^\bullet} \mid \mu^\bullet \in \mathbb{Y}^I\}$ of $\Lambda^{\otimes I}$. By definition the wreath Macdonald polynomial $\tilde{H}^w_{\mu^\bullet}$ is the unique element of $\Lambda^{\otimes I}$ satisfying

$$\mathcal{P}_{\mathrm{id}-q\chi^{-1}}(\tilde{H}^w_{\mu^\bullet}) \in \bigoplus_{\lambda^\bullet \trianglerighteq_w \mu^\bullet} \mathbb{K} s_{\lambda^\bullet} \tag{3.7}$$

$$\mathcal{P}_{\mathrm{id}-t^{-1}\chi^{-1}}(\tilde{H}^w_{\mu^\bullet}) \in \bigoplus_{\lambda^\bullet \trianglelefteq_w \mu^\bullet} \mathbb{K} s_{\lambda^\bullet} \tag{3.8}$$

$$\langle s_n[X^{(0)}], \tilde{H}^w_{\mu^\bullet}\rangle = 1 \tag{3.9}$$

where $n = |\mu^\bullet| = \sum_{i\in I} |\mu^{(i)}|$ is the total size of $\mu^\bullet$.

As in the ordinary Macdonald case ($r = 1$), the conditions (3.7) and (3.8) overdetermine the $\tilde{H}^w_{\mu^\bullet}$ and their existence therefore requires proof. Existence was conjectured by Haiman [**H2**] and first proved by Bezrukavnikov and Finkelberg [**BF**] using powerful geometric methods originating from earlier work of Bezrukavnikov and Kaledin [**BK**]; a second proof of existence was given later by Losev [**L2**]. We formulate these results in more detail in §4.3 below.

It is not difficult to show that the conditions above uniquely determine the $\tilde{H}^w_{\mu^\bullet}$. Moreover, using properties the inner product $\langle\cdot,\cdot\rangle_{q,t}$ introduced below (see §3.6), one can show that the leading terms in (3.7) and (3.8) are nonzero.

REMARK 3.14. For $r = 1$, $\mu^\bullet$ is a 1-tuple of partitions (μ) where μ is a partition, the affine Weyl group is the identity, and $\trianglerighteq_{\mathrm{id}}$ becomes dominance order. The matrix plethysm $\mathcal{P}_{\mathrm{id}-q\chi^{-1}}$ becomes the plethysm $f \mapsto f[(1-q)X]$ so that (3.7) matches (2.7), $\mathcal{P}_{\mathrm{id}-t^{-1}\chi^{-1}}$ becomes the plethysm $f \mapsto f[(1-t^{-1})X]$ so that (3.8) matches

(2.10), and (3.9) matches (2.9). Hence the wreath Macdonald polynomial $\tilde{H}^{\mathrm{id}}_{(\mu)}$ for $r=1$ is the usual modified Macdonald polynomial $\tilde{H}_\mu$.

EXAMPLE 3.15. Using the order from Example 3.10 with $w = s_2 s_1 t_{-\overline{\alpha}_1^\vee}$ we give a few wreath Macdonald polynomials and verify some of their triangularity properties using the order (3.6). The corresponding wreath Macdonalds are given explicitly and the nontrivial triangularity conditions are verified.

$$\begin{aligned}
\tilde{H}^w_{(\square,\cdot,\cdot)} &= s_1[X^{(0)}] + q^2 s_1[X^{(1)}] + q s_1[X^{(2)}] \\
\mathcal{P}_{\mathrm{id}-q\chi^{-1}} \tilde{H}^w_{(\cdot,\cdot,\square)} &= (1-q^3) s_1[X^{(0)}]
\end{aligned}$$

$$\begin{aligned}
\tilde{H}^w_{(\cdot,\square,\cdot)} &= s_1[X^{(0)}] + t s_1[X^{(1)}] + q s_1[X^{(2)}] \\
\mathcal{P}_{\mathrm{id}-q\chi^{-1}} \tilde{H}^w_{(\cdot,\square,\cdot)} &= (1-qt) s_1[X^{(0)}] + (t-q^2) s_1[X^{(1)}] \\
\mathcal{P}_{\mathrm{id}-t^{-1}\chi^{-1}} \tilde{H}^w_{(\square,\cdot,\cdot)} &= (t-qt^{-1}) s_1[X^{(1)}] + (q-t^{-1}) s_1[X^{(2)}]
\end{aligned}$$

$$\begin{aligned}
\tilde{H}^w_{(\cdot,\cdot,\square)} &= s_1[X^{(0)}] + t s_1[X^{(1)}] + t^2 s_1[X^{(2)}] \\
\mathcal{P}_{\mathrm{id}-t^{-1}\chi^{-1}} \tilde{H}^w_{(\cdot,\cdot,\square)} &= (t^2-t^{-1}) s_1[X^{(2)}]
\end{aligned}$$

3.4. Wreath Macdonald-Kostka coefficients. Define the wreath Macdonald-Kostka coefficients $\tilde{K}^w_{\lambda^\bullet,\mu^\bullet}(q,t) \in \mathbb{K}$ by

$$\tilde{H}^w_{\mu^\bullet} = \sum_{\lambda^\bullet} \tilde{K}^w_{\lambda^\bullet,\mu^\bullet}(q,t) s_{\lambda^\bullet}.$$

The wreath analog of Haiman's Theorem (Theorem 4.6 below) implies the Laurent polynomiality and positivity of wreath Macdonald-Kostka coefficients:

THEOREM 3.16 ([**BF**]). $\tilde{K}^w_{\lambda^\bullet,\mu^\bullet}(q,t) \in \mathbb{Z}_{\geq 0}[q^{\pm 1}, t^{\pm 1}]$.

Another immediate consequence is the following dimension formula for the Γ_n-isotypic components of fibers of the wreath Procesi bundle over torus fixed points.

THEOREM 3.17. *$\tilde{K}^w_{\lambda^\bullet,\mu^\bullet}(1,1)$ is the number of standard multitableaux $f^{\lambda^\bullet}$ of shape $\lambda^\bullet$.*

PROOF. Theorem 4.6 below realizes $\tilde{H}^w_{\mu^\bullet}$ as the Γ_n-Frobenius character of a wreath Procesi bundle fiber $P^w_n|_{I^w_{\mu^\bullet}}$, where $n = |\mu^\bullet|$. The important point here is that, after forgetting the $\mathbb{T}$-grading, $P^w_n|_{I^w_{\mu^\bullet}}$ affords a copy of the regular representation of Γ_n and its irreducible decomposition is given by $\sum_{|\lambda^\bullet|=n} \tilde{K}^w_{\lambda^\bullet,\mu^\bullet}(1,1)\chi^{\lambda^\bullet}$ where $\chi^{\lambda^\bullet}$ denotes the Γ_n-irreducible representation with Frobenius character $s_{\lambda^\bullet}$ (see (4.4)). Taking the coefficient of $\chi^{\lambda^\bullet}$ and using the fact that the multiplicity of an irreducible representation in the regular representation is equal to its dimension, we obtain $\tilde{K}^w_{\lambda^\bullet,\mu^\bullet}(1,1) = \dim \chi^{\lambda^\bullet}$. The dimension is given by multitableaux. $\square$

REMARK 3.18. If $|\lambda^{(i)}| = n_i$, $n = \sum_{i\in I} n_i$, and f^λ is the the number of standard tableaux of shape $\lambda \in \mathbb{Y}$, then

$$f^{\lambda^\bullet} = \binom{n}{n_0, n_1, \ldots, n_{r-1}} \prod_{i\in I} f^{\lambda^{(i)}}.$$

The notion of Γ-degree is defined in §4.1. The following says that only certain Laurent monomials in q and t may occur in a given wreath Macdonald-Kostka coefficient.

PROPOSITION 3.19 ([**H3**]). *The wreath Macdonald-Kostka coefficient* $\tilde{K}^w_{\lambda^\bullet,\mu^\bullet}(q,t)$ *is homogeneous of* Γ*-degree* $\sum_{i\in I}|\lambda^{(i)}|i$.

EXAMPLE 3.20. We see that Theorem 3.16 and Proposition 3.19 holds for the tensor Schur coefficients of the wreath Macdonald polynomials appearing in Example 3.15.

3.5. Wreath Macdonald symmetries. The wreath Macdonald polynomials satisfy a number of beautiful symmetries. Some of these symmetries generalize properties of modified Macdonald polynomials for the $r = 1$ case, while others are only visible in the wreath setting when $r \geq 2$.

For $f, g \in \Lambda^{\otimes I}$ let $f \equiv g$ mean that $g = cf$ for $c \in \mathbb{K}^\times$. In all of the following cases of wreath Macdonald symmetries presented below, the constant of proportionality is a Laurent monomial which may easily be deduced using the normalization condition (3.9). The following symmetries can be readily proved using the definitions.

3.5.1. *Symmetry swapping q and t.* Let "swap" be the $\mathbb{Q}$-algebra automorphism of $\mathbb{K}$ given by exchanging q and t. It extends to a $\mathbb{Q}$-algebra automorphism of $\Lambda^{\otimes I}$. The combination of swap and neg is compatible with Γ-degree (see §4.1) in the sense that $\mathrm{res}_\Gamma(\mathrm{swap}(f)) = \mathrm{neg}(\mathrm{res}_\Gamma(f))$ for all Γ-homogeneous $f \in R(\mathbb{T})$.

PROPOSITION 3.21. *For all* $w \in \hat{\mathfrak{S}}_I$ *and* $\mu^\bullet \in \mathbb{Y}^I$,

$$\mathrm{swap}\,\mathrm{neg}\,\tilde{H}^w_{\mu^\bullet} = \tilde{H}^{w^*}_{\mu^{\bullet *}}.$$

REMARK 3.22. For $r = 1$, $\tilde{H}_\mu[X;t,q] = \tilde{H}_{\mu^t}[X;q,t]$.

3.5.2. *Inversion symmetry.* Let "inv" be the $\mathbb{Q}$-algebra automorphism of $\mathbb{K}$ given by sending q and t to their reciprocals. It extends to a $\mathbb{Q}$-algebra automorphism of $\Lambda^{\otimes I}$ by acting on coefficients. Let ω be the $\mathbb{K}$-algebra involutive automorphism of Λ given by $s_\lambda \mapsto s_{\lambda^t}$. Consider the involutive $\mathbb{K}$-algebra automorphism of $\Lambda^{\otimes I}$ given by the I-th tensor power of ω; by abuse of notation we also call this ω. For $f \in \Lambda^{\otimes I}$ homogeneous of degree d, we have

$$\omega(f) = (-1)^d \mathcal{P}_{-\mathrm{id}}(f).$$

It follows that ω commutes with matrix plethysms:

$$\omega\,\mathcal{P}_M = \mathcal{P}_M\,\omega \qquad \text{for all } M \in M_{I\times I}(\mathbb{K}).$$

Define the operator $\downarrow$ on $\Lambda^{\otimes I}$ by

$$\downarrow = \mathrm{inv}\,\mathrm{neg}\,\omega,$$

in analogy with the operator $\downarrow$ of [**GHT**, Prop. 1.1]. It is an involution.

PROPOSITION 3.23. *For all* $w \in \hat{\mathfrak{S}}_I$ *and* $\mu^\bullet \in \mathbb{Y}^I$,

$$\downarrow \tilde{H}^w_{\mu^\bullet} \equiv \tilde{H}^{w_0 w}_{w_0\mu^\bullet}.$$

REMARK 3.24. For $r = 1$, $q^{n(\mu^t)}t^{n(\mu)}\mathrm{inv}\,\omega\,\tilde{H}_\mu[X;q,t] = \tilde{H}_\mu[X;q,t]$.

REMARK 3.25. Let c be the constant such that $\downarrow \tilde{H}^w_{\mu^\bullet} = c\tilde{H}^{w_0 w}_{w_0\mu^\bullet}$. With $n = |\mu^\bullet|$ we have

$$\begin{aligned} c &= \langle s_n[X^{(0)}], \downarrow \tilde{H}^w_{\mu^\bullet}\rangle \\ &= \mathrm{inv}\,\langle e_n[X^{(0)}], \tilde{H}^w_{\mu^\bullet}\rangle \\ &= \mathrm{inv}\, e^{(0),w}_{\mu^\bullet}, \end{aligned}$$

where $e^{(0),w}_{\mu^\bullet}$ is the eigenvalue of Haiman's operator $\nabla^{(0)}$ on $\tilde{H}^w_{\mu^\bullet}$; see §4.5. It is a Laurent monomial so $\mathrm{inv}(e^{(0),w}_{\mu^\bullet}) = 1/e^{(0),w}_{\mu^\bullet}$. It follows that

$$\downarrow (\nabla^{(0)})^{-1}\tilde{H}^w_{\mu^\bullet} = \tilde{H}^{w_0 w}_{w_0\mu^\bullet}. \tag{3.10}$$

We deduce that $\downarrow (\nabla^{(0)})^{-1}$, $\nabla^{(0)}\downarrow$, and $\downarrow \nabla^{(0)}$ are all involutions. In particular $\downarrow \nabla^{(0)} \downarrow = (\nabla^{(0)})^{-1}$; see [**GHT**, Prop. 1.1(b)].

3.5.3. *Rotational symmetry.* The following is peculiar to the wreath setting for $r \geq 2$. The quotient construction in [**Mac**, Ex. I.1.8] depends on a sufficiently large integer m. Increasing m has the effect of rotating the quotient. This leads to the following symmetry.

THEOREM 3.26. *For all $w \in \hat{\mathfrak{S}}_I$ and $\mu^\bullet \in \mathbb{Y}^I$,*

$$\chi\tilde{H}^w_{\mu^\bullet} \equiv \tilde{H}^{\chi w}_{\chi\mu^\bullet}.$$

3.5.4. *Forward rotation in matrix plethysms.* Another variant of wreath Macdonald polynomials that only appears for $r \geq 2$ is the following. Let $\hat{H}^w_{\mu^\bullet}$ be the version of wreath Macdonald polynomials defined by replacing χ^{-1} by χ in (3.7) and (3.8).

PROPOSITION 3.27. *For all $w \in \hat{\mathfrak{S}}_I$ and $\mu^\bullet \in \mathbb{Y}^I$,*

$$\hat{H}^w_{\mu^\bullet} \equiv w_0\tilde{H}^{w_0 w}_{w_0\mu^\bullet}.$$

Note how closely related this is with the inversion symmetry.

3.5.5. *Dualizing the partial order.* For $w \in \hat{\mathfrak{S}}_I$ let $\trianglerighteq^{\mathrm{opp}}_w$ be the partial order dual to $\trianglerighteq_w$ on $\mathbb{Y}^I$: $\lambda^\bullet \trianglerighteq^{\mathrm{opp}}_w \mu^\bullet$ if and only if $\mu^\bullet \trianglerighteq_w \lambda^\bullet$. Let ${}^{\mathrm{opp}}\tilde{H}^w_{\mu^\bullet}$ be the family of symmetric functions obtained by using the definition of wreath Macdonald polynomial with $\trianglerighteq^{\mathrm{opp}}_w$ instead of $\trianglerighteq_w$. Replacing $\trianglerighteq_w$ by $\trianglerighteq^{\mathrm{opp}}_w$ and sending (q,t) to (t^{-1}, q^{-1}) leaves the definition of wreath Macdonald polynomial invariant. This implies the following.

PROPOSITION 3.28. *For all $w \in \hat{\mathfrak{S}}_I$ and $\mu^\bullet \in \mathbb{Y}^I$,*

$${}^{\mathrm{opp}}\tilde{H}^w_{\mu^\bullet} = \mathrm{inv}\,\mathrm{swap}\,\tilde{H}^w_{\mu^\bullet}.$$

3.6. Orthogonality of $\tilde{H}^w_{\mu^\bullet}$. Define $A \in \mathrm{Mat}_{I\times I}(\mathbb{K})$ by

$$A = (\mathrm{id} - q\chi^{-1})^{-1}(\mathrm{id} - t\chi)^{-1}\mathrm{neg}, \tag{3.11}$$

where neg is regarded as a permutation matrix. Let swap and inv act on matrices by acting on their entries.

LEMMA 3.29.

$$\begin{aligned} A^t &= A \\ \mathrm{swap}\,\mathrm{inv}(A) &= qt\,A \\ A^{-1}\mathrm{inv}\,\mathrm{neg} &= qt\,\mathrm{inv}\,\mathrm{neg}\,A^{-1}. \end{aligned}$$

Let $\langle \cdot,\, \cdot \rangle_{q,t}$ be the pairing on $\Lambda^{\otimes I}$ defined by

$$\langle f[X^\bullet],\, g[X^\bullet] \rangle_{q,t} = \langle f[X^\bullet],\, g[A^{-1}X^\bullet] \rangle.$$

The following are straightforward consequences of the definition of wreath Macdonald polynomials and the pairing $\langle \cdot,\, \cdot \rangle_{q,t}$.

PROPOSITION 3.30.

(1) *The reproducing kernel for* $\langle \cdot,\, \cdot \rangle_{q,t}$ *is*

$$\Omega_{q,t} = \Omega[Y^{\bullet t} AX].$$

(2) *For all* $f, g \in \Lambda^{\otimes I}$, $\langle f,\, g \rangle_{q,t} = \langle g,\, f \rangle_{q,t}$.
(3) $s_{\lambda^\bullet}[AX^\bullet]$ *and* $s_{\lambda^\bullet}[X^\bullet]$ *are* $\langle \cdot,\, \cdot \rangle_{q,t}$*-dual bases.*
(4) $\mathcal{P}_{(\mathrm{id}-q\chi^{-1})^{-1}} s_{\lambda^\bullet}[X^\bullet] = s_{\lambda^\bullet}[(\mathrm{id}-q\chi)^{-1}X^\bullet]$ *and* $\mathcal{P}_{\mathrm{neg}\,(\mathrm{id}-t\chi^{-1})^{-1}} s_{\lambda^\bullet}[X^\bullet] = s_{\lambda^\bullet}[(1-t\chi)^{-1}\mathrm{neg}\, X^\bullet]$ *are* $\langle \cdot,\, \cdot \rangle_{q,t}$*-dual bases.*

PROPOSITION 3.31.

$$\langle \tilde{H}^w_{\mu^\bullet},\, \mathrm{inv}\,\mathrm{neg}\, \tilde{H}^w_{\nu^\bullet} \rangle_{q,t} = 0 \qquad \textit{if } \mu^\bullet \neq \nu^\bullet.$$

For $w \in \hat{\mathfrak{S}}_I$ and $\mu^\bullet \in \mathbb{Y}^I$ define

$$b^w_{\mu^\bullet}(q,t) = \langle \tilde{H}^w_{\mu^\bullet},\, \mathrm{inv}\,\mathrm{neg}\, \tilde{H}^w_{\nu^\bullet} \rangle_{q,t} \tag{3.12}$$

Since inv neg is an automorphism, $\{\mathrm{inv}\,\mathrm{neg}\, \tilde{H}^w_{\mu^\bullet} \mid \mu^\bullet \in \mathbb{Y}^I\}$ is a basis of $\Lambda^{\otimes I}$. We conclude that $b^w_{\mu^\bullet}(q,t) \neq 0$. It follows that $(b^w_{\mu^\bullet}(q,t))^{-1}\mathrm{inv}\,\mathrm{neg}\, \tilde{H}^w_{\mu^\bullet}$ is the $\langle \cdot,\, \cdot \rangle_{q,t}$-dual basis to $\tilde{H}^w_{\mu^\bullet}$.

THEOREM 3.32. *The value of the pairing in* (3.12) *is*

$$b^w_{\mu^\bullet}(q,t) = b_{\tau(w^{-1}(\mu^\bullet,0))}(q,t) \qquad \textit{where} \tag{3.13}$$
$$b_\mu(q,t) = \prod_{\substack{s\in\mu \\ h_\mu(s)\equiv 0 \bmod r}} (1-q^{1+a_\mu(s)}t^{-l_\mu(s)})(1-t^{1+l_\mu(s)}q^{-a_\mu(s)}).$$

REMARK 3.33. (1) We give a proof of Theorem 3.32 in §4.4 using the wreath analog of Theorem 4.6.
(2) In terms of the locus of the Hilbert scheme fixed under $\Gamma \cong \mathbb{Z}/r\mathbb{Z}$ (see §4), one has $b_\mu(q,t)^{-1} = \mathrm{ch}_{\mathbb{T}}\,\mathrm{Sym}\, T^*\mathrm{Hilb}_{|\mu|}(\mathbb{C}^2)^\Gamma|_{I_\mu}$.
(3) $(qt)^n b^w_{\mu^\bullet}(q^{-1},t^{-1}) = b^w_{\mu^\bullet}(q,t)$ for $n = |\mu^\bullet|$.

3.7. Wreath Macdonald P polynomials. As known to Haiman [**H3**] there is a wreath analogue of the Macdonald P polynomial. The analogies in this subsection are not perfect in comparison to the $r=1$ case; the variations will be pointed out.

Define $B \in \mathrm{Mat}_{I\times I}(\mathbb{K})$ by

$$B = (\mathrm{id}-q\chi^{-1})^{-1}(\mathrm{id}-t^{-1}\chi^{-1})\,\mathrm{neg}.$$

Define the pairing $\langle \cdot,\, \cdot \rangle^P_{q,t}$ on $\Lambda^{\otimes I}$ by

$$\langle f[X^\bullet],\, g[X^\bullet] \rangle^P_{q,t} = \langle f[X^\bullet],\, g[B^{-1}X^\bullet] \rangle$$

It has reproducing kernel

$$\Omega^P_{q,t} = \Omega[Y^{\bullet t} BX^\bullet].$$

LEMMA 3.34.

$$\Omega^P_{q,t} = \mathcal{P}^X_{\mathrm{id}-t^{-1}\chi^{-1}} \mathcal{P}^Y_{\mathrm{id}-t\chi} \Omega_{q,t}.$$

PROPOSITION 3.35. *For $w \in \hat{\mathfrak{S}}_I$ there is a unique basis $\{P^w_{\mu^\bullet} \mid \mu^\bullet \in \mathbb{Y}^I\}$ of $\Lambda^{\otimes I}$ called the wreath Macdonald P polynomials, such that*

$$P^w_{\mu^\bullet} \in s_{\mu^\bullet} + \bigoplus_{\lambda^\bullet \lhd_w \mu^\bullet} \mathbb{K} s_{\lambda^\bullet}$$

$$\langle P^w_{\mu^\bullet}, \operatorname{inv}\operatorname{neg} P^w_{\nu^\bullet}\rangle^P_{q,t} = 0 \qquad \text{for } \mu^\bullet \neq \nu^\bullet.$$

For $w \in \hat{\mathfrak{S}}_I$ and $\mu^\bullet \in \mathbb{Y}^I$ define the wreath Macdonald J polynomial by

$$J^w_{\mu^\bullet} = \mathcal{P}_{\mathrm{id}-t^{-1}\chi^{-1}} \tilde{H}^w_{\mu^\bullet}.$$

It follows directly from the definitions that there is a nonzero constant $c^w_{\mu^\bullet} \in \mathbb{K}$ such that

$$J^w_{\mu^\bullet} \in c^w_{\mu^\bullet} s_{\mu^\bullet} + \bigoplus_{\lambda^\bullet \lhd \mu^\bullet} \mathbb{K} s_{\lambda^\bullet}.$$

Then

$$P^w_{\mu^\bullet} = (c^w_{\mu^\bullet})^{-1} J^w_{\mu^\bullet}.$$

CONJECTURE 3.36. *For $\mu = \tau(w^{-1}(\mu^\bullet, 0))$*

$$\langle P^w_{\mu^\bullet}, \operatorname{inv}\operatorname{neg} P^w_{\mu^\bullet}\rangle^P_{q,t} = \prod_{\substack{s\in\mu \\ h(s)\equiv 0 \bmod r}} \frac{1-q^{1+a(s)}t^{-l(s)}}{1-q^{a(s)}t^{-1-l(s)}}.$$

REMARK 3.37. In $r=1$ Macdonald theory, to go from $\tilde{H}_\mu$ to J_μ one first passes to H_μ by inverting t (and rescaling) and then applies the plethystic automorphism. In the above definition there is no scaling. For $r=1$ we have $P^{\mathrm{id}}_{(\mu)} = P_\mu|_{t\mapsto t^{-1}}$.

One obtains another interesting basis of $\Lambda^{\otimes I}$ by using q instead of t^{-1} in the definition of $J^w_{\mu^\bullet}$.

CONJECTURE 3.38. *Let $f^w_{\mu^\bullet} \in \mathbb{K}$ be the sum of terms in the "diagonal" wreath Macdonald-Kostka polynomial $\tilde{K}^w_{\mu^\bullet,\mu^\bullet}(q,t)$ with maximum power of t. Then $f^w_{\mu^\bullet}$ is a single Laurent monomial in q and t and*

$$c^w_{\mu^\bullet} = f^w_{\mu^\bullet} \prod_{\substack{s\in\mu \\ h_\mu(s)\equiv 0 \bmod r}} (1-q^{a(s)}t^{-1-l(s)}),$$

where $\mu = \tau(w^{-1}(\mu^\bullet, 0))$.

PROPOSITION 3.39. *$P^w_{\mu^\bullet}[X^\bullet; q, q^{-1}] = s_{\mu^\bullet}$.*

4. Nakajima quiver varieties

In this section, we explain the geometric realization of wreath Macdonald polynomials, as formulated conjecturally by Haiman [**H2**] and later established by Bezrukavnikov and Finkelberg [**BF**] and Losev [**L1**, **L2**]. We use this realization to prove the norm formula (3.13) and to construct wreath analogs of the ∇ operator.

4.1. The cyclic group Γ. Let $\Gamma \subset \mathbb{C}^\times$ be the cyclic group of r-th roots of unity; we fix the generator $\zeta = \exp(2\pi\sqrt{-1}/r)$ for Γ. We shall use the embedding $\Gamma \to \mathbb{T} \subset GL(2,\mathbb{C})$ determined by $\zeta \mapsto \mathrm{diag}(\zeta^{-1},\zeta)$, where $\mathbb{T}$ is the torus of §2.7 regarded as a subgroup of diagonal matrices in $GL(2,\mathbb{C})$ in the standard way, via $(t_1,t_2) \mapsto \mathrm{diag}(t_1,t_2)$.

The representation ring $R(\Gamma) = \mathbb{Z}[\chi]$ is generated by the character χ sending $\zeta = \exp(2\pi\sqrt{-1}/r)$ to itself. $R(\Gamma)$ has basis χ^i for $i \in I = \mathbb{Z}/r\mathbb{Z}$. The McKay correspondence asserts a bijection between the affine Dynkin node set $I = \mathbb{Z}/r\mathbb{Z}$ (vertex set of cyclic quiver) and the irreducible characters of Γ. It is given by $i \mapsto \chi^i$ for $i \in I$.[5]

Recall that we have made the identification $R(\mathbb{T}) = \mathbb{Z}[q^{\pm 1}, t^{\pm 1}]$, with the standard coordinate functions $x(a_1,a_2) = a_1$ and $y(a_1,a_2) = a_2$ on $\mathbb{C}^2$ having $\mathbb{T}$-weights t and q, respectively. Let us call res_Γ the restriction map $R(\mathbb{T}) \to R(\Gamma)$. We have $\mathrm{res}_\Gamma(q) = \chi^{-1}$ and $\mathrm{res}_\Gamma(t) = \chi$. We say that $f \in R(\mathbb{T})$ is homogeneous of Γ-degree $i \in \mathbb{Z}/r\mathbb{Z}$ if $\mathrm{res}_\Gamma(f) \in \mathbb{Z}\chi^i$.

It is sometimes enlightening to include the Γ-grading of $R(\mathbb{T})$ explicitly in formulas. This is accomplished by applying the ring homomorphism $R(\mathbb{T}) \to R(\mathbb{T}\times\Gamma)$ given by $q \mapsto q\chi^{-1}$ and $t \mapsto t\chi$.

4.2. Quiver varieties. Let $V(\Lambda_0)$ be the irreducible integrable highest weight $\hat{\mathfrak{sl}}_r$-module of highest weight Λ_0, i.e., the basic representation. By [**Nak1**], the weight spaces of $V(\Lambda_0)$ may be realized by the top degree homology of various Nakajima quiver varieties $\mathfrak{M}$ for the cyclic quiver with one-dimensional framing space. These quiver varieties provide the geometric setting for wreath Macdonald theory [**H2**]. Their equivariant K-theory groups $K^*_{\mathbb{T}}(\mathfrak{M})$ realize the weight spaces in an irreducible quantum toroidal algebra module which we shall consider in the next section (see Remark 5.8 below).

4.2.1. *Affine root lattice and dimension vectors for cyclic quiver.* In this article we always use a one-dimensional framing at node 0 only. Given $\mathbf{v} = (v_i \mid i \in I) \in \mathbb{Z}^I_{\geq 0}$, let V_i be an v_i-dimensional vector space over $\mathbb{C}$. This dimension vector is written $\sum_{i\in I} v_i\alpha_i \in Q^+_{\mathrm{af}}$. Since the null root equals $\delta = \sum_{i\in I}\alpha_i$, the expression $n\delta$ will sometimes be used for the vector having dimension n at each quiver vertex.

4.2.2. *Stability conditions.* Let $G = \prod_{i\in I} GL(V_i)$ and let $\det_i$ be the determinant character on $GL(V_i)$ for $i \in I$. For each $\theta = (\theta_i \mid i \in I) \in \mathbb{Z}^I$, define the following character of G: $\chi^\theta = \prod_{i\in I}\det_i^{\theta_i}$.

4.2.3. *Quiver variety.* Let $\mathfrak{M}_\theta(\mathbf{v})$ be the Nakajima quiver variety for the cyclic quiver with one-dimensional framing at the node 0, quiver dimension vector $\mathbf{v}$, and stability condition given by χ^θ for $\theta \in \mathbb{Z}^I$. For the standard stability condition $\theta = (1,1,\dots,1) \in \mathbb{Z}^I$ the top degree homology of $\mathfrak{M}_\theta(\mathbf{v})$ realizes the $(\Lambda_0 - \sum_{i\in I} v_i\alpha_i)$-weight space in $V(\Lambda_0)$ [**Nak1**].

4.2.4. *Stability alcoves and the affine Weyl group.* We consider how the quiver varieties $\mathfrak{M}_\theta(\mathbf{v})$ change as the stability condition θ changes. The variety $\mathfrak{M}_{k\theta}(\mathbf{v})$ is invariant up to isomorphism for $k \in \mathbb{Z}_{>0}$. Therefore we may work with a parameter $\theta \in \Theta = \mathbb{Q}^I$, modulo scaling by $\mathbb{Q}_{>0}$. Let $X^{\mathrm{af}}_{\mathbb{Q}} = \mathbb{Q}\delta \oplus \bigoplus_{i\in I}\mathbb{Q}\Lambda_i$ be the affine weight lattice tensored with $\mathbb{Q}$, with null root δ and fundamental weights Λ_i. We use a linear isomorphism $\Theta \cong X^{\mathrm{af}}_{\mathbb{Q}}/\mathbb{Q}\delta$ defined by $\sum_{i\in I}\theta_i\epsilon_i \mapsto \sum_{i\in I}\theta_i\Lambda_i$. The level of

[5]The reason that we used the notation χ for the rotational permutation in $\mathfrak{S}_I$, was due to the fact that multiplication by χ acts by rotation on the ordered basis $\{\chi^0, \chi^1, \dots, \chi^{r-1}\}$ of $R(\Gamma)$.

$\theta \in \Theta$ is by definition $\langle \theta, c \rangle = \sum_{i \in I} \theta_i$, where c is the canonical central element. Let Θ_ℓ be the elements of Θ of level $\ell \in \mathbb{Q}$. Let us assume $\theta \neq 0$. By scaling we can assume $\theta \in \Theta_{\pm 1}$.

Let $X_{\mathbb{Q}} = \bigoplus_{i \in I \setminus \{0\}} \mathbb{Q}\omega_i$ be the finite weight lattice tensored with $\mathbb{Q}$ where ω_i is the i-th finite fundamental weight. Let $\mathrm{cl} : X_{\mathbb{Q}}^{\mathrm{af}} \to X_{\mathbb{Q}}$ be the restriction map. It has kernel $\mathbb{Q}\delta \oplus \mathbb{Q}\Lambda_0$. Consider the lift $\omega_i \mapsto \Lambda_i - \Lambda_0$ for $i \in I \setminus \{0\}$. Via this lift we have $\Theta_{\pm 1} \cong \pm\Lambda_0 \oplus X_{\mathbb{Q}}$.

Define $\theta_\pm = \pm(1/r)\sum_{i \in I} \Lambda_i$. Note that the vector $\theta_\pm$ has level ± 1 and that θ_+ is a $\mathbb{Q}_{>0}$-multiple of the standard stability condition.

There is a natural action of $\hat{\mathfrak{S}}_I$ on Θ_ℓ for each $\ell \in \mathbb{Q}$. For every $i \in I$ and $m \in \mathbb{Z}$ let $L_{i,m}$ be the hyperplane in $\mathbb{Q}^I$ consisting of the vectors β such that $\langle \beta, \alpha_i^\vee \rangle = m$. The affine hyperplane arrangement is the set of hyperplanes of the form $w(L_{i,m})$ for $w \in \hat{\mathfrak{S}}_I$, $i \in I$, and $m \in \mathbb{Z}$. Let $\Theta_{\pm 1}^{\mathrm{reg}}$ denote the subset of $\Theta_{\pm 1}$ that avoids all the hyperplanes in the affine hyperplane arrangement. The affine Weyl group $\hat{\mathfrak{S}}_I$ acts on $\Theta_{\pm 1}^{\mathrm{reg}}$, permuting the connected (for the usual topology on Euclidean space) components of $\Theta_{\pm 1}^{\mathrm{reg}}$, which are called alcoves. There are exactly two orbits of alcoves, $\hat{\mathfrak{S}}_I \cdot \theta_+$ and $\hat{\mathfrak{S}}_I \cdot \theta_-$. By [**G**] the quiver variety $\mathfrak{M}_\theta(\mathbf{v})$ is smooth and stays the same up to isomorphism as θ varies over an alcove. Therefore we have a stability condition for every pair $(w, +)$ or $(w, -)$ for $w \in \hat{\mathfrak{S}}_I$.

REMARK 4.1. Gordon [**G**] uses two parametrizations of stability condition. One is $(h, H_1, \ldots, H_{r-1}) \in \mathbb{Q}^r$, which in our notation is $-h\Lambda_0 + \sum_{i=1}^{r-1} H_i\omega_i$. The other is $\theta \in \mathbb{Q}^\ell$ which we have identified with $\sum_{i \in I} \theta_i \Lambda_i$.

4.2.5. *Weight spaces.* The nonzero weight spaces in $V(\Lambda_0)$ have weights of the form $t_{\beta^\vee}\Lambda_0 - n\delta$ for $(\beta, n) \in Q \times \mathbb{Z}_{\geq 0}$ [6]. Subtracting this weight from the highest weight we get an element $\Lambda_0 - t_{\beta^\vee}\Lambda_0 + n\delta \in \bigoplus_{i \in I} \mathbb{Z}_{\geq 0}\alpha_i$. By [**Kac**, Eq. (6.5.2)] we have

$$t_{\beta^\vee} \cdot \Lambda_0 = \Lambda_0 + \beta - \frac{(\beta, \beta)}{2}\delta.$$

Let $\mathfrak{M}_{\beta,n} = \mathfrak{M}_{\theta_+}(\Lambda_0 - t_{\beta^\vee}\Lambda_0 + n\delta) = \mathfrak{M}_{\theta_+}(-\beta + ((1/2)(\beta, \beta) + n)\delta)$.

4.2.6. *Γ-fixed locus in* Hilb_N. It is well-known (see, e.g., [**Kir**, Theorem 12.13]) that $\mathfrak{M}_{\beta,n}$ is isomorphic to the irreducible component of the Γ-fixed locus $\mathrm{Hilb}_N^\Gamma = \mathrm{Hilb}_N(\mathbb{C}^2)^\Gamma$ consisting of Γ-invariant ideals $I \subset \mathbb{C}[x, y]$ such that the equality Γ-characters $\mathrm{ch}_\Gamma \mathbb{C}[x, y]/I = \sum_{i \in I} v_i \chi^i$ holds. Here $\mathbf{v}$ is determined by $t_{\beta^\vee} \cdot \Lambda_0 - n\delta = \Lambda_0 - \sum_{i \in I} v_i\alpha_i$, $N = \sum_{i \in I} v_i$, and Γ acts on Hilb_N via the embedding $\Gamma \to \mathbb{T}$. In terms of the r-core $\gamma = \overline{\kappa}^{-1}(\beta) \in \mathcal{C}$, one has $N = |\gamma| + rn$. Moreover, this isomorphism is $\mathbb{T}$-equivariant for a natural $\mathbb{T}$-action on $\mathfrak{M}_{\beta,n}$. The $\mathbb{T}$-fixed points of $\mathfrak{M}_{\beta,n}$ correspond precisely to the monomial ideals I_λ such that $|\lambda| = N$ and $\mathrm{core}(\lambda) = \gamma$ [**H2**, Prop. 7.2.18].

4.2.7. *Correspondence between $\mathbb{T}$-fixed points.* For any $w = ut_{-\beta^\vee} \in \hat{\mathfrak{S}}_I$, where $u \in \mathfrak{S}_I$ and $\beta \in Q$, the composition of Nakajima reflection functors $R_{w^{-1}}$ [**Nak3**] defines an isomorphism

$$\mathfrak{M}_{w\theta_+}(n\delta) \to \mathfrak{M}_{\beta,n}. \tag{4.1}$$

[6] We shall use the type A isomorphism $Q \mapsto Q^\vee$ sending $\overline{\alpha}_i \mapsto \overline{\alpha}_i^\vee$ for $i \in I \setminus \{0\}$ without further mention.

By [**G**] (see also [**Nag1**, **P**]), the $\mathbb{T}$-fixed points $I^w_{\mu^\bullet}$ of $\mathfrak{M}_{w\theta_+}(n\delta)$ are naturally labeled by the set $\mathbb{Y}^I_n$ of r-multipartitions $\mu^\bullet$ of total size n, so that $I^w_{\mu^\bullet}$ corresponds to the $\mathbb{T}$-fixed point $I_{\tau(w^{-1}(\mu^\bullet,0))}$ in Hilb^Γ_N, where $\tau : \mathbb{Y}^I \times Q \to \mathbb{Y}$ is the bijection of §3.1.2.

REMARK 4.2. The combinatorial partial order $\trianglerighteq_w$ on multipartitions defined in (3.5) refines the geometric partial order defined by attracting sets with respect to the hyperbolic $\mathbb{C}^*$-action on $\mathfrak{M}_{\beta,n}$ [**G**] (see also [**BF**, §2.3] and [**P**]). This $\mathbb{C}^*$-action corresponds to the subgroup of matrices $\mathrm{diag}(t_1, t_1^{-1}) \in \mathbb{T}$ acting on Hilb^Γ_N.

4.2.8. *Small tautological bundle and bigraded reflection functors.* The following is due to Haiman [**H2**, **H3**]. Let M^w_n be the bundle of tautological quiver data over $\mathfrak{M}_{w\theta_+}(n\delta)$. We call this the small tautological bundle; it has rank rn. Using the isomorphism (4.1), we regard M^w_n as a vector bundle on $\mathfrak{M}_{\beta,n}$. In contrast the ("big") tautological bundle Taut_N over $\mathfrak{M}_{\beta,n}$, has rank $N = rn + |\overline{\kappa}^{-1}(\beta)|$.

For $\mu^\bullet \in \mathbb{Y}^I_n$ define

$$\mathcal{B}^w_{\mu^\bullet}(q,t,\chi) = \mathrm{ch}_{\Gamma\times\mathbb{T}} M^w_n|_{I^w_{\mu^\bullet}} \in R(\Gamma\times\mathbb{T}).$$

We use $R(\Gamma\times\mathbb{T}) = \mathbb{Z}[q^{\pm1}, t^{\pm1}][\chi]$; it has $R(\mathbb{T})$-basis χ^i for $i \in I$. For $i \in I$ and $f \in R(\Gamma\times\mathbb{T})$ let $[\chi^i]f$ denote the coefficient of χ^i in f.

The following operator computes the change in the $\Gamma\times\mathbb{T}$-grading of the fibers of tautological quiver data bundles over $\mathbb{T}$-fixed points under the application of a Nakajima reflection functor R_i. For $i \in I$, define the operator R^*_i on $R(\Gamma\times\mathbb{T})$ by

$$R^*_i(f) = f + \left(q^{-1}t^{-1}\delta_{i0} - [\chi^i](\mathrm{id} - q^{-1}\chi)(\mathrm{id} - t^{-1}\chi^{-1})f\right)\chi^i \quad \text{for } i \in I.$$

If $f = \sum_{i\in I} f^{(i)}\chi^i$ with $f^{(i)} \in R(\mathbb{T})$ and $g = R^*_i(f) = \sum_{i\in I} g^{(i)}\chi^i$ then for $j \in I$ we have

$$g^{(j)} = \begin{cases} f^{(j)} & \text{if } j \neq i \\ q^{-1}t^{-1}\delta_{i0} - q^{-1}t^{-1}f^{(i)} + t^{-1}f^{(i+1)} + q^{-1}f^{(i-1)} & \text{if } j = i. \end{cases}$$

THEOREM 4.3 ([**H3**]).

(1) $R^*_iR^*_{i+1}R^*_i = R^*_{i+1}R^*_iR^*_{i+1}$ *for all* $i \in I$.
(2) $(R^*_i - 1)(R^*_i + q^{-1}t^{-1}) = 0$ *for all* $i \in I$.
(3) *Let* $w \in \hat{\mathfrak{S}}_I$ *and* $n \in \mathbb{Z}_{\geq 0}$. *Then for all* $\mu^\bullet \in \mathbb{Y}^I$

$$\mathcal{B}^w_{\mu^\bullet}(q,t,\chi) = R^*_{i_1}R^*_{i_2}\cdots R^*_{i_\ell}B_\mu(q\chi^{-1}, t\chi)$$

where $w = s_{i_1}s_{i_2}\cdots s_{i_\ell}$ *is a reduced decomposition,* $\mu = \tau(w^{-1}(\mu^\bullet, 0))$, *and* $B_\mu(q,t)$ *is defined in* (2.1).

REMARK 4.4. The operators R^*_i define a representation of the affine Hecke algebra.

EXAMPLE 4.5. Let $r = 3$, $w = t_{-\overline{\alpha}_2^\vee} = s_2s_1s_0s_1$, and $\mu^\bullet = (\square, \cdot, \square)$. We have $u = \mathrm{id}$, $\beta = \overline{\alpha}_2$, $t_{\beta^\vee}\cdot\varnothing = t_{\overline{\alpha}_2^\vee}\cdot\varnothing = s_1s_0s_1s_2\cdot\varnothing = \boxminus$ and $u^{-1}\mu^\bullet = (\square, \cdot, \square)$. The partition with 3-core $\boxminus$ and 3-quotient $(\square, \cdot, \square)$ is $\mu = (4, 4)$.

$t\chi$	qt	$q^2t\chi^2$	$q^3t\chi$
1	$q\chi^2$	$q^2\chi$	q^3

Starting with $B_\mu(q\chi^{-1}, t\chi)$ we have

$$\begin{array}{cr}
(q^3+qt+1)\chi^0 + (q^3t+q^2+t)\chi^1 + (q^2t+q)\chi^2 & (\square, \cdot, \square) \\
\big\downarrow R_1^* & \\
(q^3+qt+1)\chi^0 + (q^2+t)\chi^1 + (q^2t+q)\chi^2 & (\cdot, \square, \square) \\
\big\downarrow R_0^* & \\
(qt+1)\chi^0 + (q^2+t)\chi^1 + (q^2t+q)\chi^2 & (\square, \square, \cdot) \\
\big\downarrow R_1^* & \\
(qt+1)\chi^0 + (q^2+t)\chi^1 + (q^2t+q)\chi^2 & (\square, \square, \cdot) \\
\big\downarrow R_2^* & \\
(qt+1)\chi^0 + (q^2+t)\chi^1 + (tq^{-1}+q)\chi^2 & (\square, \cdot, \square).
\end{array}$$

The result is $\mathcal{B}^w_{\mu^\bullet}$.

4.3. Wreath analog of Haiman's Positivity Theorem. We are now prepared to state the wreath analog of Theorem 2.10.

THEOREM 4.6 ([**BF**, **L2**]). *For each $\beta \in Q$, $n \in \mathbb{Z}_{\geq 0}$, and $w = ut_{-\beta^\vee} \in \hat{\mathfrak{S}}_I$, there exists an equivalence*

$$D^b(\mathrm{Coh}^{\mathbb{T}}(\mathfrak{M}_{\beta,n})) \cong D^b(\mathrm{Coh}^{\mathbb{T}\times\Gamma_n}(\mathbb{C}^{2n})) \tag{4.2}$$

between derived categories of equivariant coherent sheaves. Moreover, at the level of K-theory, this equivalence realizes the wreath Macdonald polynomial $\tilde{H}^w_{\mu^\bullet}$ as the bigraded Γ_n-Frobenius character of the fiber $P^w_n|_{I^w_{\mu^\bullet}}$ of a rank $r^n n!$ vector bundle P^w_n on $\mathfrak{M}_{\beta,n}$ whose fibers afford the regular representation of Γ_n.

The equivalence (4.2) originates from a more general result of Bezrukavnikov and Kaledin [**BK**]; the work [**BF**] connects it to wreath Macdonald theory. The vector bundle P^w_n on $\mathfrak{M}_{\beta,n}$ described above is the wreath Procesi bundle. These were first considered for arbitrary $u \in \mathfrak{S}_I$ by Losev [**L1**], who also classified Procesi bundles on symplectic resolutions of quotient singularities in greater generality. The Procesi bundle we denote by P^w_n is normalized so that

$$(P^w_n)^{\Gamma_{n-1}} = M^w_n. \tag{4.3}$$

We note that the methods of [**BF**, **L2**] provide alternative proofs of Haiman's Theorem (Theorem 2.10) in the $r = 1$ case.

As in the case of Haiman's Positivity Theorem, we pass from the setting of Theorem 4.6 to (multi-)symmetric functions by applying the K_0 functor, which gives an isomorphism

$$K^{\mathbb{T}}(\mathfrak{M}_{\beta,n}) \cong K^{\mathbb{T}\times\Gamma_n}(\mathbb{C}^{2n})$$

of $R(\mathbb{T})$-modules. We proceed as before, but replacing the $\mathfrak{S}_n$-Frobenius character by the Γ_n-Frobenius character [**Mac**, I, Appendix B] given by the map

$$\mathrm{Frob}^{\Gamma} : R(\Gamma_n) \to (\Lambda_{\mathbb{Z}}^{\otimes I})_n \tag{4.4}$$

sending an irreducible character $\chi^{\lambda^\bullet}$ of Γ_n to the tensor Schur function $s_{\lambda^\bullet}$; we denote the bigraded Γ_n-Frobenius series by $\mathrm{Frob}^{\Gamma}_{q,t}$.

Altogether we have an isomorphism

$$\Phi_n^w : K^{\mathbb{T}}(\mathfrak{M}_{\beta,n})_{\mathrm{loc}} \to (\Lambda^{\otimes I})_n$$

of $\mathbb{K}$-vector spaces, which by Theorem 4.6 satisfies

$$\Phi_n^w([I^w_{\mu^\bullet}]) = \mathrm{Frob}^{\Gamma}_{q,t}(P_n^w|_{I^w_{\mu^\bullet}}) = \tilde{H}^w_{\mu^\bullet}. \tag{4.5}$$

The Laurent integrality and positivity of wreath Macdonald-Kostka coefficients (Theorem 3.16) follows directly from (4.5).

REMARK 4.7. In contrast to the $r = 1$ case, the wreath Procesi fibers $P_n^w|_{I^w_{\mu^\bullet}}$ are not quotients of $\mathbb{C}[x_1, \dots, x_n, y_1, \dots, y_n]$ in general. For instance, it is impossible for either of the $r = 3$ examples

$$\tilde{H}^{s_1}_{(\square,\cdot,\cdot)} = s_{(\square,\cdot,\cdot)} + qt^{-1}s_{(\cdot,\square,\cdot)} + qs_{(\cdot,\cdot,\square)}$$
$$\tilde{H}^{t_{-\alpha_1^\vee-\alpha_2^\vee}}_{(\cdot,\square,\cdot)} = s_{(\square,\cdot,\cdot)} + q^{-1}s_{(\cdot,\square,\cdot)} + t^{-1}s_{(\cdot,\cdot,\square)}$$

to be a bigraded Γ_n-Frobenius character of a module $\mathbb{C}[x_1, \dots, x_n, y_1, \dots, y_n]/I$ for some Γ_n-invariant homogeneous ideal I. These examples do not have a unique minimal bi-degree, which the image of 1 would provide in such a module.

REMARK 4.8. According to [**L1**], there are exactly $2|\mathfrak{S}_I|$ non-equivalent Procesi bundles on $\mathfrak{M}_{\beta,n}$. In our formulation of Theorem 4.6, we only consider half of these. The other half is obtained by applying the dual. The resulting multi-symmetric functions are precisely the orthogonal basis $\mathrm{inv}\,\mathrm{neg}\,\tilde{H}^w_{\mu^\bullet}$ appearing in Proposition 3.31.

Let us work out the condition that (4.3) imposes on the wreath Macdonald-Kostka coefficients. Under the bigraded wreath Frobenius map, the wreath Procesi fiber maps to the wreath Macdonald polynomial: $P_n^w|_{I^w_{\mu^\bullet}} \mapsto \tilde{H}^w_{\mu^\bullet}$. Restriction to Γ_{n-1} is achieved on $\Lambda^{\otimes I}$ by the operator $\sum_{i\in I} s_1[X^{(i)}]^\perp$ where $\perp$ is the adjoint of multiplication with respect to the Hall scalar product on Λ. Taking Γ_{n-1}-invariants is accomplished by taking the coefficient of $s_{n-1}[X^{(0)}]$, the image of the trivial Γ_{n-1} character under wreath Frobenius. Thus (4.3) is equivalent to the following.

PROPOSITION 4.9 ([**H3**]). *For all $w \in \hat{\mathfrak{S}}_I$ and $\mu^\bullet \in \mathbb{Y}^I$ with $|\mu^\bullet| = n$ we have*

$$[\chi^i]\mathcal{B}^w_{\mu^\bullet} = \begin{cases} \langle s_{(n-1)}[X^{(0)}]s_1[X^{(i)}],\, \tilde{H}^w_{\mu^\bullet}\rangle & \text{for } i \in I \setminus \{0\} \\ 1 + \langle s_{(n-1,1)}[X^{(0)}],\, \tilde{H}^w_{\mu^\bullet}\rangle & \text{for } i = 0. \end{cases}$$

Here for $n = 1$ the $s_{(n-1,1)}$ should be interpreted as 0.

EXAMPLE 4.10. Let r, w, and $\mu^\bullet$ be as in Example 4.5. Using the reverse lex order on big partitions, the induced total order on multipartitions (which refines the order $\trianglerighteq_w$) is

71	62	44	422	41^4	3221	321^3	22211	1^8
$(\square\!\square,\cdot,\cdot)$	$(\cdot,\cdot,\square\!\square)$	$(\square,\cdot,\square)$	$(\substack{\square\\\square},\cdot,\cdot)$	$(\square,\square,\cdot)$	$(\cdot,\cdot,\substack{\square\\\square})$	$(\cdot,\square,\square)$	$(\cdot,\square\!\square,\cdot)$	$(\cdot,\substack{\square\\\square},\cdot)$

We have

$$\begin{aligned}\tilde{H}^w_{\mu^\bullet} &= s_{(\square\!\square,\cdot,\cdot)} + qt\, s_{(\substack{\square\\\square},\cdot,\cdot)} + (q^2+t)s_{(\square,\square,\cdot)} + (tq^{-1}+q)s_{(\square,\cdot,\square)} \\ &\quad + q\, s_{(\cdot,\square\!\square,\cdot)} + q^2t\, s_{(\cdot,\substack{\square\\\square},\cdot)} + (qt+1)s_{(\cdot,\square,\square)} + q^{-1}\, s_{(\cdot,\cdot,\square\!\square)} + t\, s_{(\cdot,\cdot,\substack{\square\\\square})}.\end{aligned}$$

We have

$$\begin{aligned}[\chi^0]\mathcal{B}^w_{\mu^\bullet} &= 1+qt = 1 + \langle s_{(\substack{\square\\\square},\cdot,\cdot)},\, \tilde{H}^w_{\mu^\bullet}\rangle \\ [\chi^1]\mathcal{B}^w_{\mu^\bullet} &= q^2+t = \langle s_{(\square,\square,\cdot)},\, \tilde{H}^w_{\mu^\bullet}\rangle \\ [\chi^2]\mathcal{B}^w_{\mu^\bullet} &= tq^{-1}+q = \langle s_{(\square,\cdot,\square)},\, \tilde{H}^w_{\mu^\bullet}\rangle.\end{aligned}$$

4.4. Geometric pairing and norms. Adapting the approach of [**H2**, §5.4.3] to the wreath setting, we can now prove formula (3.13) from Theorem 3.32 for the norms of wreath Macdonald polynomials using the isomorphisms Φ^w_n arising from Theorem 4.6.

For $\mathbb{T}$-equivariant coherent sheaves $\mathcal{F},\mathcal{G}$ on $\mathfrak{M}_{\beta,n}$, define the pairing

$$\langle\mathcal{F},\,\mathcal{G}\rangle_{\mathrm{Tor}} = \sum_{i=0}^{\infty}(-1)^i\mathrm{ch}_{\mathbb{T}}\mathrm{Tor}_i(\mathcal{F},\mathcal{G}) \in R(\mathbb{T}),$$

where $\mathrm{ch}_{\mathbb{T}}$ denotes the $\mathbb{T}$-character. This extends to a well-defined and nondegenerate $\mathbb{K}$-valued inner product on $K_{\mathbb{T}}(\mathfrak{M}_{\beta,n})_{\mathrm{loc}}$. By [**CG**, Proposition 5.4.10], it is given on the fixed point basis by

$$\begin{aligned}\langle[I^w_{\mu^\bullet}],\,[I^w_{\nu^\bullet}]\rangle_{\mathrm{Tor}} &= \delta_{\mu^\bullet\nu^\bullet}\mathrm{ch}_{\mathbb{T}}(\mathrm{Sym}(T^*\mathfrak{M}_{\beta,n}|_{I^w_{\mu^\bullet}}))^{-1} \\ &= \delta_{\mu^\bullet\nu^\bullet}\mathrm{ch}_{\mathbb{T}}(\mathrm{Sym}(T^*\mathrm{Hilb}^\Gamma_N|_{I_{\tau(w^{-1}(\mu^\bullet,0))}}))^{-1},\end{aligned}$$

where $N = |\tau(w^{-1}(\mu^\bullet,0))|$.

We also consider the pairing on $K_{\mathbb{T}}(\mathfrak{M}_{\beta,n})_{\mathrm{loc}}$ given by

$$\langle[\mathcal{F}],\,[\mathcal{G}]\rangle'_{\mathrm{Tor}} = \sum_{\mu^\bullet} c_{\mu^\bullet}(q^{-1},t^{-1})\langle[I^w_{\mu^\bullet}],\,[\mathcal{G}]\rangle_{\mathrm{Tor}} \tag{4.6}$$

for $[\mathcal{F}] = \sum_{\mu^\bullet} c_{\mu^\bullet}(q,t)[I^w_{\mu^\bullet}]$. Note that this is conjugate $\mathbb{K}$-linear in the first argument and $\mathbb{K}$-linear in the second.

PROPOSITION 4.11.

$$\langle[\mathcal{F}],\,[\mathcal{G}]\rangle'_{\mathrm{Tor}} = \langle\mathrm{inv}\,\mathrm{neg}\,\Phi^w_n([\mathcal{F}]),\,\Phi^w_n([\mathcal{G}])\rangle_{q,t}.$$

PROOF. Since both sides are conjugate linear in $[\mathcal{F}]$ and linear in $[\mathcal{G}]$, it suffices to check the assertion on $\mathbb{K}$-basis elements. We evaluate $\langle[\mathcal{F}],\,[\mathcal{G}]\rangle'_{\mathrm{Tor}}$ on the $\mathbb{K}$-bases $\{[I^w_{\mu^\bullet}]\}_{\mu^\bullet}$ and $\{[P^{w,*}_{\lambda^\bullet}]\}_{\lambda^\bullet}$ of $K_{\mathbb{T}}(\mathfrak{M}_{\beta,n})_{\mathrm{loc}}$, where

$$P^{w,*}_{\lambda^\bullet} = \mathrm{Hom}_{\Gamma_n}(V^{\lambda^\bullet *},(P^w_n)^*).$$

By [**BF**, §4.1] and [**L2**, §5.1.1], we know that $\Phi_n^w([P_{\lambda^\bullet}^{w,*}])$ coincides with the bigraded Γ_n-Frobenius character of $V^{\lambda^\bullet} \otimes \mathbb{C}[x_1,\dots,x_n,y_1,\dots,y_n]$, namely:

$$\Phi_n^w(P_{\lambda^\bullet}^{w,*}) = s_{\lambda^\bullet}[(1-q\chi)^{-1}(1-t\chi^{-1})^{-1}X^\bullet].$$

The computation then proceeds as follows:

$$\begin{aligned}
\langle [I_{\mu^\bullet}^w],\, [P_{\lambda^\bullet}^{w,*}]\rangle'_{\mathrm{Tor}} &= \langle [I_{\mu^\bullet}^w],\, [P_{\lambda^\bullet}^{w,*}]\rangle_{\mathrm{Tor}} \\
&= \mathrm{ch}_{\mathbb{T}}(P_{\lambda^\bullet}^{w,*}|_{I_{\mu^\bullet}^w}) \\
&= \mathrm{ch}_{\mathbb{T}}(\mathrm{Hom}(V^{\lambda^\bullet *}, P_{\lambda^\bullet}^{w,*}|_{I_{\mu^\bullet}^w})) \\
&= \langle \mathrm{inv}\,\mathrm{neg}\, H_{\mu^\bullet}^w,\, \mathrm{neg}\, s_{\lambda^\bullet}[X^\bullet]\rangle \\
&= \langle \mathrm{inv}\,\mathrm{neg}\, H_{\mu^\bullet}^w,\, s_{\lambda^\bullet}[\mathrm{neg}\, X^\bullet]\rangle \\
&= \langle \mathrm{inv}\,\mathrm{neg}\, H_{\mu^\bullet}^w,\, s_{\lambda^\bullet}[(1-q\chi)^{-1}(1-t\chi^{-1})^{-1}X^\bullet]\rangle_{q,t} \\
&= \langle \mathrm{inv}\,\mathrm{neg}\, \Phi_n^w([I_{\mu^\bullet}^w]),\, \Phi_n^w([P_{\lambda^\bullet}^{w,*}])\rangle_{q,t},
\end{aligned}$$

where we use that $A = (1-q\chi^{-1})^{-1}(1-t\chi)^{-1}\mathrm{neg}$ satisfies $(1-q\chi)^{-1}(1-t\chi^{-1})^{-1}A^{-1} = \mathrm{neg}$. □

We deduce (3.13):

COROLLARY 4.12.

$$\langle \mathrm{inv}\,\mathrm{neg}\, \tilde{H}_{\mu^\bullet}^w,\, \tilde{H}_{\mu^\bullet}^w\rangle_{q,t} = \mathrm{ch}_{\mathbb{T}}(\mathrm{Sym}(T^*\mathrm{Hilb}_N^\Gamma|_{I_{\tau(w^{-1}(\mu^\bullet,0))}}))^{-1}.$$

REMARK 4.13. As shown in [**H2**, §5.4.3], one can eliminate inv (and neg) from the above when $r=1$ by means of an isomorphism $P_n^* \cong P_n \otimes \varepsilon \otimes \mathcal{O}(-1)$, where ε denotes the sign representation of $\mathfrak{S}_n$ and $\mathcal{O}(-1)$ is the dual of the top exterior power of the tautological bundle (corresponding to the operator ∇^{-1} on symmetric functions).

4.5. Wreath nabla. For $w \in \hat{\mathfrak{S}}_I$, $\mu^\bullet \in \mathbb{Y}_n^I$, and $i \in I$ define the Laurent monomial

$$\begin{aligned}
e_{\mu^\bullet}^{(i),w} &= \mathrm{ch}_{\mathbb{T}} \bigwedge^n ([\chi^i]M_n^w)|_{I_{\mu^\bullet}^w} \\
&= e_n([\chi^i]\mathcal{B}_{\mu^\bullet}^w(q,t,\chi)).
\end{aligned}$$

This value is the product of the $\mathbb{T}$-weights of the χ^i component of the fiber of M_n^w over $I_{\mu^\bullet}^w$. The following result asserts that the monomial $e_{\mu^\bullet}^{(i),w}$ records the bi-degree of the unique copy of the i-th sign representation of Γ_n in $P_n^w|_{I_{\mu^\bullet}^w}$.

THEOREM 4.14 ([**H3**]). *Let $\mu^\bullet \in \mathbb{Y}^I$ have total size n. Then*

$$\langle e_n[X^{(i)}],\, \tilde{H}_{\mu^\bullet}^w\rangle = e_{\mu^\bullet}^{(i),w}.$$

EXAMPLE 4.15. With w and $\mu^\bullet$ from the previous example, we have

$$\begin{aligned}
\langle e_2[X^{(0)}],\, \tilde{H}_{\mu^\bullet}^w\rangle &= qt = e_2(qt+1) \\
\langle e_2[X^{(1)}],\, \tilde{H}_{\mu^\bullet}^w\rangle &= q^2t = e_2(q^2+t) \\
\langle e_2[X^{(2)}],\, \tilde{H}_{\mu^\bullet}^w\rangle &= t = e_2(tq^{-1}+q).
\end{aligned}$$

The input variables to the e_n on the right are the summands of the coefficients of powers of χ in $\mathcal{B}_{\mu^\bullet}^w$.

For $i \in I$ and $w \in \hat{\mathfrak{S}}_I$, the wreath nabla operator $\nabla_w^{(i)}$ is defined as the linear operator on $\Lambda^{\otimes I}$ that acts diagonally on the wreath Macdonald basis $\tilde{H}^w_{\mu^\bullet}$ with eigenvalue $e^{(i),w}_{\mu^\bullet}$. (Recall that $\nabla_w^{(0)}$ made an appearance in §3.5.2 above.)

4.6. Factorization for generic stability condition. We recall a factorization property observed by Haiman [**H3**].

Fix $n \in \mathbb{Z}_{\geq 0}$ and $u, v \in \mathfrak{S}_I$. Let $\beta^\vee \in Q^\vee$ be sufficiently antidominant with respect to n, that is, $\langle \beta^\vee, \overline{\alpha}_i\rangle \ll 0$ for all $1 \leq i \leq r-1$. Let $w = ut_{v\beta^\vee} \in \hat{\mathfrak{S}}_I$. Then

- $\mathfrak{M}_{w\theta_+}(n\delta)$ does not depend on $\beta^\vee$.
- It is isomorphic to the Hilbert scheme $\mathrm{Hilb}^n(\widehat{\mathbb{C}^2/\Gamma})$ of points on the minimal resolution $\widehat{\mathbb{C}^2/\Gamma}$ of the quotient singularity $\mathbb{C}^2/\Gamma$ [**Kuz**].

As a consequence the wreath Macdonald polynomial $\tilde{H}^w_{\mu^\bullet}$ factorizes into modified Macdonald polynomials with appropriate substitutions of variables and parameters.

In this situation we write $\tilde{H}^{u,v}_{\mu^\bullet} = \tilde{H}^w_{\mu^\bullet}$ and call it a generic wreath Macdonald polynomial.

For $i \in I$ let $\square^{(i)} \in \mathbb{Y}^I$ be the multipartition that has a single box partition in position i and empty partitions elsewhere. Let $M^{u,v} \in \mathrm{Mat}_{I\times I}(\mathbb{K})$ be the matrix

$$M^{u,v}_{ij} = \langle \tilde{H}^{u,v}_{\square^{(i)}}, s_1[X^{(j)}]\rangle. \tag{4.7}$$

EXAMPLE 4.16. Let $M^- = M^{\mathrm{id},\mathrm{id}}$. For this stability condition the partial order $\trianglerighteq_w$ on $\mathbb{Y}^I_n$ is refined by the following total order $\geq$: for distinct $\lambda^\bullet, \mu^\bullet \in \mathbb{Y}^I_n$, let $0 \leq k \leq r-1$ be minimum such that $\lambda^{(k)} \neq \mu^{(k)}$. Then $\lambda^\bullet > \mu^\bullet$ if $\lambda^{(k)}$ is reverse lex larger than $\mu^{(k)}$ (the first part that disagrees is larger in $\lambda^{(k)}$). See [**G**, Cor. 7.13] for a precise characterization of the partial order in this case. All of the generic orders are permutations of each other.

For $r = 3$ and $n = 1$ we have

$$(\square, \cdot, \cdot) \geq (\cdot, \square, \cdot) \geq (\cdot, \cdot, \square).$$

For $n = 2$

$$(\square\!\square, \cdot, \cdot) \geq (\begin{smallmatrix}\square\\\square\end{smallmatrix}, \cdot, \cdot) \geq (\square, \square, \cdot) \geq (\square, \cdot, \square) \geq (\cdot, \square\!\square, \cdot) \geq (\cdot, \begin{smallmatrix}\square\\\square\end{smallmatrix}, \cdot) \geq (\cdot, \square, \square) \geq (\cdot, \cdot, \square\!\square) \geq (\cdot, \cdot, \begin{smallmatrix}\square\\\square\end{smallmatrix}).$$

Using this for $n = 1$ one may compute

$$M^-_{ij} = \begin{cases} t^j & \text{if } i \geq j \\ q^{r-j} & \text{if } i < j. \end{cases}$$

For a symmetric function $f \in \Lambda$ and a row vector $(a_j \mid j \in I) \in \mathbb{K}^I$, we write $f[\sum_{j\in I} a_j X^{(j)}]$ to mean the image of f under the $\mathbb{K}$-algebra homomorphism $\Lambda \to \Lambda^{\otimes I}$ sending p_k to $\sum_{j \in I} a_j(q^k, t^k) p_k(X^{(j)})$. This is called a vector plethysm.

Let $X^\bullet$ be the column vector with i-th coordinate $X^{(i)}$ for $i \in I$ and ϵ_i the i-th standard basis column vector. Then

$$\epsilon_i^t M^{u,v} X^\bullet = \sum_{j \in I} M^{u,v}_{ij} X^{(j)}.$$

Define the vectors $q_\bullet, t_\bullet \in \mathbb{K}^I$ by $q_i = q^{r-i}t^{-i}$ and $t_i = q^{-i}t^{r-i}$ for $0 \leq i \leq r-1$.

PROPOSITION 4.17 ([**H3**]). *The wreath Macdonald polynomial factors:*

$$\tilde{H}^{u,v}_{\mu^\bullet} = \prod_{i \in I} \tilde{H}^{u,v}_{\chi^i \mu^{(i)}}$$

where the special notation $\chi^i\nu$ (not used elsewhere and not to be confused with the rotation of multipartitions) denotes the multipartition which has empty components except the i-th, which is $\nu \in \mathbb{Y}$. Moreover we have

$$\tilde{H}^{u,v}_{\chi^i\nu} = \tilde{H}_\nu(\epsilon_i^t M^{u,v} X^\bullet, q_{v^{-1}(i)}, t_{v^{-1}(i)})$$

where the right hand side is a usual modified Macdonald polynomial $\tilde{H}_\nu$ with parameters $(q,t) = (q_{v^{-1}(i)}, t_{v^{-1}(i)})$ to which is applied the given vector plethysm.

EXAMPLE 4.18. Let $r = 3$, $i = 2$, $\nu = (2)$, $u = v = \mathrm{id}$. We have $\tilde{H}_{(2)} = s_2 + q s_{11}$. First the parameters are substituted: $q \mapsto q_2 = q/t^2$ and $t \mapsto t_2 = t^3$, giving $s_2 + (q/t^2)s_{11}$. The vector plethysms of s_2 and s_{11} using $Z = X^{(0)} + tX^{(1)} + t^2X^{(2)}$ are:

$$\begin{aligned}
s_2[Z] &= s_2[X^{(0)}] + t^2 s_2[X^{(1)}] + t^4 s_2[X^{(2)}] \\
&\quad + t s_1[X^{(0)}]s_1[X^{(1)}] + t^2 s_1[X^{(0)}]s_1[X^{(2)}] + t^3 s_1[X^{(1)}]s_1[X^{(2)}] \\
s_{11}[Z] &= s_{11}X^{(0)}] + t^2 s_{11}[X^{(1)}] + t^4 s_{11}[X^{(2)}] \\
&\quad + t s_1[X^{(0)}]s_1[X^{(1)}] + t^2 s_1[X^{(0)}]s_1[X^{(2)}] + t^3 s_1[X^{(1)}]s_1[X^{(2)}].
\end{aligned}$$

One may verify that

$$\tilde{H}^{\mathrm{id},\mathrm{id}}_{(\cdot,\cdot,(2))} = s_2[Z] + (q/t^2)s_{11}[Z].$$

5. Quantum toroidal algebras and wreath Macdonald eigenoperators

Throughout this section, we assume $r \geq 3$. Our goal is to explain how the quantum toroidal algebra provides a wreath generalization of the Macdonald eigenoperators from Corollary 2.9 for wreath Macdonald functions $\tilde{H}^w_{\mu^\bullet}$ with $w = t_{-\beta^\vee}$ a translation element. See Remark 5.20 for a brief discussion of the $r = 2$ case and the connection to classical Macdonald theory for $r = 1$.

5.1. Quantum toroidal algebras. Quantum toroidal algebras were introduced in [**GKV**] for arbitrary simply-laced Lie types. Here we consider only the type A case and follow the conventions of [**T, FJMM2**]. In this case there is a remarkable symmetry of the quantum toroidal algebra due to Miki [**M1**], which plays a key role in the construction of wreath Macdonald eigenoperators.

5.1.1. *Setup.* We enlarge the base field to $\mathbb{F} = \mathbb{Q}(s,u)$, with $\mathbb{K} = \mathbb{Q}(q,t)$ embedded into $\mathbb{F}$ via $q = s^2u^2$ and $t = s^2u^{-2}$. We set $p = s^2$ and $d = u^2$.

Let $A = (a_{ij})_{i,j\in I}$ be the Cartan matrix of $\hat{\mathfrak{sl}}_r$ and let $M = (m_{ij})_{i,j\in I}$ be the skew-symmetric matrix given by:

$$m_{ij} = \begin{cases} \pm 1 & \text{if } i = j \pm 1 \\ 0 & \text{otherwise.} \end{cases}$$

5.1.2. *Definition.* The quantum toroidal algebra $U_r^{\mathrm{tor}} = U_{s,u}^{\mathrm{tor}}(\mathfrak{gl}_r)$ is the associative $\mathbb{F}$-algebra generated by

$$C_h^{\pm 1/2}, C_v^{\pm 1/2}, e_{i,k}, f_{i,k}, \psi^{\pm}_{i,l} \qquad \text{for } i \in I, k \in \mathbb{Z}, \pm l \in \mathbb{Z}_{\geq 0}$$

subject to the relations given below.[7] To state these, it is convenient to collect the generators into currents:

$$e_i(z) = \sum_{k\in\mathbb{Z}} e_{i,k} z^{-k}, \quad f_i(z) = \sum_{k\in\mathbb{Z}} f_{i,k} z^{-k}, \quad \psi_i^{\pm}(z) = \sum_{\pm l\geq 0} \psi_{i,l}^{\pm} z^{-l}.$$

The defining relations of U_r^{tor} are then as follows for all $i, j \in I$:

$$\begin{aligned}
&C_h^{1/2}, C_v^{1/2} \text{ are central}\\
&\psi_{i,0}^{+}\psi_{i,0}^{-} = \psi_{i,0}^{-}\psi_{i,0}^{+} = 1\\
&\psi_i^{\pm}(z)\psi_j^{\pm}(w) = \psi_j^{\pm}(w)\psi_i^{\pm}(z)\\
&C_h = (C_h^{1/2})^2 = \prod_{i\in I}\psi_{i,0}\\
&\theta_{a_{ij}}(C_v^{-1} d^{m_{ij}} z/w)\psi_i^{+}(z)\psi_j^{-}(w) = \theta_{a_{ij}}(C_v d^{m_{ij}} z/w)\psi_j^{-}(w)\psi_i^{+}(z)\\
&\psi_i^{\pm}(z)e_j(w) = \theta_{a_{ij}}(C_v^{\pm\frac{1}{2}} d^{m_{ij}} z/w)e_j(w)\psi_i^{\pm}(z)\\
&\psi_i^{\pm}(z)f_j(w) = \theta_{a_{ij}}(C_v^{\mp\frac{1}{2}} d^{m_{ij}} z/w)^{-1} f_j(w)\psi_i^{\pm}(z)\\
&e_i(z)e_j(w) = \theta_{a_{ij}}(d^{m_{ij}} z/w)e_j(w)e_i(z)\\
&f_i(z)f_j(w) = \theta_{a_{ij}}(d^{m_{ij}} z/w)^{-1} f_j(w)f_i(z)\\
&[e_i(z), f_j(w)] = \frac{\delta_{ij}}{p-p^{-1}}\left(\delta(C_v w/z)\psi_i^{+}(C_v^{\frac{1}{2}} w) - \delta(C_v z/w)\psi_i^{-}(C_v^{\frac{1}{2}} z)\right)\\
&[e_i(z_1), [e_i(z_2), e_{i\pm 1}(w)]_p]_{p^{-1}} + [e_i(z_2), [e_i(z_1), e_{i\pm 1}(w)]_p]_{p^{-1}} = 0\\
&[f_i(z_1), [f_i(z_2), f_{i\pm 1}(w)]_p]_{p^{-1}} + [f_i(z_2), [f_i(z_1), f_{i\pm 1}(w)]_p]_{p^{-1}} = 0
\end{aligned}$$

where $\theta_m(z) = \frac{p^m z - 1}{z - p^m}$ for any $m \in \mathbb{Z}$, $\delta(z) = \sum_{n\in\mathbb{Z}} z^n$, and $[x, y]_c = xy - cyx$.

Instead of the generators $\psi_{i,l}^{\pm}$ for $i \in I$ and $\pm l \geq 0$, we may use

(5.1) $$\psi_{i,0}^{\pm},\ h_{i,l} \qquad \text{for } i \in I \text{ and } l \in \mathbb{Z}_{\neq 0}$$

where the $h_{i,l}$ are defined by

(5.2) $$\psi_i^{\pm}(z) = \psi_{i,0}^{\pm} \exp\left(\pm(p - p^{-1}) \sum_{\pm l > 0} h_{i,l} z^{-l}\right).$$

[7] For the purposes of our discussion, we ignore the additional generators q^{d_1}, q^{d_2}.

5.1.3. *Mode form of relations.* In terms of individual generators, the relations of U_r^{tor} can be fully unpacked as follows:

$$C_h^{1/2}, C_v^{1/2} \text{ are central}$$
$$\psi^+_{i,0}\psi^-_{i,0} = \psi^-_{i,0}\psi^+_{i,0} = 1$$
$$\psi^\pm_{i,k}\psi^\pm_{j,l} = \psi^\pm_{j,l}\psi^\pm_{i,k}$$
$$C_h = (C_h^{1/2})^2 = \prod_{i\in I}\psi_{i,0}$$
$$[h_{i,k}, h_{j,-l}] = \delta_{k,l}\frac{d^{-km_{ij}}}{k}\frac{(p^{ka_{ij}} - p^{-ka_{ij}})(C_v^k - C_v^{-k})}{(p-p^{-1})^2} \qquad (k,l\neq 0) \tag{5.3}$$
$$\psi_{i,0}e_{j,k} = p^{a_{ij}}e_{j,k}\psi_{i,0}$$
$$\psi_{i,0}f_{j,k} = p^{-a_{ij}}f_{j,k}\psi_{i,0}$$
$$[h_{i,k}, e_{j,l}] = \frac{C_v^{-|k|/2}d^{-m_{ij}k}}{k}\frac{p^{ka_{ij}} - p^{-ka_{ij}}}{p-p^{-1}}e_{j,k+l} \quad (k\neq 0)$$
$$[h_{i,k}, f_{j,l}] = -\frac{C_v^{|k|/2}d^{-m_{ij}k}}{k}\frac{p^{ka_{ij}} - p^{-ka_{ij}}}{p-p^{-1}}f_{j,k+l} \quad (k\neq 0)$$
$$d^{m_{ij}}[e_{i,k+1}, e_{j,l}]_{p^{a_{ij}}} + [e_{j,l+1}, e_{i,k}]_{p^{a_{ij}}} = 0$$
$$[f_{i,k}, f_{j,l+1}]_{p^{a_{ij}}} + d^{m_{ij}}[f_{j,l}, f_{i,k+1}]_{p^{a_{ij}}} = 0$$
$$[e_{i,k}, f_{j,l}] = \delta_{ij}\frac{C_v^{(k-l)/2}\psi^+_{i,k+l} - C_v^{-(l-k)/2}\psi^-_{i,k+l}}{p-p^{-1}}$$
$$[e_{i,k_1}, [e_{i,k_2}, e_{i\pm 1,l}]_p]_{p^{-1}} + [e_{i,k_2}, [e_{i,k_1}, e_{i\pm 1,l}]_p]_{p^{-1}} = 0$$
$$[f_{i,k_1}, [f_{i,k_2}, f_{i\pm 1,l}]_p]_{p^{-1}} + [f_{i,k_2}, [f_{i,k_1}, f_{i\pm 1,l}]_p]_{p^{-1}} = 0$$

where $i,j\in I$ and $k,l\in\mathbb{Z}$ and we set $\psi^+_{i,l} = \psi^-_{i,-l} = 0$ for all $i\in I$ and $l<0$.

5.1.4. *Involution on a quotient.* Let $U_{r,0}^{\mathrm{tor}} = U_r^{\mathrm{tor}}/\langle C_v^{1/2} - 1\rangle$. Observe that there exists a unique $\mathbb{Q}(u)$-linear involutive automorphism $\dagger$ of U_0^{tor} extending the assignment:

$$s \mapsto s^{-1}$$
$$e_i(z) \mapsto f_i(z)$$
$$f_i(z) \mapsto e_i(z)$$
$$\psi_i^\pm(z) \mapsto \psi_i^\pm(z)$$

for all $i\in I$. To see this, one uses that $\theta_m(x)^{-1} = \theta_{-m}(x)$.

Observe that $q^\dagger = t^{-1}$ and $t^\dagger = q^{-1}$.

5.1.5. *Quantum affine subalgebras and the Miki automorphism.* The quantum toroidal algebra U_r^{tor} contains two distinguished copies of the quantum affine algebra $U_r^{\mathrm{af}} = U_p'(\hat{\mathfrak{sl}}_r)$.

The *horizontal* quantum affine subalgebra $U^h\subset U_r^{\mathrm{tor}}$ is the $\mathbb{Q}(p)$-subalgebra generated by $e_{i,0}, f_{i,0}$, and $\psi^{\pm 1}_{i,0}$ for all $i\in I$. An isomorphism $h: U_r^{\mathrm{af}}\to U^h$ of $\mathbb{Q}(p)$-algebras is given by identifying these generators with the Chevalley generators in the Drinfeld-Jimbo presentation of U_r^{af}.

The *vertical* quantum affine subalgebra $U^v\subset U_r^{\mathrm{tor}}$ is the $\mathbb{Q}(p)$-subalgebra generated by C_v and $d^{ik}e_{i,k}, d^{ik}f_{i,k}$, and $d^{ik}C_v^{k/2}\psi^\pm_{i,k}$ for $1\le i\le r-1$ and $k\in\mathbb{Z}$. An isomorphism $v : U_r^{\mathrm{af}} \to U^h$ of $\mathbb{Q}(p)$-algebras is given by identifying these generators with the generators in the Drinfeld loop presentation of U_r^{af}.

THEOREM 5.1 ([**M1**]). *There exists an automorphism $\varsigma : U_r^{\mathrm{tor}} \to U_r^{\mathrm{tor}}$ such that*

$$\varsigma(U^h) = U^v \quad \textit{and} \quad \varsigma(U^v) = U^h. \tag{5.4}$$

Moreover, ς is uniquely determined by compatibility with the isomorphisms $h : U_r^{\mathrm{af}} \to U^h$ and $v : U_r^{\mathrm{af}} \to U^v$.

We call ς the *Miki automorphism.* We note that ς is not uniquely determined by (5.4) alone (for instance, $\varsigma \neq \varsigma^{-1}$). The compatibility of ς with the isomorphisms $h : U_r^{\mathrm{af}} \to U^h$ and $v : U_r^{\mathrm{af}} \to U^v$, which we will not state explicitly, is asymmetric with respect to h and v; see [**M1**, Theorem 1] or [**FJMM2**, Theorem 2.2] for details.

It is in general difficult to describe the action of ς on elements of U_r^{tor}. For our purposes, the following explicit formulas will suffice to characterize ς:

PROPOSITION 5.2 ([**T**, **FJMM2**]). *The Miki automorphism $\varsigma : U_r^{\mathrm{tor}} \to U_r^{\mathrm{tor}}$ is uniquely determined by the following formulas, where $1 \leq i \leq r-1$:*

$$\begin{aligned}
\varsigma(C_v^{1/2}) &= C_h^{-1/2}\\
\varsigma(C_h^{1/2}) &= C_v^{1/2}\\
\varsigma(\psi_{i,0}) &= \psi_{i,0}\\
\varsigma(\psi_{0,0}) &= \psi_{0,0}C_h^{-1}C_v\\
\varsigma(e_{i,0}) &= e_{i,0}\\
\varsigma(f_{i,0}) &= f_{i,0}\\
\varsigma(e_{0,-1}) &= (-1)^r u^{2r} e_{0,1}\\
\varsigma(f_{0,1}) &= (-1)^r u^{-2r} f_{0,-1}
\end{aligned}$$

and

$$\varsigma(h_{i,1}) = (-1)^{i+1}u^{-2i}C_h^{-1/2} f_{0,0} \cdot \begin{bmatrix} f_{r-1,0} & \cdots & f_{i+1,0} & f_{1,0} & \cdots & f_{i-1,0} & f_{i,0} \\ p & \cdots & p & p & \cdots & p & p^2 \end{bmatrix}$$

$$\varsigma(h_{i,-1}) = (-1)^{i+1}u^{2i}C_h^{1/2} \begin{bmatrix} e_{i,0} & e_{i-1,0} & \cdots & e_{1,0} & e_{i+1,0} & \cdots & e_{r-1,0} \\ p^{-2} & p^{-1} & \cdots & p^{-1} & p^{-1} & \cdots & p^{-1} \end{bmatrix} \cdot e_{0,0}$$

$$\varsigma(h_{0,1}) = (-1)^r u^{2-2r} C_h^{-1/2} f_{1,1} \cdot \begin{bmatrix} f_{2,0} & \cdots & f_{r-1,0} & f_{0,-1} \\ p & \cdots & p & p^2 \end{bmatrix}$$

$$\varsigma(h_{0,-1}) = (-1)^r u^{2r-2} C_h^{1/2} \begin{bmatrix} e_{0,1} & e_{r-1,0} & \cdots & e_{2,0} \\ p^{-2} & p^{-1} & \cdots & p^{-1} \end{bmatrix} \cdot e_{1,-1},$$

in terms of the iterated commutators:

$$x \cdot \begin{bmatrix} y_1 & \cdots & y_m \\ c_1 & \cdots & c_m \end{bmatrix} = [\ldots[[x, y_1]_{c_1}, y_2]_{c_2} \cdots, y_m]_{c_m}$$

$$\begin{bmatrix} x_m & \cdots & x_1 \\ c_m & \cdots & c_1 \end{bmatrix} \cdot y = [x_m, \ldots [x_2, [x_1, y]_{c_1}]_{c_2} \cdots]_{c_m}.$$

REMARK 5.3. We extracted the formulas above from **[FJMM2]** via the following identification:

	ς	C_h	C_v	$e_{i,k}$	$f_{i,k}$	$\psi^{\pm}_{i,0}$	$h_{i,l}$	p	d
[FJMM2]	θ^{-1}	κ	q^c	$E_{i,k}$	$F_{i,k}$	$K_i^{\pm 1}$	$q^{-lc/2}H_{i,l}$	q	d

5.1.6. *Heisenberg subalgebras.* Let $\mathfrak{H}_v \subset U_r^{\mathrm{tor}}$ be the $\mathbb{F}$-span of $C_v^{\pm 1}$ and $h_{i,l}$ for $i \in I$ and $l \neq 0$. The subspace $\mathfrak{H}_v$ is closed under the Lie bracket and is isomorphic to a p-deformed Heisenberg Lie algebra with relations given by (5.3). We call $\mathfrak{H}_v$ the *vertical* Heisenberg (Lie) subalgebra of U_r^{tor}.

The *horizontal* Heisenberg subalgebra of U_r^{tor} is the Lie subalgebra $\mathfrak{H}_h = \varsigma(\mathfrak{H}_v)$. As a vector space, $\mathfrak{H}_h$ is spanned by $C_h^{\pm 1}$ and $\varsigma(h_{i,l})$ for $i \in I$ and $l \neq 0$. The horizontal Heisenberg elements $\varsigma(h_{i,\pm 1})$ are given explicitly by Proposition 5.2.

Acting in the vertex representation $\mathcal{V}$ of U_r^{tor} introduced below, the elements $\varsigma(h_{i,l})$ will lead to wreath Macdonald eigenoperators.

5.2. Fock representation. The eigenvalues of wreath Macdonald eigenoperators are found in the Fock representation of U^{tor}.

5.2.1. *Lowest weight Fock representation.* The Fock space $\mathcal{F}_\xi = \bigoplus_{\mu\in\mathbb{Y}} \mathbb{F}|\mu\rangle$ affords an action of U_r^{tor} depending on a parameter $\xi \in \mathbb{F}^\times$ **[VV1, STU]**, which simultaneously extends the well-known level 1 action of U_r^{af} on this space **[KMS]** and the level 0 action of U_r^{af} due to **[TU]** (via the horizontal and vertical embeddings, respectively). Here we follow the conventions of **[FJMM2, T]**.

To describe this action explicitly, let $\mathcal{F}_\xi^\vee = \oplus_{\mu\in\mathbb{Y}}\mathbb{F}\langle\mu|$ be the finite dual space, where $\langle\mu|\nu\rangle = \delta_{\mu\nu}$, and define the rational function $\phi(z) = \frac{p-p^{-1}z}{1-z}$.

THEOREM 5.4 (**[FJMM2**, Proposition 3.3**]**). *The space $\mathcal{F}_\xi$ affords an action of the quantum toroidal algebra U_r^{tor} whose nonzero matrix coefficients are as follows:*

$$\begin{aligned}
\langle\mu|C_h^{1/2}|\mu\rangle &= s^{-1}\\
\langle\mu|C_v^{1/2}|\mu\rangle &= 1\\
\langle\nu|e_i(z)|\mu\rangle &= \prod_{\substack{b<y\\(a,b)\in A_i(\mu)}} \phi(t^{a-x}q^{b-y}) \prod_{\substack{b<y\\(a,b)\in R_i(\mu)}} \phi(t^{x-a}q^{y-b})\delta(t^{-x}q^{-y}\xi/z)\\
\langle\mu|f_i(z)|\nu\rangle &= \prod_{\substack{b>y\\(a,b)\in A_i(\mu)}} \phi(t^{a-x}q^{b-y}) \prod_{\substack{b>y\\(a,b)\in R_i(\mu)}} \phi(t^{x-a}q^{y-b})\delta(t^{-x}q^{-y}\xi/z)\\
\langle\mu|\psi_i^{\pm}(z)|\mu\rangle &= \prod_{(a,b)\in A_i(\mu)} \phi(t^{-a}q^{-b}qt\xi/z)^{-1} \prod_{(a,b)\in R_i(\mu)} \phi(t^{-a}q^{-b}\xi/z)
\end{aligned}$$

for all $\mu \in \mathbb{Y}$, $i \in I$, and all $\nu \in \mathbb{Y}$ such that ν/μ is a single cell (x, y) satisfying $y - x \equiv i \pmod r$.

In particular, $\mathcal{F}_\xi$ is an irreducible lowest weight module over $U_{r,0}^{\mathrm{tor}}$, with lowest weight vector $|\varnothing\rangle$ and lowest weight $(\phi(z/\xi)^{\delta_{i0}})_{i\in I}$.

Here the matrix coefficients for $\psi_i^{\pm}(z)$ are obtained by expanding the rational functions in $z^{\mp 1}$, respectively. The lowest weight condition in $\mathcal{F}_\xi$ reads as follows:

$$\begin{aligned}\psi_i^{\pm}(z)|\varnothing\rangle &= \phi(z/\xi)^{\delta_{i0}}|\varnothing\rangle\\ f_i(z)|\varnothing\rangle &= 0 \quad \text{for all } i \in I.\end{aligned}$$

REMARK 5.5. More generally, there is a Fock representation of U_r^{tor} for each $p \in I$, corresponding to a fundamental weight of $\hat{\mathfrak{sl}}_r$. We only consider the $p = 0$ case, which matches the choice of vertex supporting the one-dimensional framing space in the quiver varieties $\mathfrak{M}_{\beta,n}$ from the previous section.

5.2.2. *Twisting.* To obtain a highest weight $U_{r,0}^{\mathrm{tor}}$-module $\mathcal{F}_\xi^\dagger$, we twist the Fock module by the involutive automorphism $\dagger$ of $U_{r,0}^{\mathrm{tor}}$ from §5.1.4. More precisely, let $\rho_{\mathcal{F}_\xi} : U_{r,0}^{\mathrm{tor}} \to \mathrm{End}_{\mathbb{F}}(\mathcal{F}_\xi) \cong \mathrm{Mat}_{\mathbb{Y}}(\mathbb{F})$ be the $\mathbb{F}$-algebra homomorphism determined by the Fock representation $\mathcal{F}_\xi$, where $\mathrm{Mat}_{\mathbb{Y}}(\mathbb{F})$ is the algebra of $\mathbb{Y} \times \mathbb{Y}$-matrices over $\mathbb{F}$ whose columns have finite support. We identify $\mathrm{End}_{\mathbb{F}}(\mathcal{F}_\xi) \cong \mathrm{Mat}_{\mathbb{Y}}(\mathbb{F})$ by means of the basis $\{|\mu\rangle\}_{\mu\in\mathbb{Y}}$.

Then, we define $\rho_{\mathcal{F}_\xi^\dagger}$ as the composition:

$$\rho_{\mathcal{F}_\xi^\dagger} : U_{r,0}^{\mathrm{tor}} \xrightarrow{\dagger} U_{r,0}^{\mathrm{tor}} \xrightarrow{\rho_{\mathcal{F}_\xi}} \mathrm{Mat}_{\mathbb{Y}}(\mathbb{F}) \xrightarrow{\dagger} \mathrm{Mat}_{\mathbb{Y}}(\mathbb{F})$$

where, by an abuse of notation, the map $\mathrm{Mat}_{\mathbb{Y}}(\mathbb{F}) \xrightarrow{\dagger} \mathrm{Mat}_{\mathbb{Y}}(\mathbb{F})$ is given by entry-wise application of the automorphism $a(s,u) \mapsto a(s^{-1},u)$ of $\mathbb{F}$. We use the same vector space $\mathcal{F}_\xi^\dagger = \oplus_{\mu\in\mathbb{Y}}\mathbb{F}|\mu\rangle$ and regard the images of $\rho_{\mathcal{F}_\xi^\dagger}$ as matrices with respect to the basis $\{|\mu\rangle\}_{\mu\in\mathbb{Y}}$.

COROLLARY 5.6. *The homomorphism $\rho_{\mathcal{F}_\xi^\dagger}$ determines an action of U_r^{tor} on $\mathcal{F}_\xi^\dagger$ whose nonzero matrix coefficients are as follows:*

$$\begin{aligned}
\langle\mu|C_h^{1/2}|\mu\rangle &= s\\
\langle\mu|C_v^{1/2}|\mu\rangle &= 1\\
\langle\mu|e_i(z)|\nu\rangle &= \prod_{\substack{b>y\\(a,b)\in A_i(\mu)}} \phi(q^{a-x}t^{b-y}) \prod_{\substack{b>y\\(a,b)\in R_i(\mu)}} \phi(q^{x-a}t^{y-b})\delta(q^x t^y \xi^\dagger/z)\\
\langle\nu|f_i(z)|\mu\rangle &= \prod_{\substack{b<y\\(a,b)\in A_i(\mu)}} \phi(q^{a-x}t^{b-y}) \prod_{\substack{b<y\\(a,b)\in R_i(\mu)}} \phi(q^{x-a}t^{y-b})\delta(q^x t^y \xi^\dagger/z)
\end{aligned}$$

$$\langle\mu|\psi_i^{\pm}(z)|\mu\rangle = \prod_{(a,b)\in A_i(\mu)} \phi(q^{-a}t^{-b}qtz/\xi^\dagger)^{-1} \prod_{(a,b)\in R_i(\mu)} \phi(q^{-a}t^{-b}z/\xi^\dagger) \tag{5.5}$$

for all $\mu \in \mathbb{Y}$, $i \in I$, and all $\nu \in \mathbb{Y}$ such that ν/μ is a single cell (x,y) satisfying $y - x \equiv i \pmod r$.

In particular, $\mathcal{F}_\xi$ is an irreducible highest weight module over $U_{r,0}^{\mathrm{tor}}$, with highest weight vector $|\varnothing\rangle$ and highest weight $(\phi(\xi^\dagger/z)^{\delta_{i0}})_{i\in I}$.

To compute these matrix coefficients, we use that $q^\dagger = t^{-1}$, $t^\dagger = q^{-1}$, and

$$\phi(c/z)^\dagger = \left(\frac{p - p^{-1}c/z}{1 - c/z}\right)^\dagger = \frac{p^{-1} - pc^\dagger/z}{1 - c^\dagger/z} = \phi(z/c^\dagger)$$

for any $c \in \mathbb{F}$.

The highest weight condition in $\mathcal{F}_\xi^\dagger$ reads as follows:

$$\psi_i^\pm(z)|\varnothing\rangle = \phi(\xi^\dagger/z)^{\delta_{i0}}|\varnothing\rangle$$
$$e_i(z)|\varnothing\rangle = 0 \quad \text{for all } i \in I.$$

From (5.2) and (5.5), one computes the following finer description of the joint eigenvalues of $\mathfrak{H}^v$ in $\mathcal{F}_\xi^\dagger$:

$$\langle\mu|h_{i,\pm l}|\mu\rangle = A_\mu^{(i)}(q^{\pm l}, t^{\pm l})\frac{p^l - p^{-l}}{p - p^{-1}}\frac{(p^{-1}\xi^\dagger)^{\pm l}}{l} \tag{5.6}$$

for all $\mu \in \mathbb{Y}$, $i \in I$, and $l > 0$, where $A_\mu^{(i)}$ denotes the coefficient of χ^i in $A_\mu(q\chi^{-1}, t\chi)$. At this point, it is instructive to recall that the quantities $A_\mu(q^{\pm 1}, t^{\pm 1})$ from (2.2) were precisely the eigenvalues of Macdonald eigenoperators in Corollary 2.9. As we will see below, (5.6) determines the eigenvalues of wreath Macdonald eigenoperators.

REMARK 5.7. In [**T**], it is asserted that the dual $\mathcal{F}_\xi^\vee$ can be made into a U_r^{tor}-module via the antipode of a topological Hopf algebra structure on U_r^{tor}. Here we achieve the same desired effect—namely, converting a lowest weight module to a highest weight module—by using the automorphism $\dagger$.

REMARK 5.8. By [**VV2, Nak2**] (see also [**Nag2**]), the space

$$\mathcal{K} = \bigoplus_{\substack{\beta\in Q\\ n\geq 0}} K^{\mathbb{T}}(\mathfrak{M}_{\beta,n}) \otimes_{R(\mathbb{T})} \mathbb{F} = \bigoplus_{N\geq 0} K^{\mathbb{T}}(\mathrm{Hilb}_N^\Gamma) \otimes_{R(\mathbb{T})} \mathbb{F} \tag{5.7}$$

has the structure of an irreducible highest weight $U_{r,0}^{\mathrm{tor}}$-module with highest weight $(\phi(z^{-1})^{\delta_{i0}})_{i\in I}$. This action is defined geometrically, in fashion similar to the action appearing in Theorem 2.12, with the currents $e_i(z)$ and $f_i(z)$ acting via correspondences, and the Cartan currents $\psi_i^\pm(z)$ acting by K-theory multiplication with tautological classes.

Recall that we identify the Nakajima quiver varieites $\mathfrak{M}_{\beta,n}$ for $\beta \in Q$ and $n \in \mathbb{Z}_{\geq 0}$ with the irreducible components of Hilb_N^Γ for $N \in \mathbb{Z}_{\geq 0}$. The space $\mathcal{K}$ has a basis given by the classes $[I_\mu]^\Gamma$ of $\mathbb{T}$-fixed points $I_\mu \in \mathrm{Hilb}_{|\mu|}^\Gamma$ for $\mu \in \mathbb{Y}$. These classes form a weight basis of $\mathcal{K}$ under the U_r^{tor} action:

$$\psi_i^\pm(z)[I_\mu]^\Gamma = \prod_{(a,b)\in A_i(\mu)} \phi(q^{-a}t^{-b}qtz)^{-1} \prod_{(a,b)\in R_i(\mu)} \phi(q^{-a}t^{-b}z)[I_\mu]^\Gamma. \tag{5.8}$$

In particular, $[I_\varnothing]^\Gamma$ is a highest weight vector in $\mathcal{K}$.

By the uniqueness of irreducible highest weight U_r^{tor}-modules (cf. [**FJMM1**]), the modules $\mathcal{K}$ and $\mathcal{F}_1^\dagger$ must be isomorphic. Moreover, the joint spectrum $(\psi_i^\pm(z))_{i\in I}$ is simple in these modules, as the explicit formula (5.5) (or (5.8)) shows. Thus, for any isomorphism $\Phi : \mathcal{K} \to \mathcal{F}_1^\dagger$ of U_r^{tor}-modules, it must be true that

$$\Phi([I_\mu]^\Gamma) \in \mathbb{F}^\times|\mu\rangle \subset \mathcal{F}_1^\dagger, \quad \text{for } \mu \in \mathbb{Y}. \tag{5.9}$$

5.3. Vertex representation. Wreath Macdonald eigenoperators act in the vertex reresentation of U_r^{tor} [**S**], as we now describe.

Recall that Q denotes the root lattice of $\mathfrak{sl}_r$ and that $\overline{\alpha}_i = \mathrm{cl}(\alpha_i)$ for all $i \in I$. Let $\langle\cdot,\cdot\rangle : Q^\vee \times Q \to \mathbb{Z}$ denote the canonical pairing.

5.3.1. *Skew group algebra.* Let $\mathbb{Z}\{Q\} = \bigoplus_{\alpha\in Q}\mathbb{Z}e^{\alpha}$. We equip $\mathbb{Z}\{Q\}$ with a noncommutative product as follows. For $\alpha = \sum_{i=1}^{r-1} m_i\alpha_i$ and $\beta = \sum_{j=1}^{r-1} n_j\alpha_j$, define

$$e^{\alpha}e^{\beta} = \mathrm{s}(\alpha,\beta)e^{\alpha+\beta} \tag{5.10}$$

where

$$\mathrm{s}(\alpha,\beta) = (-1)^{\sum_{j=1}^{r-2} m_{j+1}n_j}. \tag{5.11}$$

One verifies directly that (5.10), extended by linearity, gives $\mathbb{Z}\{Q\}$ the structure of a ring with $1 = e^0$.

The following are easily verified using the above definition:

LEMMA 5.9. *For all $i,j\in I$, we have $e^{\alpha_i}e^{\alpha_j} = (-1)^{a_{ij}}e^{\alpha_j}e^{\alpha_i}$.*

LEMMA 5.10. *For any $\alpha\in Q$, we have $(e^{\alpha})^{-1} = \mathrm{s}(\alpha)e^{-\alpha}$ where $\mathrm{s}(\alpha) := \mathrm{s}(\alpha,\alpha)$. We have $\mathrm{s}(\overline{\alpha}_i) = 1$ for $i\neq 0$, and $\mathrm{s}(\overline{\alpha}_0) = (-1)^r$.*

Let $\mathbb{F}\{Q\} = \mathbb{Z}\{Q\}\otimes_{\mathbb{Z}}\mathbb{F}$.

5.3.2. *Vertex representation.* The vertex representation of U_r^{tor} has underlying vector space

$$\mathcal{V} = \Lambda_{\mathbb{F}}^{\otimes I}\otimes_{\mathbb{F}}\mathbb{F}\{Q\}$$

where $\Lambda_{\mathbb{F}}^{\otimes I}$ is the I-fold tensor product of $\Lambda_{\mathbb{F}}$ with itself over $\mathbb{F}$. In preparation for the definition of the U_r^{tor}-action on $\mathcal{V}$, we define linear maps $L_\alpha, p^{\partial_{\alpha_i}} : \mathbb{F}\{Q\}\to\mathbb{F}\{Q\}$ and $z^{H_{i,0}} : \mathbb{F}\{Q\}\to\mathbb{F}\{Q\}[z^{\pm1}]$ as follows:

$$\begin{aligned} L_\alpha e^{\beta} &= e^{\alpha}e^{\beta}\\ p^{\partial_{\alpha_i}}e^{\beta} &= s^{2\langle\overline{\alpha}_i^\vee,\beta\rangle}e^{\beta}\\ z^{H_{i,0}}e^{\beta} &= z^{\langle\overline{\alpha}_i^\vee,\beta\rangle}u^{\langle\overline{\alpha}_i^\vee,M_i\beta\rangle}e^{\beta}\end{aligned}$$

where $\alpha,\beta\in Q$, $i\in I$, and for $\beta = \sum_{j=1}^{r-1} n_j\alpha_j$:

$$M_i\beta = n_{i-1}\overline{\alpha}_{i-1} - n_{i+1}\overline{\alpha}_{i+1}$$

with $n_0 = 0$ and, as usual, subscripts taken mod r.

We also introduce the vertex operators $E^{(i)}(z), F^{(i)}(z)\in\mathrm{End}(\Lambda^{\otimes I})[[z^{\pm1}]]$ given by:

$$\begin{aligned} E^{(i)}(z) &= \sum_{k\in\mathbb{Z}} E_k^{(i)}z^{-k} = \Omega[s^{-1}zX^{(i)}]\Omega[-sz^{-1}Y^{(i)}(X,s,u)]^{\perp}\\ F^{(i)}(z) &= \sum_{k\in\mathbb{Z}} F_k^{(i)}z^{-k} = \Omega[-szX^{(i)}]\Omega[s^{-1}z^{-1}Y^{(i)}(X,s^{-1},u)]^{\perp}\end{aligned}$$

where

$$\begin{aligned} Y^{(i)}(X,s,u) &= (1+s^{-4})X^{(i)} - s^{-2}u^2X^{(i+1)} - s^{-2}u^{-2}X^{(i-1)}\\ &= (1+q^{-1}t^{-1})X^{(i)} - t^{-1}X^{(i+1)} - q^{-1}X^{(i-1)}\\ Y^{(i)}(X,s^{-1},u) &= (1+s^4)X^{(i)} - s^2u^2X^{(i+1)} - s^2u^{-2}X^{(i-1)}\\ &= (1+qt)X^{(i)} - qX^{(i+1)} - tX^{(i-1)}.\end{aligned}$$

REMARK 5.11. The vertex operators $E^{(i)}(z)$ and $F^{(i)}(z)$ are natural wreath analogs of $\tilde{D}^*(z)$ and $\tilde{D}(z)$ from (2.13) and (2.12). However, to obtain wreath Macdonald eigenoperators, it is necessary to combine various coefficients of $E^{(i)}(z)$ and $F^{(i)}(z)$ for *all* $i \in I$ (see Theorem 5.18 and Example 5.16 below).

Now we are ready to describe Saito's vertex representation of U_r^{tor}:

THEOREM 5.12 ([**S**]). *The following formulas define an action of U_r^{tor} on $\mathcal{V}$:*

$$\begin{aligned}
C_v^{1/2} &\mapsto s \\
C_h^{1/2} &\mapsto 1 \\
e_i(z) &\mapsto E^{(i)}(z) \otimes L_{\overline{\alpha}_i} z^{1+H_{i,0}} \\
f_i(z) &\mapsto F^{(i)}(z) \otimes \mathrm{s}(\overline{\alpha}_i) L_{-\overline{\alpha}_i} z^{1-H_{i,0}} \\
\psi_i^-(z) &\mapsto \Omega[-z(p-p^{-1})X^{(i)}] \otimes p^{-\partial_{\alpha_i}} \\
\psi_i^+(z) &\mapsto \Omega[z^{-1}(p^2-1)Y^{(i)}(X,s,u)]^\perp \otimes p^{\partial_{\alpha_i}}
\end{aligned}$$

for all $i \in I$. Moreover, $\mathcal{V}$ is an irreducible U_r^{tor}-module.

On the subspaces $\mathcal{V}_\alpha = \Lambda_{\mathbb{F}}^{\otimes I} \otimes \mathbb{F}e^\alpha$, the action of U_r^{tor} is given more explicitly as follows:

$$\begin{aligned}
e_i(z)\Big|_{\mathcal{V}_\alpha} &= \left(z^{1+\langle \overline{\alpha}_i^\vee, \alpha\rangle} u^{\langle \overline{\alpha}_i^\vee, M_i\alpha\rangle} E^{(i)}(z) \otimes L_{\overline{\alpha}_i}\right)\Big|_{\mathcal{V}_\alpha} \\
e_{i,k}\Big|_{\mathcal{V}_\alpha} &= \left(u^{\langle \overline{\alpha}_i^\vee, M_i\alpha\rangle} E^{(i)}_{k+1+\langle \overline{\alpha}_i^\vee, \alpha\rangle} \otimes L_{\overline{\alpha}_i}\right)\Big|_{\mathcal{V}_\alpha} \\
f_i(z)\Big|_{\mathcal{V}_\alpha} &= \left(z^{1-\langle \overline{\alpha}_i^\vee, \alpha\rangle} u^{-\langle \overline{\alpha}_i^\vee, M_i\alpha\rangle} F^{(i)}(z) \otimes \mathrm{s}(\overline{\alpha}_i) L_{-\overline{\alpha}_i}\right)\Big|_{\mathcal{V}_\alpha} \\
f_{i,k}\Big|_{\mathcal{V}_\alpha} &= \left(u^{-\langle \overline{\alpha}_i^\vee, M_i\alpha\rangle} F^{(i)}_{k+1-\langle \overline{\alpha}_i^\vee, \alpha\rangle} \otimes \mathrm{s}(\overline{\alpha}_i) L_{-\overline{\alpha}_i}\right)\Big|_{\mathcal{V}_\alpha} \\
h_{i,-k}\Big|_{\mathcal{V}_\alpha} &= s^{-2\langle \overline{\alpha}_i^\vee, \alpha\rangle} \frac{p_k[X^{(i)}(s^2-s^{-2})]}{k(s^2-s^{-2})} \qquad (k>0) \\
h_{i,k}\Big|_{\mathcal{V}_\alpha} &= s^{2\langle \overline{\alpha}_i^\vee, \alpha\rangle} \frac{p_k[Y^{(i)}(X,s,u)(s^4-1)]^\perp}{k(s^2-s^{-2})} \qquad (k>0).
\end{aligned}$$

5.3.3. *Miki twist.* The vertex representation is not a highest (or lowest) weight representation in the ordinary sense. It is a miraculous fact that its twist by the Miki automorphism is a highest weight representation.

More precisely, let $\rho_{\mathcal{V}} : U_r^{\mathrm{tor}} \to \mathrm{End}(\mathcal{V})$ be Saito's vertex representation. Its Miki twist $\mathcal{V}^\varsigma$ is given on the same representation space $\mathcal{V}^\varsigma = \mathcal{V}$ but with the action of U^{tor} determined by the homomorphism $\rho_{\mathcal{V}^\varsigma} = \rho_{\mathcal{V}} \circ \varsigma$.

THEOREM 5.13 ([**T**]). *The Miki-twisted vertex representation $\mathcal{V}^\varsigma$ is an irreducible highest weight U_r^{tor}-module with highest weight vector $1 \otimes e^0$ and highest weight $(\phi(\xi_\circ^\dagger/z)^{\delta_{i0}})_{i\in I}$ where $\xi_\circ^\dagger = (-1)^r s^2 u^{-r}$.*

REMARK 5.14. The power of u in our constant $\xi^\dagger$ comes from a correction to [**T**] made by [**W1**]. The sign in $\xi_\circ^\dagger$ is slightly different from that of [**T**, **W1**] and it results from our choice of 2-cocycle (5.11) used to define the product in $\mathbb{F}\{Q\}$.

REMARK 5.15. A precursor to Theorem 5.13 was given in [**STU**], where it was shown that the U^v-action in $\mathcal{V}$ agrees with the U^h-action in $\mathcal{F}_\xi^\dagger$ (which is independent of ξ).

Recalling now the highest weight Fock representation $\mathcal{F}_\xi^\dagger$ and the uniqueness of irreducible highest weight modules, we deduce that $\mathcal{V}^\varsigma \cong \mathcal{F}_{\xi_\circ}^\dagger$ as $U_{r,0}^{\mathrm{tor}}$-modules. In particular, taking into account the action of $\mathfrak{H}^v$ in $\mathcal{F}_{\xi_\circ}^\dagger$ given by Corollary 5.6, Theorem 5.13 implies the following remarkable fact: the horizontal Heisenberg subalgebra $\mathfrak{H}^h = \varsigma(\mathfrak{H}^v)$ acts diagonally in Saito's vertex representation $\mathcal{V}$, with one-dimensional joint eigenspaces and eigenvalues given by (5.6).

5.3.4. *Eigenoperators.* We now consider operators on $\mathcal{V}$ arising from the action of particular elements of $\mathfrak{H}^h$, namely the following scalar multiples of $\rho_{\mathcal{V}^\varsigma}(h_{i,\pm 1}) = \rho_{\mathcal{V}}(\varsigma(h_{i,\pm 1}))$ (cf. Proposition 5.2):

$$\begin{aligned}\tilde{D}_1^{(i)} &= c_i(u)\hat{D}_1^{(i)}\\ \tilde{D}_1^{(i)*} &= c_i^*(u)\hat{D}_1^{(i)*},\end{aligned}$$

where

$$\hat{D}_1^{(i)} = \begin{cases} \rho_{\mathcal{V}}\left(f_{0,0}\cdot\begin{bmatrix} f_{r-1,0} & \cdots & f_{i+1,0} & f_{1,0} & \cdots & f_{i-1,0} & f_{i,0}\\ p & \cdots & p & p & \cdots & p & p^2\end{bmatrix}\right) & \text{if } i\neq 0\\[2ex] \rho_{\mathcal{V}}\left(f_{1,1}\cdot\begin{bmatrix} f_{2,0} & \cdots & f_{r-1,0} & f_{0,-1}\\ p & \cdots & p & p^2\end{bmatrix}\right) & \text{if } i=0\end{cases}$$

$$\hat{D}_1^{(i)*} = \begin{cases} \rho_{\mathcal{V}}\left(\begin{bmatrix} e_{i,0} & e_{i-1,0} & \cdots & e_{1,0} & e_{i+1,0} & \cdots & e_{r-1,0}\\ p^{-2} & p^{-1} & \cdots & p^{-1} & p^{-1} & \cdots & p^{-1}\end{bmatrix}\cdot e_{0,0}\right) & \text{if } i\neq 0\\[2ex] \rho_{\mathcal{V}}\left(\begin{bmatrix} e_{0,1} & e_{r-1,0} & \cdots & e_{2,0}\\ p^{-2} & p^{-1} & \cdots & p^{-1}\end{bmatrix}\cdot e_{1,-1}\right) & \text{if } i=0\end{cases}$$

and

$$c_i(u) = \begin{cases} (-1)^{r-i-1}u^{r-2i} & \text{if } i\neq 0\\ u^{2-r} & \text{if } i=0\end{cases}$$

$$c_i^*(u) = \begin{cases} (-1)^{r-i-1}u^{2i-r} & \text{if } i\neq 0\\ u^{r-2} & \text{if } i=0,\end{cases}$$

and where we take $0\leq i\leq r-1$ throughout. Inspecting the eigenvalues given by (5.6), one sees that joint eigenspaces of $\{D_1^{(i)}\}_{i\in I}$ (or $\{D_1^{(i)*}\}_{i\in I}$) alone are one-dimensional, i.e., these sets of operators are sufficient to uniquely characterize the joint eigenfunctions of $\mathfrak{H}^h$ in $\mathcal{V}$.

EXAMPLE 5.16. For $r=3$, we have

$$D_1^{(1)*} = -u^{-1}\rho_{\mathcal{V}}([e_{1,0},[e_{2,0},e_{0,0}]_{p^{-1}}]_{p^{-2}}).$$

The iterated commutator expands into

$$e_{1,0}e_{2,0}e_{0,0} - p^{-1}e_{1,0}e_{0,0}e_{2,0} - p^{-2}e_{2,0}e_{0,0}e_{1,0} + p^{-3}e_{0,0}e_{2,0}e_{1,0}$$

and its summands act, for instance, as follows in $\mathcal{V}$:

$$\begin{aligned}&e_{1,0}e_{2,0}e_{0,0}\cdot(f\otimes e^\alpha)\\ &= u^{m_{1,2,0}(\alpha)}(E^{(1)}_{1+\langle\overline{\alpha}_1^\vee,\overline{\alpha}_2+\overline{\alpha}_0+\alpha\rangle}E^{(2)}_{1+\langle\overline{\alpha}_2^\vee,\overline{\alpha}_0+\alpha\rangle}E^{(0)}_{1+\langle\overline{\alpha}_0^\vee,\alpha\rangle}\cdot f)\otimes e^{\overline{\alpha}_1}e^{\overline{\alpha}_2}e^{\overline{\alpha}_0}e^\alpha,\end{aligned}$$

where $m_{1,2,0}(\alpha) = \langle \overline{\alpha}_1^\vee, M_1(\overline{\alpha}_2 + \overline{\alpha}_0 + \alpha)\rangle + \langle \overline{\alpha}_2^\vee, M_2(\overline{\alpha}_0 + \alpha)\rangle + \langle \overline{\alpha}_0^\vee, M_0\alpha\rangle$ and the product in $\mathbb{F}\{Q\}$ simplifies as $e^{\overline{\alpha}_1}e^{\overline{\alpha}_2}e^{\overline{\alpha}_0}e^{\alpha} = -e^{\alpha}$ since $\mathrm{s}(\alpha_2, \alpha_0) = -1$.

5.3.5. *Eigenfunctions are wreath Macdonald polynomials.* By analogy with Theorem 2.12, it is natural to hope that the joint eigenfunctions of $\mathfrak{H}^h$ in $\mathcal{V}$ are related to wreath Macdonald polynomials. Wen [**W1**] proved that this is indeed the case, giving a spectacular extension of the Macdonald eigenoperators from Corollary 2.9 to the wreath setting.

For any $\mu \in \mathbb{Y}$, define a vector $\tilde{H}_\mu \in \mathcal{V}$ by

$$\tilde{H}_\mu = \tilde{H}_{\mu^\bullet}^{t_{-\beta^\vee}} \otimes e^\beta$$

where $\mu = \tau(\mu^\bullet, \beta)$.

EXAMPLE 5.17. Let $r = 3$ and $\mu = (3,3,2,2)$. We have $\overline{\kappa}(\mu) = \beta = -\overline{\alpha}_2$, $\mathrm{quot}_3(\mu) = (\boxminus, \cdot, \cdot)$. So $\tilde{H}_\mu = \tilde{H}^{t_{\overline{\alpha}_2}}_{(\boxminus,\cdot,\cdot)} \otimes e^{-\overline{\alpha}_2}$.

THEOREM 5.18 ([**W1**]). *The vectors $\{\tilde{H}_\mu\}_{\mu\in\mathbb{Y}}$ form a joint eigenbasis for the horizontal Heisenberg algebra acting in $\mathcal{V}$. In particular, we have $\mathcal{V} = \bigoplus_{\mu\in\mathbb{Y}} \mathbb{F}\tilde{H}_\mu$ and*

$$\tilde{D}_1^{(i)} \cdot \tilde{H}_\mu = A_\mu^{(i)}(q^{-1}, t^{-1})\tilde{H}_\mu \tag{5.12}$$

$$\tilde{D}_1^{(i)*} \cdot \tilde{H}_\mu = A_\mu^{(i)}(q, t)\tilde{H}_\mu \tag{5.13}$$

for all $i \in I$ and $\mu \in \mathbb{Y}$.

REMARK 5.19. Just as in (5.9), Theorem 5.18 implies that the isomorphism $\mathcal{F}^\dagger_{\xi_\circ} \cong \mathcal{V}^\varsigma$ maps the joint eigenspace $\mathbb{K}|\mu\rangle$ isomorphically to $\mathbb{K}\tilde{H}_\mu$. However, it does not send $|\mu\rangle$ to $\tilde{H}_\mu$ in general. The isomorphism $\mathcal{F}^\dagger_{\xi_\circ} \cong \mathcal{V}^\varsigma$ is unique up to a single nonzero scalar, while the images of $\{|\mu\rangle\}_{\mu\in\mathbb{Y}}$ differ from $\{\tilde{H}_\mu\}_{\mu\in\mathbb{Y}}$ by a nonconstant function of μ which appears to be quite subtle.

REMARK 5.20. We have made the assumption $r \geq 3$ throughout this section for simplicity of exposition and in order to match the existing literature. However, we have verified by computation that (5.13) and (5.12) from Wen's Theorem continue to hold as stated in the case $r = 2$. In the $r = 1$ case, one can make sense of the above formulas by taking $Q = 0$ and $\mathbb{F}\{Q\} = \mathbb{F}e^0$, and then the images of $e_{0,-1}$ and $f_{0,-1}$ under $\rho_{\mathcal{V}}$ are exactly the operators $\tilde{D}_0^*$ and $\tilde{D}_0^*$ from Corollary 2.9.

5.4. Further implications and related results. The toolkit for working with wreath Macdonald polynomials provided by Theorem 5.18 goes beyond the eigenoperators of (5.12) and (5.13).

5.4.1. *Pieri formulas.* The vertical Heisenberg algebra $\mathfrak{H}^v$ acts in $\mathcal{V}$ by symmetric function multiplication and skewing operators. Thus, in light of the isomorphism $\mathcal{F}^\dagger_{\xi_\circ} \cong \mathcal{V}^\varsigma$, one may use the action of $\varsigma^{-1}(\mathfrak{H}^v)$ in $\mathcal{F}^\dagger_{\xi_\circ}$ to obtain Pieri formulas for the $\tilde{H}_\mu$. However, a major technical challenge is that, while the isomorphism $\mathcal{F}^\dagger_{\xi_\circ} \cong \mathcal{V}^\varsigma$ identifies the eigenspaces $\mathbb{K}|\mu\rangle \cong \mathbb{K}\tilde{H}_\mu$, the explicit scalars relating these two bases are not known (see Remark 5.19). For more detail and recent progress in this direction, see [**W2**].

5.4.2. *Wreath Macdonald difference operators.* One may consider the wreath Macdonald polynomials specialized to finitely many variables. In [**OS**, **OSW**], we use the shuffle algebra realization of U_r^{tor} [**Neg2**] to construct q-difference operators acting diagonally on such a specialization of the wreath Macdonald polynomials $P_{\mu^\bullet}^{t_{\beta^\vee}}$. These operators directly generalize the Macdonald difference operator M_N from (2.3). A key point is that the shuffle algebra provides access to the higher degree Miki images $\varsigma(h_{i,\pm l})$ and their action in $\mathcal{V}$ [**T**, **W1**].

5.4.3. *Dual eigenoperators.* As explained to us by J. Wen, one can show that the twist of $\mathcal{V}$ by ς^{-1} is a *lowest* weight representation of U_r^{tor} isomorphic to $\mathcal{F}_\xi$ for a particular ξ. From this point of view, one obtains eigenoperators for the functions $\operatorname{swap}\operatorname{inv}\omega\tilde{H}_{\mu^\bullet}^{t_{-\beta^\vee}}$ by considering the action of the Heisenberg algebra $\varsigma^{-1}(\mathfrak{H}^v)$ (which is different from $\mathfrak{H}^h$) in $\mathcal{V}$.

Acknowledgements

We thank Mark Haiman for helpful discussions and for sharing his unpublished results on wreath Macdonald polynomials [**H3**]. We also thank Joshua Wen for many fruitful discussions and related collaborations. D.O. gratefully acknowledges support from the Simons Foundation (Collaboration Grant for Mathematicians, 638577) and the Max Planck Institute for Mathematics (MPIM Bonn).

References

[BF] R. Bezrukavnikov and M. Finkelberg, *Wreath Macdonald polynomials and the categorical McKay correspondence*, Camb. J. Math. **2** (2014), no. 2, 163–190, DOI 10.4310/CJM.2014.v2.n2.a1. With an appendix by Vadim Vologodsky. MR3295916

[BK] R. V. Bezrukavnikov and D. B. Kaledin, *McKay equivalence for symplectic resolutions of quotient singularities* (Russian, with Russian summary), Tr. Mat. Inst. Steklova **246** (2004), no. Algebr. Geom. Metody, Svyazi i Prilozh., 20–42; English transl., Proc. Steklov Inst. Math. **3(246)** (2004), 13–33. MR2101282

[BHMPS] J. Blasiak, M. Haiman, J. Morse, A. Pun, G. Seelinger, *Dens, nests and the Loehr-Warrington conjecture*, arXiv:2112.07070

[BS] I. Burban and O. Schiffmann, *On the Hall algebra of an elliptic curve, I*, Duke Math. J. **161** (2012), no. 7, 1171–1231, DOI 10.1215/00127094-1593263. MR2922373

[C] I. Cherednik, *Double affine Hecke algebras*, London Mathematical Society Lecture Note Series, vol. 319, Cambridge University Press, Cambridge, 2005, DOI 10.1017/CBO9780511546501. MR2133033

[CG] N. Chriss and V. Ginzburg, *Representation theory and complex geometry*, Birkhäuser Boston, Inc., Boston, MA, 1997. MR1433132

[FJMM1] B. Feigin, M. Jimbo, T. Miwa, and E. Mukhin, *Representations of quantum toroidal* $\mathfrak{gl}_n$, J. Algebra **380** (2013), 78–108, DOI 10.1016/j.jalgebra.2012.12.029. MR3023228

[FJMM2] B. Feigin, M. Jimbo, T. Miwa, and E. Mukhin, *Branching rules for quantum toroidal* $\mathfrak{gl}_n$, Adv. Math. **300** (2016), 229–274, DOI 10.1016/j.aim.2016.03.019. MR3534833

[FT] B. L. Feigin and A. I. Tsymbaliuk, *Equivariant K-theory of Hilbert schemes via shuffle algebra*, Kyoto J. Math. **51** (2011), no. 4, 831–854, DOI 10.1215/21562261-1424875. MR2854154

[GH] A. M. Garsia and M. Haiman, *A remarkable q, t-Catalan sequence and q-Lagrange inversion*, J. Algebraic Combin. **5** (1996), no. 3, 191–244, DOI 10.1023/A:1022476211638. MR1394305

[GHT] A. M. Garsia, M. Haiman, and G. Tesler, *Explicit plethystic formulas for Macdonald q, t-Kostka coefficients*, Sém. Lothar. Combin. **42** (1999), Art. B42m, 45 pp. The Andrews Festschrift (Maratea, 1998). MR1701592

[GT] A. M. Garsia and G. Tesler, *Plethystic formulas for Macdonald q, t-Kostka coefficients*, Adv. Math. **123** (1996), no. 2, 144–222, DOI 10.1006/aima.1996.0071. MR1420484

[GKV] V. Ginzburg, M. Kapranov, and É. Vasserot, *Langlands reciprocity for algebraic surfaces*, Math. Res. Lett. **2** (1995), no. 2, 147–160, DOI 10.4310/MRL.1995.v2.n2.a4. MR1324698

[G] I. G. Gordon, *Quiver varieties, category $\mathcal{O}$ for rational Cherednik algebras, and Hecke algebras*, Int. Math. Res. Pap. IMRP **3** (2008), Art. ID rpn006, 69 pp. MR2457847

[GN] E. Gorsky and A. Neguţ, *Refined knot invariants and Hilbert schemes* (English, with English and French summaries), J. Math. Pures Appl. (9) **104** (2015), no. 3, 403–435, DOI 10.1016/j.matpur.2015.03.003. MR3383172

[H1] M. Haiman, *Hilbert schemes, polygraphs and the Macdonald positivity conjecture*, J. Amer. Math. Soc. **14** (2001), no. 4, 941–1006, DOI 10.1090/S0894-0347-01-00373-3. MR1839919

[H2] M. Haiman, *Combinatorics, symmetric functions, and Hilbert schemes*, Current developments in mathematics, 2002, Int. Press, Somerville, MA, 2003, pp. 39–111. MR2051783

[H3] M. Haiman. Personal communication, 2019.

[Kac] V. G. Kac, *Infinite-dimensional Lie algebras*, 3rd ed., Cambridge University Press, Cambridge, 1990, DOI 10.1017/CBO9780511626234. MR1104219

[Kir] A. Kirillov Jr., *Quiver representations and quiver varieties*, Graduate Studies in Mathematics, vol. 174, American Mathematical Society, Providence, RI, 2016, DOI 10.1090/gsm/174. MR3526103

[Kuz] A. Kuznetsov, *Quiver varieties and Hilbert schemes* (English, with English and Russian summaries), Mosc. Math. J. **7** (2007), no. 4, 673–697, 767, DOI 10.17323/1609-4514-2007-7-4-673-697. MR2372209

[KMS] M. Kashiwara, T. Miwa, and E. Stern, *Decomposition of q-deformed Fock spaces*, Selecta Math. (N.S.) **1** (1995), no. 4, 787–805, DOI 10.1007/BF01587910. MR1383585

[LR] N. A. Loehr and J. B. Remmel, *A computational and combinatorial exposé of plethystic calculus*, J. Algebraic Combin. **33** (2011), no. 2, 163–198, DOI 10.1007/s10801-010-0238-4. MR2765321

[L1] I. Losev, *On Procesi bundles*, Math. Ann. **359** (2014), no. 3-4, 729–744, DOI 10.1007/s00208-014-1017-7. MR3231013

[L2] I. Losev, *Procesi bundles and symplectic reflection algebras*, Algebraic and analytic microlocal analysis, Springer Proc. Math. Stat., vol. 269, Springer, Cham, 2018, pp. 3–61, DOI 10.1007/978-3-030-01588-6_1. MR3903316

[LLT] A. Lascoux, B. Leclerc, and J.-Y. Thibon, *Hecke algebras at roots of unity and crystal bases of quantum affine algebras*, Comm. Math. Phys. **181** (1996), no. 1, 205–263. MR1410572

[Mac] I. G. Macdonald, *Symmetric functions and Hall polynomials*, 2nd ed., Oxford Classic Texts in the Physical Sciences, The Clarendon Press, Oxford University Press, New York, 2015. With contribution by A. V. Zelevinsky and a foreword by Richard Stanley; Reprint of the 2008 paperback edition [MR1354144]. MR3443860

[M1] K. Miki, *Toroidal braid group action and an automorphism of toroidal algebra $U_q(\mathrm{sl}_{n+1,\mathrm{tor}})$ $(n \geq 2)$*, Lett. Math. Phys. **47** (1999), no. 4, 365–378, DOI 10.1023/A:1007556926350. MR1693755

[M2] K. Miki, *A (q,γ) analog of the $W_{1+\infty}$ algebra*, J. Math. Phys. **48** (2007), no. 12, 123520, 35, DOI 10.1063/1.2823979. MR2377852

[MM] K. Misra and T. Miwa, *Crystal base for the basic representation of $U_q(\mathfrak{sl}(n))$*, Comm. Math. Phys. **134** (1990), no. 1, 79–88. MR1079801

[Nag1] K. Nagao, *Quiver varieties and Frenkel-Kac construction*, J. Algebra **321** (2009), no. 12, 3764–3789, DOI 10.1016/j.jalgebra.2009.03.012. MR2517812

[Nag2] K. Nagao, *K-theory of quiver varieties, q-Fock space and nonsymmetric Macdonald polynomials*, Osaka J. Math. **46** (2009), no. 3, 877–907. MR2583334

[Nak1] H. Nakajima, *Quiver varieties and Kac-Moody algebras*, Duke Math. J. **91** (1998), no. 3, 515–560, DOI 10.1215/S0012-7094-98-09120-7. MR1604167

[Nak2] H. Nakajima, *Quiver varieties and finite-dimensional representations of quantum affine algebras*, J. Amer. Math. Soc. **14** (2001), no. 1, 145–238, DOI 10.1090/S0894-0347-00-00353-2. MR1808477

[Nak3] H. Nakajima, *Reflection functors for quiver varieties and Weyl group actions*, Math. Ann. **327** (2003), no. 4, 671–721, DOI 10.1007/s00208-003-0467-0. MR2023313

[Neg1] A. Negut, *The shuffle algebra revisited*, Int. Math. Res. Not. IMRN **22** (2014), 6242–6275, DOI 10.1093/imrn/rnt156. MR3283004

[Neg2] A. Neguţ, *Quantum toroidal and shuffle algebras*, Adv. Math. **372** (2020), 107288, 60, DOI 10.1016/j.aim.2020.107288. MR4128572

[OS] D. Orr and M. Shimozono, *Difference operators for wreath Macdonald polynomials*, arXiv:2110.08808

[OSW] D. Orr, M. Shimozono, and J. Wen, *Wreath Macdonald operators*, arXiv:2211.03851

[P] T. Przeździecki, *The combinatorics of $\mathbb{C}^*$-fixed points in generalized Calogero-Moser spaces and Hilbert schemes*, J. Algebra **556** (2020), 936–992, DOI 10.1016/j.jalgebra.2020.04.003. MR4089565

[S] Y. Saito, *Quantum toroidal algebras and their vertex representations*, Publ. Res. Inst. Math. Sci. **34** (1998), no. 2, 155–177, DOI 10.2977/prims/1195144759. MR1617066

[STU] Y. Saito, K. Takemura, and D. Uglov, *Toroidal actions on level 1 modules of $U_q(\widehat{\mathrm{sl}}_n)$*, Transform. Groups **3** (1998), no. 1, 75–102, DOI 10.1007/BF01237841. MR1603798

[SV1] O. Schiffmann and E. Vasserot, *The elliptic Hall algebra, Cherednik Hecke algebras and Macdonald polynomials*, Compos. Math. **147** (2011), no. 1, 188–234, DOI 10.1112/S0010437X10004872. MR2771130

[SV2] O. Schiffmann and E. Vasserot, *The elliptic Hall algebra and the K-theory of the Hilbert scheme of $\mathbb{A}^2$*, Duke Math. J. **162** (2013), no. 2, 279–366, DOI 10.1215/00127094-1961849. MR3018956

[TU] K. Takemura and D. Uglov, *Level-0 action of $U_q(\widehat{\mathfrak{sl}}_n)$ on the q-deformed Fock spaces*, Comm. Math. Phys. **190** (1998), no. 3, 549–583, DOI 10.1007/s002200050252. MR1600311

[T] A. Tsymbaliuk, *Several realizations of Fock modules for toroidal $\ddot{U}_{q,d}(\mathfrak{sl}_n)$*, Algebr. Represent. Theory **22** (2019), no. 1, 177–209, DOI 10.1007/s10468-017-9761-5. MR3908899

[VV1] M. Varagnolo and E. Vasserot, *Double-loop algebras and the Fock space*, Invent. Math. **133** (1998), no. 1, 133–159, DOI 10.1007/s002220050242. MR1626481

[VV2] M. Varagnolo and E. Vasserot, *On the K-theory of the cyclic quiver variety*, Internat. Math. Res. Notices **18** (1999), 1005–1028, DOI 10.1155/S1073792899000525. MR1722361

[W1] B. Ion, *Macdonald polynomials, Demazure modules and positivity*, ProQuest LLC, Ann Arbor, MI, 2002. Thesis (Ph.D.)–Princeton University. MR2703107

[W2] J. Wen. Shuffle approach to wreath Pieri operators. In preparation.

VIRGINIA TECH, DEPARTMENT OF MATHEMATICS, BLACKSBURG, VA USA
Email address: dorr@math.vt.edu

VIRGINIA TECH, DEPARTMENT OF MATHEMATICS, BLACKSBURG, VA USA
Email address: mshimo@math.vt.edu

Contemporary Mathematics
Volume **804**, 2024
https://doi.org/10.1090/conm/804/16114

Meromorphic connections on the projective line with specified local behavior

Daniel S. Sage

ABSTRACT. A meromorphic connection on the complex projective line induces formal connections at each singular point, and these formal connections constitute the local behavior at the singularities. In this primarily expository paper, we discuss the extent to which specified local behavior at singular points determines the global connection. In particular, given a finite set of points and a collection of "formal types" at these points, does there exist a moduli space of meromorphic connections with this local behavior, and if so, when is this moduli space nonempty or a singleton? In this paper, we discuss variants of these problems (for example, the Deligne–Simpson and rigidity problems) as the allowed singularities get progressively more complicated: first connections with only regular singularities, next connections with additional unramified irregular singularities allowed, and finally the general case.

CONTENTS

1. Introduction
2. Connections with only regular singularities
3. Connections with unramified irregular singular points
4. Connections with ramified singular points
5. Meromorphic G-connections
References

1. Introduction

Consider a system of first-order linear ordinary differential equations on the Riemann sphere with rational coefficients:

$$X' + A(z)X = 0, \tag{1.1}$$

where $A(z)$ is an $n \times n$ matrix whose entries are rational functions. If the matrix $A(z)$ is not constant, then this differential equation will have singularities at the finite collection of points where these matrix entries have poles. By expanding the

2020 *Mathematics Subject Classification.* Primary 34M50, 14D05; Secondary 22E67, 34M35, 14D24, 20G25.

Key words and phrases. Deligne–Simpson problem, meromorphic connections, Fuchsian connections, irregular singularities, moduli spaces, parahoric subgroups, fundamental strata, toral connections, rigid connections.

The author received support from Simons Collaboration Grant 637367.

rational functions in Laurent series at each singular point, one obtains a collection of "local" systems of differential equations which encode the local behavior of the original system. A natural question then arises: To what extent does this local behavior determine the global differential equation?

In order to make this question more precise, it is convenient to give a geometric reformulation in terms of the equivalent language of *meromorphic connections.* Let V be a rank n vector bundle over the Riemann sphere $\mathbb{P}^1$ endowed with a meromorphic connection ∇, i.e., a $\mathbb{C}$-linear derivation $V \to V \otimes \Omega^1_{\mathbb{C}(z)/\mathbb{C}}$ taking sections of the vector bundle to sections of the corresponding bundle of one-forms. We will always assume that V is trivializable. Upon fixing a trivialization, the connection may be expressed as $\nabla = d + A(z)dz$, where $A(z) \in \mathfrak{gl}_n(\mathbb{C}(z))$ is a matrix of rational functions. This is the connection corresponding to our original differential equation.

Given $p \in \mathbb{P}^1$, the local behavior of ∇ is governed by the corresponding *formal connection* $\widehat{\nabla}_p$. Let $F = \mathbb{C}((z))$ denote the formal Laurent series, so that $\Delta^\times = \operatorname{Spec}(F)$ is the formal punctured disk. A vector bundle over $\Delta^\times$ is just an F-vector space $\hat{V}$, and a formal connection $\widehat{\nabla}$ is a derivation $\hat{V} \to \hat{V} \otimes \Omega^1_{F/\mathbb{C}}$. One gets the formal connection $\widehat{\nabla}_p$ by expanding the entries of the matrix $A(z)$ as Laurent series in terms of a local parameter at p: $z - p$ if $p \in \mathbb{C}$ and z^{-1} if $p = \infty$.

As usual, we are only interested in connections (and differential equations) up to isomorphism, which is given here by *gauge equivalence.* For ODE's, this means that we do not distinguish between equations which are related by a change of coordinates $Y = gX$, where g is an invertible $n \times n$ matrix whose coefficients are regular functions on the base space, namely $\mathbb{P}^1$ or $\Delta^\times$. Accordingly, g is in $\mathrm{GL}_n(\mathbb{C})$ or $\mathrm{GL}_n(F)$, depending on whether we are in the global or local case. It is an easy calculation to show that our original ODE (1.1) becomes

$$Y' + (gA(z)g^{-1} - g'g^{-1})Y = 0 \tag{1.2}$$

in the new variables. For meromorphic ODE's, the nonlinear gauge term disappears, so that changing coordinates just amounts to conjugating the original matrix $A(z)$.

In terms of connections, we first observe that the set of trivializations of a rank n vector bundle over $\Delta^\times$ is a $\mathrm{GL}_n(F)$-torsor. If $\widehat{\nabla} = d + A\,dz$ with respect to the trivialization ϕ, then the connection operator for the new trivialization $g\phi$ is obtained by conjugating the original operator:

$$g(d + A\,dz)g^{-1} = d + (gA(z)g^{-1} - g'g^{-1})\,dz. \tag{1.3}$$

Similar considerations apply to meromorphic connections.

As we have seen, a meromorphic connection ∇ on $\mathbb{P}^1$ gives rise to a finite set of singular points and associated formal connections at these points. We view the isomorphism classes of these formal connections as the local behavior at the singularities. It is now natural to consider the fiber of this map from global connections to collections of local connections. In particular, we are interested in the following basic questions.

THREE QUESTIONS. Fix points $a_1, \dots, a_k \in \mathbb{P}^1$ and formal connections $\widehat{\nabla}_1, \dots, \widehat{\nabla}_k$.

(1) (Existence) Does there exists a meromorphic connection ∇, regular on $\mathbb{P}^1 \setminus \{a_1, \dots, a_k\}$, such that for all i, the associated formal connection $\widehat{\nabla}_{a_i}$ is isomorphic to $\widehat{\nabla}_i$?

(2) (Uniqueness/Rigidity) If such a meromorphic connection exists, is it unique? If so, it is called *physically rigid*.
(3) (Moduli spaces) Is there a moduli space of meromorphic connections with this specified local behavior?

In this paper, we will discuss variants of these problems as the allowed singularities get progressively more complicated. A formal connection is called *regular singular* if there exists a trivialization for which the connection matrix has a simple pole. Otherwise, the connection is called *irregular singular*.

In §2, we consider meromorphic connections which only have regular singularities. In particular, we describe Crawley–Boevey's solution of the Deligne–Simpson problem for Fuchsian connections. In the following section, we turn to a discussion of connections which also have "unramified" irregular singularities. We explain the construction of moduli spaces of "unramified framable connections" due to Boalch, Hiroe, and Yamakawa [**Boa01**, **HY14**] and Hiroe's solution of the unramified Deligne–Simpson problem. In §4, we discuss ramified irregular singular points with an emphasis on "toral singularities". We construct the moduli space of framable connections where all ramified singular points are toral and formulate the *toral Deligne–Simpson problem*. We then restrict to a class of meromorphic connection on $\mathbb{G}_m$ called *Coxeter connections*–certain generalizations of the well-known Frenkel–Gross and Airy connections. We describe the solution of the Deligne–Simpson problem for Coxeter connections; under additional hypotheses, we determine when they are rigid. In the final section, we discuss generalization of these problems to meromorphic G-connections for complex reductive groups G.

It is a great pleasure to thank Neal Livesay for many helpful comments and suggestions.

2. Connections with only regular singularities

2.1. Monodromy of regular singular connections. It is well-known that the isomorphism class of a regular singular formal connection is determined by its *monodromy*. The monodromy is easy to compute as long as one puts a very mild restriction on the leading term of the connection. Consider the formal connection $\widehat{\nabla} = d + (B_0 + B_1 z + \dots)\frac{dz}{z}$, where $B_k \in \mathfrak{gl}_n(\mathbb{C})$. It is called *nonresonant* if no two eigenvalues of B_0 differ by a nonzero integer.

LEMMA 2.1. If the connection $\widehat{\nabla}$ is nonresonant, then it is formally gauge equivalent to $d + B_0\frac{dz}{z}$.

We briefly recall the proof. It suffices to find $g = g_0 + g_1 z + g_2 z^2 + \dots \in \mathrm{GL}_n(\mathbb{C}[\![z]\!])$ with $g_i \in \mathfrak{gl}_n(\mathbb{C})$ such that $B_0/z = g(\sum_{k=0}^{\infty} B_k z^{k-1})g^{-1} - g'g^{-1}$. The power series g is invertible as long as $g_0 \in \mathrm{GL}_n(\mathbb{C})$. Multiplying through by zg on the right and rearranging gives the following differential equation for g:

$$zg' = g(\sum_{k=0}^{\infty} B_k z^k) - B_0 g. \tag{2.1}$$

If we set $g_0 = I$, the z^0 coefficient on both sides is 0. Equating the coefficients of z^k gives

$$(B_0 + kI)g_k - g_k B_0 = \sum_{i=0}^{k-1} g_i B_{k-i}. \tag{2.2}$$

We now consider the linear map $f_{C,D} : \mathfrak{gl}_n(\mathbb{C}) \to \mathfrak{gl}_n(\mathbb{C})$ given by $X \mapsto CX - XD$, where $C, D \in \mathfrak{gl}_n(\mathbb{C})$. It is a standard fact from matrix theory that this map is an isomorphism if and only if C and D do not have a common eigenvalue [**Was76**, Theorem 4.1]. Applying this result to $C = B_0 + kI$, $D = B_0$, and using the fact that B_0 is nonresonant, we see that we can solve the equations (2.2) recursively to obtain the desired gauge change.

In the nonresonant case, we are now reduced to computing the monodromy of $d + B_0 \frac{dz}{z}$ or equivalently, of the ODE $X' + z^{-1}B_0 X = 0$. A fundamental matrix of this equation is given by a branch of $z^{-B_0} = e^{-B_0 \log z}$. Analytically continuing a local solution at 1 along a counterclockwise loop around 0 gives a new fundamental solution which differs from the first by the monodromy matrix $e^{-2\pi i B_0}$. The monodromy is only determined up to conjugation by a constant matrix; indeed, a constant gauge change of $d + B_0 \frac{dz}{z}$ acts by conjugation on B_0 and hence on the monodromy. Thus, the isomorphism class is given by the conjugacy class of $e^{-2\pi i B_0}$ in $\mathrm{GL}_n(\mathbb{C})$. In particular, the adjoint orbit of the leading term determines the isomorphism class of the formal connection, though different orbits can lead to isomorphic formal connections. We will view this adjoint orbit (or any of its representatives) as a *formal type* of $\widehat{\nabla}$.

2.2. Fuchsian connections. It is difficult to study our three questions for arbitrary meromorphic connections with only regular singularities. Instead, we will restrict attention to *Fuchsian connections*. These are meromorphic connections which have a global representation in which all singularities are manifestly simple poles. We will assume without loss of generality that ∞ is a regular point.

DEFINITION 2.2. A meromorphic connection ∇ is called *Fuchsian* with singular points $a_1, \dots, a_k$ if there exists a trivialization in which

$$\nabla = d + \sum_{i=1}^{k} \frac{A_i}{z - a_i} dz, \tag{2.3}$$

where $A_i \in \mathfrak{gl}_n(\mathbb{C})$.

Since we are assuming that ∞ is not a singular point, the residue theorem implies that $\sum_{i=1}^k A_i = 0$.

The local behavior of a Fuchsian connection is determined by a collection of nonresonant adjoint orbits $\mathcal{O}_1, \dots, \mathcal{O}_k$, where $\mathcal{O}_i$ is the formal type at a_i. A Fuchsian connection has local behavior determined by these formal types if $A_i \in \mathcal{O}_i$ for all i; we let $\mathcal{M}(\mathcal{O}_1, \dots \mathcal{O}_k)$ denote the corresponding moduli space. (The moduli space does not depend on the specific k points in $\mathbb{P}^1$, so we omit the points from the notation.)

PROPOSITION 2.3. Given nonresonant adjoint orbits $\mathcal{O}_1, \dots, \mathcal{O}_k \subset \mathfrak{gl}_n(\mathbb{C})$, the corresponding moduli space of Fuchsian connections is given by

$$\mathcal{M}(\mathcal{O}_1, \dots \mathcal{O}_k) = \{(A_1, \dots, A_k) \in \mathcal{O}_1 \times \dots \times \mathcal{O}_k \mid \sum_{i=1}^{k} A_i = 0\} / \mathrm{GL}_n(\mathbb{C}). \tag{2.4}$$

We remark that the form of the moduli space is very natural from the point of view of symplectic geometry. Adjoint orbits can be viewed as symplectic manifolds via the identification with coadjoint orbits given by the trace form. The action of

$\mathrm{GL}_n(\mathbb{C})$ on the symplectic manifold $\mathscr{O}_1 \times \cdots \times \mathscr{O}_k$ is Hamiltonian with moment map $\mu : (A_1, \ldots, A_k) \mapsto \sum_{i=1}^k A_i$. The moduli space is thus the Hamiltonian reduction

$$(\mathscr{O}_1 \times \cdots \times \mathscr{O}_k) /\!\!/_0 \mathrm{GL}_n(\mathbb{C}) := \mu^{-1}(0)/\mathrm{GL}_n(\mathbb{C}). \tag{2.5}$$

2.3. The Deligne–Simpson problem. While it is easy to find the explicit form for the moduli space (2.4), it is more challenging to understand its geometry. In fact, it is quite difficult to determine when the moduli space is nonempty except for very small values of k (or $n = 1$). If there is only one singular point, then the moduli space is a singleton (consisting of the trivial connection d) if $\mathscr{O}_1 = \{0\}$ and is empty otherwise. If $k = 2$, then the moduli space is empty unless $\mathscr{O}_2 = -\mathscr{O}_1$, in which case it is again a singleton.

To go further, we restrict attention to the *stable moduli space* consisting of irreducible Fuchsian connections. Explicitly, this means that the residue matrices do not have any nonzero proper invariant subspaces in common. Note that for $k = 2$ (and $n \geq 2$), the stable moduli space is always empty. We can now state the (additive) Deligne–Simpson problem:

> *Given adjoint orbits $\mathscr{O}_1, \ldots, \mathscr{O}_k$, determine whether there exists an irreducible k-tuple $(A_1, \ldots, A_k)$ with $A_i \in \mathscr{O}_i$ satisfying $\sum_{i=0}^k A_i = 0$. Equivalently, determine whether there exists an irreducible Fuchsian connection with residues in the prescribed orbits* [**Kos03**].

The original Deligne–Simpson problem was a multiplicative version involving the equation $M_1 \ldots M_k = 1$ with the matrices M_i lying in fixed conjugacy classes of $\mathrm{GL}_n(\mathbb{C})$ [**Sim91**]. In terms of connections, this is the existence problem for irreducible Fuchsian connections with given monodromy classes, and a solution for the additive problem gives rise to a solution for the multiplicative one. The additive Deligne–Simpson problem was formulated by Kostov, who solved it under various genericity hypotheses [**Kos02, Kos03, Kos04**]. A complete solution was given by Crawley-Boevey by reinterpreting the problem in terms of representations of quivers [**CB03**].

We now sketch Crawley-Boevey's solution.

STEP 1. One begins by associating three objects to the collections of orbits: a quiver Q, say with vertex set I, a dimension vector $\alpha \in \mathbb{Z}^I_{\geq 0}$, and a deformation vector $\lambda \in \mathbb{C}^I$. The quiver only depends on the degrees of the minimal polynomials of the orbits while the definition of the dimension vector also involves the sizes of the Jordan blocks. Only the deformation vector depends on the specific eigenvalues appearing in the orbits.

Let $p_i(x) = \prod_{j=1}^{d_i}(x - \eta_{ij})$ be the minimal polynomial of $\mathscr{O}_i$. The quiver Q has a sink labeled by 0 and a path of length $d_i - 1$ leading into it for each orbit. The vertices on the ith path are labeled by $[i, j]$ for $1 \leq j < d_i$.[1]

[1] If $\mathscr{O}_i$ is a scalar matrix, then the path has length 0, and the orbit does not contribute any additional vertices to the quiver.

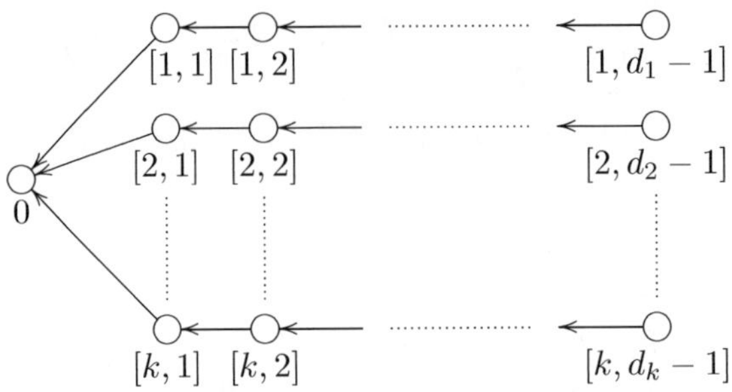

To define the dimension vector α, we take $C_i \in \mathscr{O}_i$, and let r_{ij} be the rank of $\prod_{\ell=1}^{j}(C_i - \eta_{i\ell})$ for $0 \leq j \leq d_i$. We then set $\alpha_0 = n$ and $\alpha_{[i,j]} = r_{i,j-1} - r_{ij}$. Finally, the deformation vector is defined by $\lambda_0 = -\sum_{i=1}^{k} \eta_{i1}$ and $\lambda_{[i,j]} = \eta_{ij} - \eta_{i,j+1}$.

Step 2. Let $\overline{Q}$ be the doubled quiver of Q. This is a quiver on the same vertex set I, but for each arrow a from i to i' in Q, there is a new arrow a^* from i' to i. One can now consider the representations $(V_i)_{i\in I}$ of $\overline{Q}$ with dimension vector α (i.e., $\dim V_i = \alpha_i$) and which satisfy the relations

$$\sum_{h(a)=i} aa^* - \sum_{t(a)=i} a^*a = \lambda_i 1_{V_i}, \tag{2.6}$$

where $h(a)$ and $t(a)$ denote the head and tail of the arrow a. Equivalently, these are the representations of dimension α of the deformed preprojective algebra consisting of the path algebra $\mathbb{C}\overline{Q}$ modulo the ideal generated by the given relations.

Crawley-Boevey shows that there is a bijection between irreducible solutions of the equation $\sum_{i=1}^{k} A_i = 0$ with $A_i \in \mathscr{O}_i$ and irreducible representations of $\overline{Q}$ with dimension vector α and satisfying the deformation conditions (2.6). In fact, this bijection induces a symplectic isomorphism between the corresponding moduli spaces.

Step 3. It remains to determine when such representations of $\overline{Q}$ exist. The solution to this problem is given in terms of the Kac–Moody Lie algebra $\mathfrak{g}_Q$ associated to the quiver Q. A symmetric matrix with integer coefficients C is called a *symmetric generalized Cartan matrix* if its diagonal entries are all 2 and its off-diagonal entries are nonpositive. Such a matrix gives rise to a corresponding Kac–Moody Lie algebra. In particular, one gets $\mathfrak{g}_Q$ from the matrix C_Q defined by

$$(C_Q)_{ij} = 2\delta_{ij} - \#\{\text{edges from } i \text{ to } j\}. \tag{2.7}$$

The root system R of $\mathfrak{g}_Q$ is a certain subset of $\mathbb{Z}^I$. A root is called positive if it lies in $\mathbb{Z}^I_{\geq 0}$, and $R = R_+ \sqcup -R_+$, where R_+ is the set of positive roots.

Let $R_+^\lambda = \{\beta \in R_+ \mid \beta \cdot \lambda = 0\}$. Let $p : \mathbb{Z}^I \to \mathbb{Z}$ be the function defined by $p(\beta) = 1 - (1/2)\beta^t C_Q \beta$. If β is a root, then $p(\beta) \geq 0$. We now let Σ^λ be the set of $\beta \in R_+^\lambda$ such that if $\beta = \sum_{j=1}^{\ell} \gamma_j$ with $\gamma_j \in R_+^\lambda$ and $\ell \geq 2$, then $p(\beta) > \sum_{j=1}^{\ell} p(\gamma_j)$.

Crawley-Boevey showed that there exists an irreducible representation of $\overline{Q}$ of dimension α satisfying (2.6) if and only if $\alpha \in \Sigma^\lambda$. In particular:

Theorem 2.4 ([**CB03**]). *There is an irreducible Fuchsian connection with residues in the $\mathscr{O}_i$'s if and only if $\alpha \in \Sigma^\lambda$.*

2.4. Rigidity. There are several different notions of rigidity for irreducible Fuchsian connections. One can ask whether such a connection is physically rigid, i.e., uniquely determined by the formal isomorphism classes at its singular points. One can also ask whether the moduli space of Fuchsian connections with the given formal types is a singleton. In this section, we will call an irreducible Fuchsian connection rigid if the corresponding stable moduli space is a singleton. Note that a rigid Fuchsian connection may not be physically rigid; for example, there could be a reducible Fuchsian connection with the same formal types.

In [**CB03**], Crawley-Boevey also determined when an irreducible Fuchsian connection is rigid. To explain this, we need to recall a bit more about the root system R of $\mathfrak{g}_Q$.

The coordinate vectors $e_i \in \mathbb{Z}^I$ lie in R; they are called the simple roots of R. The symmetric matrix C_Q defines a bilinear form on $\mathbb{Z}^I$, and so one can define the simple reflection $s_i : \mathbb{Z}^I \to \mathbb{Z}^I$ in the root e_i given by $s_i(\beta) = \beta - (\beta^t C_Q e_i)e_i$. The Weyl group W of $\mathfrak{g}_Q$ is the group generated by the s_i's. A root β is called *real* if $\beta = we_i$ for some $w \in W$ and $i \in I$; otherwise, it is called *imaginary*. For example, if the quiver Q is a Dynkin quiver (i.e., the underlying undirected graph is a simply-laced Dynkin diagram), then all roots are real. The real roots can also be characterized as the roots satisfying $p(\beta) = 0$.

It turns out that if there exists an irreducible Fuchsian connection for the adjoint orbits $\mathscr{O}_1, \dots, \mathscr{O}_k$, then it is rigid if and only if α is a real root. In fact, one can say more.

THEOREM 2.5 ([**CB03**]). Suppose that $\alpha \in \Sigma^\lambda$. Then, the stable moduli space $\mathscr{M}^{\mathrm{st}}(\mathscr{O}_1, \dots, \mathscr{O}_k)$ is a singleton if α is real and is infinite if α is imaginary.

If α is real, then it is more straightforward to check whether $\alpha \in \Sigma^\lambda$: if α is a real positive root, then $\alpha \in \Sigma^\lambda$ if and only if $\alpha \cdot \lambda = 0$ and for any nontrivial decomposition of α into a sum of positive roots $\alpha = \sum_{j=1}^{\ell} \beta_j$, $\beta_k \cdot \lambda \neq 0$ for some k.

2.5. Examples.

EXAMPLE 2.6. Suppose that $k = 2$ (with $n \geq 2$ arbitrary), so that there are two singular points. In this case, the underlying graph of the quiver is the Dynkin quiver $A_{d_1+d_2-1}$. Every positive root of this quiver is a sum of consecutive simple roots, all with multiplicity 1. However, e_0 appears in α with multiplicity $n \geq 2$, so we recover the elementary fact that the stable moduli space is empty in this case.

EXAMPLE 2.7. We now consider the case of three singular points on a rank 2 connection, i.e., $k = 3$, $n = 2$ [**CB03**]. We will show the following:

An irreducible solution exists (and is unique) $\iff$ (a) $\displaystyle\sum_{i=1}^{3} \mathrm{Tr}(\mathscr{O}_i) = 0$; (b) the sum of 3 eigenvalues, one from each $\mathscr{O}_i$, is nonzero.

We denote the eigenvalues of the (nonscalar) adjoint orbits $\mathscr{O}_i$ by η_{11}, η_{12}. Since the minimal polynomial of each orbit has degree 2, the corresponding quiver has its underlying graph given by the D_4 Dynkin diagram. The dimension vector is $\alpha = (2, 1, 1, 1)$, so the quiver representations are given by the diagram below. The deformation vector is

$$\lambda = (-(\eta_{11} + \eta_{21} + \eta_{31}), \eta_{11} - \eta_{12}, \eta_{21} - \eta_{22}, \eta_{31} - \eta_{32}).$$

The D_4 root system has 12 positive roots: e_0, e_1, e_2, e_3, e_0+e_1, e_0+e_2,e_0+e_3, $e_0+e_1+e_2$, $e_0+e_1+e_3$, $e_0+e_2+e_3$, $e_0+e_1+e_2+e_3$, and $2e_0+e_1+e_2+e_3$. Thus, α is a (real) root. The condition that $\alpha\cdot\lambda = 0$ is just the trivial necessary condition that if there exist matrices $A_i \in \mathscr{O}_i$ summing to 0, then their traces sum to 0. Given $J \subset \{1, 2, 3\}$, let $\beta_J = e_0 + \sum_{i\in J} e_i$. Note that $\beta_J \cdot \lambda = -(\sum_{i\in J} \eta_{i2} + \sum_{i\notin J} \eta_{i1})$, so the stated condition on the sum of 3 eigenvalues means precisely that $\beta_J \cdot \lambda \neq 0$ for all J.

Every decomposition of α into the sum of two positive roots is of the form $\alpha = \beta_J + \beta_{J^c}$, so for an irreducible solution to exist, we need either $\beta_J \cdot \lambda \neq 0$ or $\beta_{J^c} \cdot \lambda \neq 0$. However, under the assumption that $\alpha \cdot \lambda = 0$, these are equivalent. Thus, $\alpha \in \Sigma^\lambda$ implies that $\beta_J \cdot \lambda \neq 0$ for all J. On the other hand, if this condition on the β_J's is satisfied, then $\alpha \in \Sigma^\lambda$. Indeed, any nontrivial decomposition of α into a sum of positive roots involves at least one β_J.

3. Connections with unramified irregular singular points

3.1. The Levelt–Turrittin normal form. We now shift our attention from connections with only regular singularities to connections where irregular singular points are allowed. First, we need to understand isomorphism classes of irregular singular formal connections and to get an appropriate notion of formal type. To do this, we recall that there is an analogue of Jordan canonical form for formal connections called the *Levelt–Turrittin (LT) normal form.* This allows one to "upper-triangularize" a formal connection at the cost of possibly introducing roots of the formal parameter z.

THEOREM 3.1 (Levelt–Turrittin normal form [**Tur55, Lev75**]). *Any formal connection is isomorphic to a connection of the form*

$$d + (D_r z^{-r/b} + D_{r-1} z^{-(r-1)/b} + \cdots + D_1 z^{-1} + D_0 + N)\frac{dz}{z}, \tag{3.1}$$

where the D_i's are diagonal, $D_r \neq 0$, N is strictly upper triangular and commutes with each D_i, and $b \in \mathbb{N}$.

Note that the LT normal form does not contain any terms in degrees above the residue term, and it is diagonal except possibly for a nilpotent component to the residue. The rational number r/b is an invariant of the formal connection called its *slope.* It turns out that $\widehat{\nabla}$ is irregular singular if and only if its slope is positive. Moreover, the denominator of the slope in lowest form is at most n. (This does not imply that one can always take $b \leq n$.)

There is a more geometric formulation of the Levelt–Turritin form as the *slope decomposition* of a formal connection. Let $\Delta_b^\times = \operatorname{Spec}(\mathbb{C}((z^{1/b})))$; it is a degree b ramified cover of the formal punctured disk $\Delta^\times$. It follows from this theorem that after passing to this ramified cover, the pullback connection is isomorphic to a finite direct sum

$$\bigoplus (L_i \otimes M_i, \widehat{\nabla}_{L_i} \otimes \widehat{\nabla}_{M_i}), \tag{3.2}$$

where L_i is a line bundle and $\widehat{\nabla}_{M_i}$ is regular singular. One can now consider the collection of nonnegative rational numbers $\operatorname{slope}(\widehat{\nabla}_{L_i})$, each taken with multiplicity $\dim M_i$; these are called the *slopes* of $\widehat{\nabla}$. More concretely, they are the negatives of the valuations of the eigenvalues of the matrix of the LT form.[2] These are also invariants of $\widehat{\nabla}$, and the slope of $\widehat{\nabla}$ is the maximum of the slopes, i.e., $\operatorname{slope}(\widehat{\nabla}) = \max_i(\operatorname{slope}(\widehat{\nabla}_{L_i}))$. The *irregularity* of $\widehat{\nabla}$ is the sum of the slopes; it is a nonnegative integer which is 0 if and only if $\widehat{\nabla}$ is regular singular [**Kat87**].

DEFINITION 3.2.

(1) A formal connection $\widehat{\nabla}$ is called *unramified* if the LT form (or the slope decomposition) may be obtained without passing to a ramified cover (i.e. if one may take $b = 1$). In particular, all of the slopes are integers.
(2) The set of *unramified formal types of slope* r is the collection of matrix one-forms $(D_{-r}z^{-r} + \cdots + D_0 + N)\frac{dz}{z}$ with the D_i's diagonal, $D_{-r} \neq 0$, and N strictly upper triangular and commuting with each D_i. If $r = 0$, we further assume that D_0 is nonresonant.

A regular singular formal connection is an unramified formal connection of slope (or irregularity) 0. The notion of a regular singular unramified formal type is slightly different, but equivalent to the regular singular formal types used in the previous section. We will continue to use the orbit notation $\mathscr{O}$ for regular singular formal types.

It will be helpful in our discussion of the unramified Deligne–Simpson problem below to express unramified formal types in terms of the simultaneous eigenspaces of the purely irregular part. If $\mathcal{D}$ is an unramified formal type of rank n, then there exists a decomposition $V = V_1 \oplus \cdots \oplus V_\ell$ such that

$$\mathcal{D} = \left(\sum_{j=1}^{\ell} (q_j(z)I_{V_j} + R_j)\right)\frac{dz}{z}, \tag{3.3}$$

where $q_j(z) \in z^{-1}\mathbb{C}[z^{-1}]$ with $q_j(z) \neq q_{j'}(z)$ for $j \neq j'$ and $R_j \in \mathfrak{gl}(V_j)$. Note that if $\mathcal{D}$ is regular singular, then $\ell = 1$ and $q_1 = 0$.

3.2. Unramified framable connections. As in the regular singular case, we will not consider all meromorphic connections with specified unramified behavior, but rather will limit ourselves to an irregular version of Fuchsian connections called *framable connections.*

Fix singular points $a_1, \ldots, a_k$ and unramified formal types $\mathcal{D}_1, \ldots, \mathcal{D}_k$ with slopes $r_1, \ldots, r_k$. For convenience, we will assume that ∞ is a regular point. We

[2]The slopes corresponding to 0 eigenvalues are taken to be 0.

will restrict attention to connections of the form

$$d+\left(\sum_{i=1}^{k}\sum_{\nu=0}^{r_i}\frac{A_\nu^{(i)}}{(z-a_i)^\nu}\right)\frac{dz}{z} \tag{3.4}$$

for which $A_{r_i}^{(i)}$ is nonnilpotent if $r_i>0$. We call these connections *framable*. Since ∞ is regular, the residue theorem implies that $\sum_{i=1}^k A_0^{(i)}=0$. Such connections clearly have a singular point of slope r_i at a_i and have no other singular points.

We need to clarify what it means for a framable connection to have unramified formal type $\mathcal{D}_i$ at a_i. Recall that for Fuchsian connections, the principal part reduces to the residue term, and we require the residue to be in the specified adjoint orbit. The analogue of this for irregular unramified formal types is that the principal part of ∇ at a_i must lie in the orbit of $\mathcal{D}_i$ under a certain action of the group $\mathrm{GL}_n([\![z-a_i]\!])$.

Let $\mathcal{B}_r=\{(B_{-r}z^{-r}+\cdots+B_0)\frac{dz}{z}\mid B_i\in\mathfrak{gl}_n(\mathbb{C})\}$ denote the space of principal parts of order at most r. If we let $\mathfrak{o}=\mathbb{C}[\![z]\!]$ be the ring of formal power series, then the group $\mathrm{GL}_n(\mathfrak{o})$ acts on $\mathcal{B}_r$ by conjugation followed by truncation at the residue term. Note that this action factors through the finite-dimensional group $\mathrm{GL}_n(\mathfrak{o}/z^{r+1}\mathfrak{o})$.

One can interpret truncated conjugation as a coadjoint action. Let $\mathfrak{gl}_n(F)^\vee$ be the space of smooth $\mathbb{C}$-linear functionals on $\mathfrak{gl}_n(F)$. Here, smooth means that the functional vanishes on a nonempty bounded open subgroup. Any one-form $B\frac{dz}{z}$ with $B\in\mathfrak{gl}_n(F)$ defines a smooth functional on $\mathfrak{gl}_n(F)$ via $A\mapsto\operatorname{Res}\operatorname{Tr}(AB\frac{dz}{z})$. In fact, if we identify $B\frac{dz}{z}$ with the corresponding functional, the map $B\to B\frac{dz}{z}$ is an isomorphism $\mathfrak{gl}_n(F)\to\mathfrak{gl}_n(F)^\vee$. It is easy to see that $\mathfrak{gl}_n(\mathfrak{o})^\perp=z\,\mathfrak{gl}_n(\mathfrak{o})\frac{dz}{z}$, so $\mathfrak{gl}_n(\mathfrak{o})^\vee=\mathfrak{gl}_n(\mathbb{C}[z^{-1}])\frac{dz}{z}$. Thus, $\mathrm{GL}_n(\mathfrak{o})$ acts on $\mathfrak{gl}_n(\mathbb{C}[z^{-1}])\frac{dz}{z}$ via the coadjoint action with $\mathcal{B}_r$ as a subrepresentation, and this action corresponds to truncated conjugation.

If $\mathcal{D}$ is an unramified formal type of slope r, we call the orbit $\mathscr{O}_\mathcal{D}\subset\mathcal{B}_r$ under this action the *truncated orbit* of $\mathcal{D}$. Since it is a coadjoint orbit, it has a natural symplectic structure. Moreover, the $\mathrm{GL}_n(\mathbb{C})$-action on $\mathscr{O}_\mathcal{D}$ is Hamiltonian with moment map given by the residue. If $\mathcal{D}$ has slope 0, $\mathscr{O}_\mathcal{D}$ may be identified with the usual adjoint orbit of $\mathcal{D}/\frac{dz}{z}$.

The framable connection (3.4) has formal type $\mathcal{D}_i$ at a_i if $\sum_{\nu=0}^{r_i}A_\nu^{(i)}z^{-\nu}\frac{dz}{z}\in\mathscr{O}_{\mathcal{D}_i}$. Accordingly, we get the following expression for the moduli space of framable connections with the given unramified formal types.

THEOREM 3.3 ([**Boa01, HY14**]). The moduli space of framable connections with unramified formal types $\mathcal{D}_i$ at a_i is given by

$$\mathscr{M}(\mathcal{D}_1,\dots,\mathcal{D}_k)=\{(A_1,\dots,A_k)\in\mathscr{O}_{\mathcal{D}_1}\times\cdots\times\mathscr{O}_{\mathcal{D}_k}\mid\sum_{i=1}^k\operatorname{Res}A_i=0\}/\mathrm{GL}_n(\mathbb{C}). \tag{3.5}$$

Equivalently, it is given by Hamiltonian reduction:

$$(\mathscr{O}_{\mathcal{D}_1}\times\cdots\times\mathscr{O}_{\mathcal{D}_k})/\!\!/_0\,\mathrm{GL}_n(\mathbb{C}). \tag{3.6}$$

Note that if all the singular points are regular, we recover the moduli space of Fuchsian connections with specified formal types given in Theorem 2.3. Just as in the Fuchsian case, the location of the singular points is irrelevant.

REMARK 3.4. Boalch constructed this moduli space under the additional hypothesis that the leading term of every irregular unramified formal type is regular semisimple [**Boa01**] while Hiroe and Yamakawa proved the theorem in general [**HY14**].

We end this section by working out this moduli space in the simplest non-Fuchsian case. This is a generalization of an example that appeared in [**Sag17**].

EXAMPLE 3.5. Consider the case of two singular points on a rank 2 connection, i.e., $k = 2$, $n = 2$, with formal types $\mathcal{D}$ unramified of slope 1 and $\mathcal{O}$ regular singular. There are two possible forms for $\mathcal{D}$, depending on whether or not it is diagonalizable:

$$\mathcal{D} = \begin{cases} \left(\operatorname{diag}(a,b)z^{-1} + \operatorname{diag}(c,d)\right)\frac{dz}{z} & a, b \text{ not both } 0; \\ \left((az^{-1}+c)I + e_{12}\right)\frac{dz}{z} & a \neq 0. \end{cases} \tag{3.7}$$

The moduli space is given by

$$\mathscr{M}(\mathcal{D},\mathcal{O}) = \{(\alpha, Y) \in \operatorname{Ad}^*(\mathrm{GL}_2(\mathfrak{o}))(\mathcal{D}) \times \mathcal{O} \mid \operatorname{Res}(\alpha) + Y = 0\}/\operatorname{GL}_2(\mathbb{C}). \tag{3.8}$$

In order to compute the truncated orbit, we will use the fact that $\mathrm{GL}_2(\mathfrak{o}) = \mathrm{GL}_2(\mathbb{C}) \ltimes (1 + t\,\mathfrak{gl}_2(\mathfrak{o}))$.

Case 1: $a \neq b$. We first observe that $\operatorname{Ad}^*(1+z\,\mathfrak{gl}_2(\mathfrak{o}))(\mathcal{D}) = \left\{\begin{pmatrix} az^{-1}+c & u \\ v & bz^{-1}+d\end{pmatrix}\frac{dz}{z}\right\}$ with $u, v \in \mathbb{C}$ arbitrary. Indeed, if $X, X' \in \mathfrak{gl}_2(\mathbb{C})$, then $(1+zX)X'(1+zX)^{-1} \equiv X' + z[X,X'] \pmod{z^2}$, and the claim follows since $\operatorname{ad}(\operatorname{diag}(a,b))(\mathfrak{gl}_2(\mathbb{C}))$ consists of the off-diagonal matrices. Thus,

$$\operatorname{Ad}^*(\mathrm{GL}_2(\mathfrak{o}))(\mathcal{D}) = \operatorname{Ad}^*(\mathrm{GL}_2(\mathbb{C}))\left\{\begin{pmatrix} az^{-1}+c & u \\ v & bz^{-1}+d\end{pmatrix}\frac{dz}{z}\,\middle|\, u, v \in \mathbb{C}\right\}. \tag{3.9}$$

It follows from (3.9) that every $\mathrm{GL}_2(\mathbb{C})$-orbit of pairs (α, Y) has a representative with α in the standard form $\begin{pmatrix} az^{-1}+c & u \\ v & bz^{-1}+d\end{pmatrix}\frac{dz}{z}$ for some $u, v \in \mathbb{C}$. We view the pair (u, v) as coordinates on the set of standard α's. Since the diagonal torus T is the stabilizer of the diagonal component, hence preserves the set of standard forms, it follows that the moduli space is the set of T-orbits of pairs (α, Y) with α in standard form. We claim that

$$|\mathscr{M}(\mathcal{D},\mathcal{O})| = \begin{cases} 3, & c \neq d,\ \det(\mathcal{O}) = cd \text{ and } \operatorname{Tr}(\mathcal{O}) = -c-d; \\ 2, & c = d,\ \det(\mathcal{O}) = cd \text{ and } \operatorname{Tr}(\mathcal{O}) = -c-d; \\ 1, & \det(\mathcal{O}) \neq cd \text{ and } \operatorname{Tr}(\mathcal{O}) = -c-d; \\ 1, & c = d \text{ and } \mathcal{O} = \{-cI\}; \\ 0, & \operatorname{Tr}(\mathcal{O}) \neq -c-d. \end{cases} \tag{3.10}$$

The torus T acts on the set of standard α's via $\operatorname{diag}(s,t)\cdot(u,v) = (st^{-1}u, ts^{-1}v)$. Thus, the T-orbits are parameterized by the pairs $(u, 1)$ with $u \in \mathbb{C}$, $(1, 0)$, and $(0, 0)$. The residue of a standard α can have minimal polynomial of degree 1 only if $c = d$, and then one needs $(u, v) = (0, 0)$. Accordingly, we may now assume that the minimal polynomial of $\mathcal{O}$ is the characteristic polynomial. Since $\operatorname{Res}(\alpha) + Y = 0$, their traces also sum to zero, implying that the moduli space is empty unless $\operatorname{Tr}(\mathcal{O}) = -c - d$. If $\det(\mathcal{O}) \neq cd$, then there is a unique T-orbit representative which equates $\det(\operatorname{Res}(\alpha))$ and $\det(\mathcal{O})$, namely $(cd - \det(\mathcal{O}), 1)$. Thus, in this case, the moduli space is a singleton. On the other hand, if $\det(\mathcal{O}) = cd$, the three orbit representatives $(0, 1)$, $(1, 0)$, and $(0, 0)$ all give rise to a standard α with

$\det(\operatorname{Res}(\alpha)) = \det(\mathscr{O})$. If $c \neq d$, these give the 3 points in the moduli space. If $c = d$, then only the first two give rise to a nonscalar adjoint orbit, so there are only 2 points in the moduli space.

Case 2: $a = b$. In this case, $1 + z\,\mathfrak{gl}_2(\mathfrak{o})$ fixes $\mathcal{D}$, so $\operatorname{Ad}^*(\mathrm{GL}_2(\mathfrak{o}))(\mathcal{D}) = \operatorname{Ad}^*(\mathrm{GL}_2(\mathbb{C}))(\mathcal{D})$ is a single $\mathrm{GL}_2(\mathbb{C})$-orbit. It follows that $\mathscr{M}(\mathcal{D}, \mathscr{O})$ is empty unless $\operatorname{Res}(\mathcal{D}) \in -\mathscr{O}$, in which case the moduli space is a singleton.

3.3. The unramified Deligne–Simpson problem. As in the Fuchsian case, it is not easy to understand the moduli space (3.5). It can be quite badly behaved; in fact, it does not even have to be Hausdorff [**Boa01**]. However, as in the Fuchsian case, the moduli space is much nicer if one restricts to irreducible framable connections. The framable connection (3.4) is irreducible if the collection of matrices $\{A_\nu^{(i)}\}$ have no common nontrivial proper invariant subspace. Hiroe has shown that the stable moduli space consisting of the irreducible framable connections is a connected complex manifold if it nonempty [**Hir17**]. The unramified Deligne–Simpson problem is simply the determination of when this stable moduli space is nonempty.

EXAMPLE 3.6. We return to the situation of Example 3.5. If $a \neq b$, then $\mathscr{M}^{\mathrm{st}}(\mathcal{D}, \mathscr{O})$ is empty except when $\det(\mathscr{O}) \neq cd$ and $\operatorname{Tr}(\mathscr{O}) = -c-d$, in which case it is a singleton. Indeed, in all the cases where $\det(\mathscr{O}) = cd$, there is a representative (α, Y) of each solution with $\alpha = \mathcal{D}$ (hence diagonal) and Y either upper or lower triangular. If $\det(\mathscr{O}) \neq cd$, then the unique solution has a representative with α having leading term $\operatorname{diag}(a,b)z^{-1}\frac{dz}{z}$ and $Y = -\left(\begin{smallmatrix} c & u \\ 1 & d \end{smallmatrix}\right)$ with $u \neq 0$. This means that the two eigenspaces of $\operatorname{diag}(a,b)$ are not invariant under Y. Finally, if $a = b$, then it is trivial that the stable moduli space is empty as the 3 relevant matrices are aI, $\operatorname{Res}(\alpha)$ and $Y = -\operatorname{Res}(\alpha)$.

Hiroe has given a complete solution of the unramified Deligne–Simpson problem, building on earlier work of Boalch [**Boa01**] and Hiroe and Yamakawa [**HY14**]. Just as in the Fuchsian case, Hiroe shows that the existence of an irreducible framable connection with the given formal types is equivalent to the existence of an irreducible quiver representation with certain properties.

The quiver associated to a collection $\boldsymbol{\mathcal{D}} = (\mathcal{D}_1, \ldots, \mathcal{D}_k)$ of unramified types is rather complicated. The set of vertices is partitioned into "base points" and "path points": $I = I_{\mathrm{base}} \sqcup I_{\mathrm{path}}$. The path points arise from the residue terms appearing in the formal types as in (3.3). The base points come from the purely irregular part of (some of) the formal types.

As a warm-up, we describe the quiver $Q^{\mathcal{D}}_{\mathrm{base}}$ for a single purely irregular formal type $\mathcal{D}$ as in (3.3). The vertices are labeled by the simultaneous eigenspaces of $\mathcal{D}$, so $I^{\mathcal{D}}_{\mathrm{base}} = \{1, \ldots, \ell\}$. There are only arrows from j to j' if $j < j'$, in which case we draw $\deg_{z^{-1}}(q_j(z) - q_{j'}(z)) - 1$ arrows.

EXAMPLE 3.7. If $\mathcal{D} = \left(\operatorname{diag}(a,b,b)z^{-3} + \operatorname{diag}(0,c,d)z^{-2}\right)\frac{dz}{z}$ with $a, b, c \neq 0$, $a \neq b$, and $c \neq d$, then the corresponding quiver is

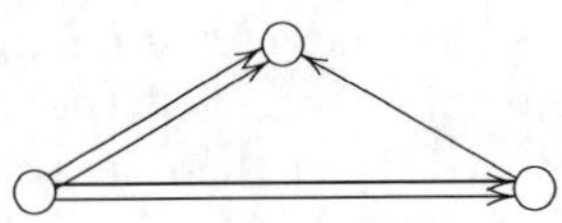

The following theorem explains the significance of this quiver in studying moduli spaces of connections.

THEOREM 3.8 ([**HY14**]). Let $\mathcal{D}$ be a purely irregular formal type, and let $V = (V_1, \dots, V_\ell)$ be the eigenspace decomposition in (3.3). Then there is a symplectic isomorphism between $\mathrm{Ad}^*(1 + z\,\mathfrak{gl}(\mathfrak{o}))(\mathcal{D})$ and $\mathrm{Rep}_{\overline{Q}_{\mathrm{base}}}(V)$.

We now return to constructing Q for $\boldsymbol{\mathcal{D}} = (\mathcal{D}_1, \dots, \mathcal{D}_k)$. In what follows, we call the index i irregular (resp. regular) if $\mathcal{D}_i$ is, and we assume that 0 is irregular. Each $\mathcal{D}_i$ has a block diagonal decomposition, say with ℓ_i blocks, as in (3.3). We now let $I_{\mathrm{base}} = \{[i,j] | i = 0 \text{ or } \ell_i \geq 2,\ 1 \leq j \leq \ell_i\}$ be the disjoint union of the $I_{\mathrm{base}}^{\mathcal{D}_i}$'s where either $i = 0$ or i is irregular with at least 2 blocks. We include all arrows in $Q_{\mathrm{base}}^{\mathcal{D}_i}$ and also draw arrows $[0,j] \to [i,j']$ for every $i \in I_{\mathrm{base}} \setminus \{0\}$, $1 \leq j \leq \ell_0$, $1 \leq j' \leq \ell_i$.

Each block in the decomposition of $\mathcal{D}_i$ gives rise to a residue term, so we obtain a collection of residues R_{ij}. As in the Fuchsian case, each residue R_{ij} has an associated path. If d_{ij} is the degree of its minimal polynomial, then this path consists of the $d_{ij} - 1$ vertices $\{[i,j,k]\}$ with arrows $[i,j,k] \to [i,j,k-1]$. We thus set $I_{\mathrm{path}} = \{[i,j,k]\}$. In the Fuchsian case, the path endpoints are all connected to the single base point 0. Here, the $[i,j,1]$'s will be connected to one or more vertices in I_{base}. If either $i = 0$ or $\ell_i \geq 2$, then there is an arrow $[i,j,1] \to [i,j]$; otherwise, there are arrows $[i,1,1] \to [0,j]$ for $1 \leq j \leq \ell_0$.

The dimension vector α is defined just as in the Fuchsian case in terms of the Jordan block structure of the R_{ij}'s. Accordingly, $\alpha_{[i,j]}$ is the size of the jth block for $\mathcal{D}_i$ while the $\alpha_{[i,j,k]}$'s are defined in terms of partial products of the minimal polynomial of R_{ij}. The deformation vector λ is again defined in terms of the eigenvalues η_{ij}^k of the R_{ij}. For path vertices, it is just the same: $\lambda_{[i,j,k]} = \eta_{ij}^k - \eta_{ij}^{k+1}$. For base vertices, we set $\lambda_{[i,j]} = -\eta_{ij}^1$ if $i \neq 0$ while $\lambda_{[0,j]} = -\eta_{0j}^1 - \sum_{i \neq 0, \ell_i = 1} \eta_{i1}^1$.

We are finally ready to state Hiroe's theorem. Let $L \subset \mathbb{Z}^I$ be the sublattice defined by

$$L = \{\beta \in \mathbb{Z}^I \mid \sum_{j=1}^{\ell_0} \beta_{[0,j]} = \sum_{j=1}^{\ell_i} \beta_{[i,j]} \text{ if } \ell_i \geq 2\}. \tag{3.11}$$

Recall that $p(\beta) = 1 - (1/2)\beta^t C_Q \beta$.

THEOREM 3.9 ([**Hir17**]). There is an irreducible framable connection with formal types $\boldsymbol{\mathcal{D}}$ at the singular points $a_1, \dots, a_k$ if and only if

(1) α is a positive root of Q, and $\alpha \cdot \lambda = 0$;

(2) if $\alpha = \sum_{j=1}^{\ell} \beta_j$ with $\ell > 2$ and $\beta_j \in L \cap \mathbb{Z}^I_{\geq 0}$, then $p(\alpha) > \sum_{j=1}^{\ell} p(\beta_j)$.

In the Fuchsian case, Crawley-Boevey actually shows that there is an isomorphism between $\mathscr{M}^{\mathrm{st}}$ and the moduli space $\mathscr{M}_\lambda^{\mathrm{reg}}(Q, \alpha)$ of irreducible representations of Q with dimension vector α and satisfying the deformation conditions (2.6). This is not true in the general unramified case. Indeed, Boalch has constructed an example where the moduli space of irreducible framable connections is not isomorphic to any quiver variety [**Boa08**]. However, Hiroe show that the stable moduli space is always isomorphic to an open subset of a quiver variety.

THEOREM 3.10 ([**Hir17**]). There exists a vector $\lambda' \in \mathbb{C}^I$, agreeing with λ on components in I_{path}, for which there is an open embedding $\mathscr{M}^{\mathrm{st}} \hookrightarrow \mathscr{M}_{\lambda'}^{\mathrm{reg}}(Q, \alpha)$. If there is at most one irregular unramified type, then $\lambda' = \lambda$ and the open embedding is an isomorphism.

REMARK 3.11. Crawley-Boevey proved that the open embedding is an isomorphism when all formal types are regular singular [**CB03**]. The analogue for one irregular formal type is due to Hiroe and Yamakawa [**HY14**].

4. Connections with ramified singular points

4.1. The Airy and Frenkel–Gross connections. In order to take ramified singularities into account, we need a notion of *ramified formal type*. The LT normal form is not suitable as it is not defined over the ground field F. Instead, we need a *rational canonical form* for formal connections.

Recall that if $\mathcal{D}$ is an unramified formal type of slope r, then $\mathcal{D}$ is an element of $\mathcal{B}_r$, the principal parts of bounded degree r. The space $\mathcal{B}_r$ is defined using the natural degree filtration on the loop algebra and indeed can be identified with $(z^{-r}\,\mathfrak{gl}_n(\mathfrak{o})/z\,\mathfrak{gl}_n(\mathfrak{o}))\,\frac{dz}{z}$. The polynomial loop group $\mathrm{GL}_n(\mathfrak{o})$ acts on $\mathcal{B}_r$, and one then considers the truncated orbit of $\mathcal{D}$ in $\mathcal{B}_r$.

In order to discuss our basic problems in the ramified case, we need analogues of these objects. Thus, we need to associate the following data to an arbitrary singularity:

- a formal type $\mathcal{F}$ giving the rational canonical form $d + \mathcal{F}$;
- a subgroup $P \subset \mathrm{GL}_n(\mathfrak{o})$ with Lie algebra $\mathfrak{p} \subset \mathfrak{gl}_n(\mathfrak{o})$;
- a filtration $(\mathfrak{p}^i)_{i\in\mathbb{Z}}$ of $\mathfrak{gl}_n(F)$ with $\mathfrak{p}^0 = \mathfrak{p}$ which is preserved by conjugation by P;
- a level r in the filtration for which $\mathcal{F}$ may be viewed as an element of $\mathfrak{p}^{-r}/\mathfrak{p}^1\,\frac{dz}{z}$; and
- the orbit of $\mathcal{F}$ in $\mathfrak{p}^{-r}/\mathfrak{p}^1\,\frac{dz}{z}$ under truncated conjugation by P (or equivalently, the P-coadjoint orbit).

We begin by describing two important motivating examples. First, consider the classical Airy equation $y'' = xy$. This is the simplest equation which has a turning point, namely a point where the solutions (the *Airy functions*) change from being oscillatory to exponential. For example, the solution of Schrödinger's equation for a particle confined to a triangular potential well involves Airy functions, reflecting the fact that the wave function of the particle oscillates within the well, but then exhibits quantum tunneling into the wall as it decays exponentially. More generally, one can consider the order n version of the Airy equation: $y^{(n)} = xy$. The Airy equation is nonsingular except at ∞. It will be convenient to make the change of variables $z = x^{-1}$. After converting the equation into matrix form, one sees that the corresponding connection is given by

$$\nabla_{\mathrm{Airy}} = d + z^{-1}\begin{pmatrix} 0 & \cdots & 0 & z^{-1} \\ 1 & 0 & \cdots & 0 \\ \vdots & \ddots & \ddots & \vdots \\ 0 & \cdots & 1 & 0 \end{pmatrix}\frac{dz}{z}. \tag{4.1}$$

Although the matrix of this connection has a pole of order 3, the matrix in the leading term $z^{-2}e_{1n}\frac{dz}{z}$ is nilpotent, so one cannot conclude that the slope at $z = 0$ is 2. In fact, the LT normal form is $d + z^{-\frac{n+1}{n}}\mathrm{diag}(1, \zeta, \dots, \zeta^{n-1})\frac{dz}{z}$, where $\zeta = e^{2\pi i/n}$, and the slope is the common value $\frac{n+1}{n}$ of all the slopes. Thus, ∇_{Airy} is ramified; in fact, since the slope has largest possible denominator (in lowest form), we say that it is *maximally ramified.*

The Airy connection suggests how to construct a meromorphic connection with a maximally ramified singularity of slope $1/n$. Let $\omega_n = \sum_{i=1}^{n-1} e_{i,i+1} + ze_{n,1}$, i.e., ω_n is the matrix with 1's along the superdiagonal, z in the lower-left entry, and 0's elsewhere. Note that $\omega_n^n = z$. With this notation, $\nabla_{\text{Airy}} = d + z^{-1}\omega_n^{-1}\frac{dz}{z}$. We can now define the rank n Frenkel–Gross (or Kloosterman) connection by

$$\nabla_{\text{FG}} = d + \omega_n^{-1}\frac{dz}{z}. \tag{4.2}$$

The corresponding differential equation is hypergeometric and was extensively studied by Katz [**Kat88, Kat90**] (see also Deligne [**Del77**]); it is a geometric incarnation of a certain exponential sum called a Kloosterman sum which arises in number theory. Unlike the Airy connection, it has two singular points, one regular singular and the other irregular. The irregular singular point has slope $1/n$ and LT normal form $d + z^{-1/n}\text{diag}(1, \zeta, \ldots, \zeta^{n-1})\frac{dz}{z}$ while the regular singular point has principal unipotent monodromy. As we will see later, both the Frenkel–Gross and Airy connections are *rigid*, i.e., uniquely determined by their local behavior.

For these two connections, we will view the matrix of the connection in (4.1) and (4.2) as the ramified formal type of the irregular singularity. The group associated to these formal types is the "standard Iwahori subgroup" I, consisting of those $g \in \mathrm{GL}_n(\mathfrak{o})$ which are upper triangular modulo z. The corresponding Iwahori filtration $\{\mathfrak{i}^s\}$ will be defined below, but it will satisfy $\omega_n^s \in \mathfrak{i}^s \setminus \mathfrak{i}^{s+1}$. This means that $\omega^{-1}\frac{dz}{z}$ may be viewed as the leading term of the formal Frenkel–Gross connection $\widehat{\nabla}_{\text{FG}}$ with respect to this filtration, and similarly for the formal Airy connection. The Iwahori filtration is periodic with period n in the sense that $\mathfrak{i}^{s+n} = z\mathfrak{i}^s$, and as we will see, these facts show that the irregular singularities of ∇_{Airy} and ∇_{FG} have slopes $\frac{n+1}{n}$ and $1/n$ respectively, without needing to compute the LT forms.

The matrix ω_n is regular semisimple (though not diagonalizable without extending scalars), and this means that the leading terms of $\widehat{\nabla}_{\text{Airy}}$ and $\widehat{\nabla}_{\text{FG}}$ are regular semisimple as well. Formal connections with "regular semisimple leading term" with respect to an appropriate filtration are called *toral.* To avoid technicalities, we focus here on toral formal types. However, a rational canonical form for arbitrary ramified connections may be obtained from Sabbah's refined Levelt–Turrittin decomposition [**Sab08**]; we will return to this in a future paper.

4.2. Lattice chain filtrations and parahoric subgroups. As we have seen, ramified formal types should be associated to certain filtrations of the loop algebra. For simplicity, we will restrict attention to filtrations arising from "lattice chains"; this will be enough to discuss toral formal types [**BS13, BS20**]. It will also suffice to define a notion of the "leading term" of a formal connection when the naive leading term with respect to the usual valuation is not useful.

An $\mathfrak{o}$-*lattice* L in F^n is a finitely generated $\mathfrak{o}$-module satisfying $L \otimes_{\mathfrak{o}} F \cong F^n$. Concretely, it is the $\mathfrak{o}$-span of some basis of F^n. A *lattice chain* in F^n is a collection $(L^i)_{i\in\mathbb{Z}}$ of lattices satisfying the following properties:

(1) $L^i \supsetneq L^{i+1}$ for all i; and
(2) there exists a positive integer e, called the *period*, such that $L^{i+e} = zL^i$ for all i.

A *parahoric subgroup* $P \subset \mathrm{GL}_n(F)$ is the stabilizer of a lattice chain, i.e., $P = \{g \in \mathrm{GL}_n(F) \mid gL^i = L^i \text{ for all } i\}$. We let e_P denote the period of the lattice chain stabilized by P.

We will be particular interested in parahoric subgroups associated to lattice chains with nice uniformity properties.

DEFINITION 4.1. A lattice chain $(L^i)_{i\in\mathbb{Z}}$ (and the corresponding parahoric subgroup) is called

(1) *uniform* if $e|n$ and $\dim L^i/L^{i-1} = n/e$ for all i;
(2) *almost uniform* if $e|(n-1)$, $e \neq 1$, $\dim L^i/L^{i-1} = 1 + (n-1)/e$ for i in a single residue class modulo e, and $\dim L^i/L^{i-1} = (n-1)/e$ otherwise;
(3) *complete* if its period is as large as possible, i.e., $e = n$.

Note that a complete lattice chain is necessarily uniform. The parahoric subgroups associated to complete lattice chains are called *Iwahori subgroups.*

The group $\mathrm{GL}_n(F)$ acts on the set of lattice chains with finitely many orbits parametrized by subsets of $\mathbb{Z}/n\mathbb{Z}$ modulo the translation action. To see this, first note that any lattice L has a valuation $\nu(L) \in \mathbb{Z}$: if L has an $\mathfrak{o}$-basis $f_1, \dots, f_n$, then the valuation of the determinant of the matrix whose columns are the f_i's is independent of the choice of basis. Since $\nu(zL) = z^n\nu(L)$, it is clear that a lattice chain consists of lattices with valuations determined by some subset $J \subset \mathbb{Z}/n\mathbb{Z}$. If we let X_J be the set of all such lattice chains, then $\mathrm{SL}_n(F)$ acts transitively on X_J. (These are the (partial) affine flag varieties for SL_n [**Sag00**].) On the other hand, the action of $g \in \mathrm{GL}_n(F)$ will change the type of the lattice chain by translating J by $\nu(\det(g))$.

If $0 \in J$, one can choose a lattice chain in X_J whose corresponding parahoric subgroup is block-upper triangular modulo z. Let $0 = k_0 < k_1 < \cdots < k_{e-1}$ be the representatives of J in $[0, n-1]$; we set $k_e = n$. Let $L^0 = \mathfrak{o}^n$, and let L^j be the lattice with basis $\{ze_i \mid i > n - k_j\} \cup \{e_i \mid i \leq n - k_j\}$ for $j > 0$. These lattices generate a lattice chain $\mathcal{L}_J \in X_J$; we denote its stabilizer by P_J and call it the standard parabolic subgroup of type J. Let $Q_J \subset \mathrm{GL}_n(\mathbb{C})$ be the block-upper triangular parabolic subgroup with blocks of sizes $k_j - k_{j-1}$ arranged in order of decreasing j. Then $P_J = \pi^{-1}(Q_J)$, where $\pi : \mathrm{GL}_n(\mathfrak{o}) \to \mathrm{GL}_n(\mathbb{C})$ is the homomorphism induced by the "evaluation at 0" map $z \mapsto 0$.

Uniform standard parahoric subgroups correspond to subgroups of $\mathbb{Z}/n\mathbb{Z}$; all of their blocks have the same size. If $e|n$, we let P_e denote the uniform standard parahoric subgroup of period e. In particular, the trivial subgroup corresponds to the maximal parahoric subgroup $P_1 = \mathrm{GL}_n(\mathfrak{o})$ while $J = \mathbb{Z}/n\mathbb{Z}$ gives the standard Iwahori subgroup $P_n = I$: the pullback of the standard Borel subgroup of upper triangular matrices. If $e|(n-1)$, but $e \neq 1$, we will similarly let P_e denote the almost uniform standard parabolic subgroup with J given by the residues of $\{j\frac{n-1}{e} \mid 0 \leq j < e\}$.

Each lattice chain $\{L^i\}_{i\in\mathbb{Z}}$—say with corresponding parahoric subgroup P—determines a filtration $\{\mathfrak{p}^s\}_{s\in\mathbb{Z}}$ of $\mathfrak{gl}_n(F)$:

$$\mathfrak{p}^s = \{X \in \mathfrak{gl}_n(F) \mid XL^i \subset L^{i+s} \text{ for all } i\}; \tag{4.3}$$

in particular, $\mathfrak{p}^0 = \mathfrak{p} := \mathrm{Lie}(P)$. One also gets a filtration $\{P^s\}_{s\in\mathbb{Z}_{\geq 0}}$ of P defined by $P^0 = P$ and $P^s = 1 + \mathfrak{p}^s$ for $s > 0$. In the special case that each lattice in the lattice chain stabilized by P admits an $\mathfrak{o}$-basis of the form $\{z^{k_j}e_j\}_{j=1}^n$, the corresponding filtration $\{\mathfrak{p}^s\}_i$ is induced by a grading $\mathfrak{gl}_n(\mathbb{C}[z,z^{-1}]) = \bigoplus_s \mathfrak{p}(s)$. In particular, this is true for the standard lattice chains $\mathcal{L}_J$.

When P is uniform, this filtration has a particular simple description. Indeed, there exists an element $\omega_P \in \mathrm{GL}_n(F)$ such that $\mathfrak{p}^s = \omega_P^s \mathfrak{p} = \mathfrak{p}\omega_P^s$ [**Bus87**]. In the standard case $P = P_J$, one can choose $\omega_J = \omega_{P_J}$ so that it generates the grading as well. This means that if ℓ_J is the usual block-diagonal Levi subalgebra of $\mathfrak{q}_J = \mathrm{Lie}(Q_J)$, then $\mathfrak{p}(s) = \ell_J\omega_J^s = \omega_J^s\ell_J$. In particular, if I is the standard Iwahori subgroup, then $\omega = \omega_n$ from the previous section has the desired property: $\mathfrak{i}^s = \omega^s\mathfrak{i} = \mathfrak{i}\omega^s$ and $\mathfrak{i}(s) = \omega^s\mathfrak{t} = \mathfrak{t}\omega^s$, where $\mathfrak{t}$ is the diagonal Cartan subalgebra. In the same way that ω only has nonzero entries on the "first circulant superdiagonal", in the general standard uniform case, the element ω_P will have nonzero entries on the first circulant block superdiagonal.

EXAMPLE 4.2. Let $n = 3$. Then

$$\omega_3 = \begin{pmatrix} 0 & 1 & 0 \\ 0 & 0 & 1 \\ z & 0 & 0 \end{pmatrix} \quad \text{and} \quad I = \begin{pmatrix} \mathfrak{o}^* & \mathfrak{o} & \mathfrak{o} \\ z\mathfrak{o} & \mathfrak{o}^* & \mathfrak{o} \\ z\mathfrak{o} & z\mathfrak{o} & \mathfrak{o}^* \end{pmatrix}.$$

Some steps in the standard Iwahori filtration on $\mathfrak{gl}_n(F)$ are shown below:

$$\mathfrak{i}^{-2} = \begin{pmatrix} \mathfrak{o} & z^{-1}\mathfrak{o} & z^{-1}\mathfrak{o} \\ \mathfrak{o} & \mathfrak{o} & z^{-1}\mathfrak{o} \\ \mathfrak{o} & \mathfrak{o} & \mathfrak{o} \end{pmatrix} \subsetneq \mathfrak{i}^{-1} = \begin{pmatrix} \mathfrak{o} & \mathfrak{o} & z^{-1}\mathfrak{o} \\ \mathfrak{o} & \mathfrak{o} & \mathfrak{o} \\ z\mathfrak{o} & \mathfrak{o} & \mathfrak{o} \end{pmatrix} \subsetneq \mathfrak{i}^{0} = \begin{pmatrix} \mathfrak{o} & \mathfrak{o} & \mathfrak{o} \\ z\mathfrak{o} & \mathfrak{o} & \mathfrak{o} \\ z\mathfrak{o} & z\mathfrak{o} & \mathfrak{o} \end{pmatrix}.$$

4.3. Fundamental strata. Let $\widehat{\nabla} = d + (M_{-r}z^{-r} + M_{-r+1}z^{-r+1} + \dots)\frac{dz}{z}$ be a formal connection with $r > 0$. If M_{-r} is non-nilpotent, then the slope of $\widehat{\nabla}$ is r. However, if the slope of $\widehat{\nabla}$ is not an integer, then M_{-r} will be nilpotent regardless of the trivialization for the underlying vector bundle. In this case, the native leading term $M_{-r}z^{-r}\frac{dz}{z}$ of the connection is not useful.

A better notion of the "leading term" of $\widehat{\nabla}$ was introduced by Bremer and Sage in [**BS13**] in terms of a geometric theory of fundamental strata, motivated by the analogous p-adic theory of Bushnell and Kutzko [**Bus87, BK93**]. A GL_n-*stratum* is a triple (P, s, β) consisting of a parahoric subgroup, a nonnegative integer s, and a functional $\beta \in (\mathfrak{p}^s/\mathfrak{p}^{s+1})^\vee \cong \mathfrak{p}^{-s}/\mathfrak{p}^{-s+1}\frac{dz}{z}$. We say that the stratum has depth s/e_P. An element $\tilde{\beta} \in \mathfrak{p}^{-s}$ inducing β is called a *representative* of β. If the filtration associated to P comes from a grading (for example, if P is a standard parahoric subgroup), then there is a unique homogeneous representative $\beta^\flat \in \mathfrak{p}(-s)$. The stratum is called *fundamental* if every representative is nonnilpotent or equivalently, if $\beta^\flat$ is nonnilpotent (when it is defined).

Given a formal connection $(\hat{V}, \widehat{\nabla})$, we let $[\widehat{\nabla}]_\phi$ denote the matrix of the connection with respect to the trivialization ϕ. In other words, $\widehat{\nabla} = d + [\widehat{\nabla}]_\phi$ in this gauge. We say that $\widehat{\nabla}$ contains a positive depth stratum with respect to ϕ if $[\widehat{\nabla}]_\phi/\frac{dz}{z} \in \mathfrak{p}^{-s}$ and is a representative of β. (See [**BS13, BS18**] for the general definition.) Intuitively, this means that β is the leading term of the connection with respect to the filtration $\{\mathfrak{p}^j\}$.

THEOREM 4.3 ([**BS13**], Theorem 4.10). Any formal connection $(\hat{V}, \widehat{\nabla})$ contains a fundamental stratum (P, r, β) with P a standard parahoric subgroup. The depth is positive if and only if $\widehat{\nabla}$ is irregular singular, and $\mathrm{slope}(\widehat{\nabla}) = r/e_P$. Moreover,

(1) If $\widehat{\nabla}$ contains the stratum (P', r', β'), then $r'/e_{P'} \geq \mathrm{slope}(\widehat{\nabla})$.

(2) If $\widehat{\nabla}$ is irregular singular, then a stratum (P', r', β') contained in $\widehat{\nabla}$ is fundamental if and only if $r'/e_{P'} = \text{slope}(\widehat{\nabla})$.

For example, an unramified connection of slope r in Levelt–Turrittin form (3.1) contains the fundamental stratum $(\mathrm{GL}_n(\mathfrak{o}), r, D_r z^{-r}\frac{dz}{z})$. The Frenkel–Gross and Airy formal connections contain the fundamental strata $(I, 1, \omega^{-1}\frac{dz}{z})$ and $(I, 1, \omega^{-(n+1)}\frac{dz}{z})$ respectively. This gives another proof that their slopes are $1/n$ and $(n+1)/n$.

4.4. Regular strata and toral connections. Classically, much more information is available about a formal connection when its (naive) leading term is regular semisimple and not just nonnilpotent. The analogue for strata involves the notion of a *regular stratum.* A stratum (P, r, β) is called regular if every representative $\tilde{\beta}$ is regular semisimple. If this is the case, the centralizers of the various representatives are maximal tori in the loop group which are conjugate under P^1. We abuse notation and call the stratum S-regular if S is one of these maximal tori. If the homogeneous representative $\beta^\flat$ is defined, then the stratum is regular if and only if $\beta^\flat$ is regular semisimple. We call a formal connection S*-toral* if it contains an S-regular stratum. For example, an unramified formal connection is $T(F)$-toral (with T the standard diagonal maximal torus) if and only if the leading term of the LT normal form is regular semisimple.

Most classes of maximal tori do not give rise to S-toral connections. Recall that there is a bijection between the conjugacy classes of maximal tori in $\mathrm{GL}_n(F)$ and conjugacy classes in the symmetric group $\mathfrak{S}_n$, hence to partitions of n [**KL88**]. If $\mu = (\mu_1, \dots, \mu_k)$ is a partition, then the corresponding maximal tori are those isomorphic to $\prod_{i=1}^k F[z^{1/\mu_i}]^\times$. Such a maximal torus may be realized explicitly as the block-diagonal embedding $\prod_{i=1}^k \mathbb{C}((\omega_{\mu_i}))^\times$, where we set $\omega_1 = z$. We call a maximal torus *uniform* if each block has the same size e (so $n = ek$). It is *almost uniform* if it splits as $S' \times F^\times$ where S' is a uniform maximal torus in $\mathrm{GL}_{n-1}(F)$. It turns out that there is an S-regular stratum (P, r, β) when (1) S is uniform with e blocks and $\gcd(r, e) = 1$; or (2) $S = S' \times F^\times$, S' is a uniform maximal torus in $\mathrm{GL}_{n-1}(F)$ with e blocks, and $\gcd(r, e) = 1$. Moreover, P can be taken conjugate to a standard parahoric subgroup whose block structure corresponds to the partition dual to μ. (For example, one can take P uniform with period e in the first case.) In particular, we conclude that toral connections necessarily have slope r/e with $\gcd(r, e) = 1$ and e dividing n or $n-1$.

We will be particularly interested in the case of maximally ramified toral connections, i.e., toral connections of slope r/n with $\gcd(r, n) = 1$. There is a unique class of maximal tori in $\mathrm{GL}_n(F)$ that are anisotropic modulo the center, meaning that they have no non-central rational cocharacters. Concretely, such tori are as far from being split as possible. They are called *Coxeter maximal tori,* since they correspond to the Coxeter class in $\mathfrak{S}_n$ consisting of the n-cycles. A specific representative of this class is the standard Coxeter torus $\mathcal{C} = \mathbb{C}((\omega_n))^\times$ with Lie algebra $\mathfrak{c}$. The formal Frenkel–Gross and Airy connections are both $\mathcal{C}$-toral.

In order to discuss toral formal types, we now introduce representatives of the classes of uniform and almost uniform Cartan subalgebras that are well-behaved with respect to the parahoric subgroups P_e defined in section 4.2. If $\mathfrak{s}$ is a uniform Cartan subalgebra isomorphic to $F[z^{1/e}]^{n/e}$, it is endowed with a filtration $\mathfrak{s}^j$ obtained by assigning degree one to $z^{1/e}$. (There is also an associated grading with graded pieces $\mathfrak{s}(j)$.) We say that $\mathfrak{s}$ is *compatible* to $\mathfrak{p}$ if $\mathfrak{s}^j = \mathfrak{p}^j \cap \mathfrak{s}$ for all j. In the case that $\mathfrak{p}$ is standard, we say that $\mathfrak{s}$ is *graded compatible* to $\mathfrak{p}$ if $\mathfrak{s}(j) = \mathfrak{p}(j) \cap \mathfrak{s}$ for all j.

The block-diagonal uniform Cartan subalgebra with blocks of size e is not compatible with $\mathfrak{p}_e$. Instead, set $V_j = \operatorname{span}\{e_i | i \equiv j \pmod e\}$, so $F^n = V_1 \oplus \cdots \oplus V_{n/e}$. Let ω_{ej} be our usual matrix ω_e with respect to the given graded basis on V_j. We now set $\mathfrak{s}_e = \bigoplus_{j=1}^{n/e} \mathbb{C}((\omega_{ej}))$; it is graded compatible with $\mathfrak{p}_e$. Note that if $\gcd(r,e) = 1$, then an element of $\mathfrak{s}_e(r)$ is regular semisimple if it is of the form $(c_j \omega_{ej}^r)_j$, where the c_j's are nonzero constants such that $c_j/c_{j'}$ is not an eth root of 1 for $j \neq j'$.

If $e|(n-1)$, one can similar define $\mathfrak{s}_e$ graded compatible with $\mathfrak{p}_e$; it is just the direct sum of a uniform Cartan subalgebra in $\mathfrak{gl}_{n-1}(F)$ with Fe_{nn}.

The crucial feature of toral connections is that they can be "diagonalized" into a Cartan subalgebra. If $(r,e) = 1$, then the space of S_e-formal types of slope r/e is the subset $\mathcal{A}(S_e, r) \subset (\mathfrak{s}(-r) + \cdots + \mathfrak{s}(0)) \frac{dz}{z}$ consisting of matrices with regular semisimple component in degree $-r$. It may also be identified with a Zariski-open subset of $\mathfrak{s}^{-r}/\mathfrak{s}^1 \frac{dz}{z}$.

THEOREM 4.4. **[BS13]**,**[BS20]** *If $\widehat{\nabla} = d + [\widehat{\nabla}] \frac{dz}{z}$ contains an S_e-regular stratum (S_e, r, β), then $[\widehat{\nabla}] \frac{dz}{z}$ is P_e^1-gauge equivalent to a unique element of $\mathcal{A}(S_e, r)$.*

Thus, we get a rational canonical form for toral connections.

When $e = 1$, so $S_e = T(F)$, this is just the familiar fact that if $\nabla = d + (D_{-r} z^{-r} + M) \frac{dz}{z}$ with $D_{-r} \in \mathfrak{gl}_n(\mathbb{C})$ regular semisimple and $M \in z^{1-r} \mathfrak{gl}_n(\mathfrak{o})$, then it is gauge equivalent to a diagonal unramified LT normal form. At the opposite extreme when $e = n$, then $S_1 = \mathcal{C}$ and

$$\mathcal{A}(\mathcal{C}, r) = \{p(\omega_n^{-1}) \mid p \in \mathbb{C}[x], \deg p = r\}. \tag{4.4}$$

In fact, one can show that all formal connections of slope r/n with $\gcd(r,n) = 1$ are $\mathcal{C}$-toral, so we have obtained a rational canonical form for maximally ramified connections.

THEOREM 4.5. **[KLM+22]** *Let $\widehat{\nabla}$ be a maximally ramified connection of slope r/n with $\gcd(r,n) = 1$. Then $\widehat{\nabla}$ is formally gauge equivalent to a connection of the form $d + p(\omega^{-1}) \frac{dz}{z}$ with p a polynomial of degree r.*

REMARK 4.6. Let $S_e^0 = \prod_{j=1}^{n/e} \mathbb{C}[\![\omega_{ej}]\!]^\times$; it is the unique maximal bounded subgroup of S_e. Let $W_{S_e}^{\text{aff}} = N(S_e)/S_e^0$ be the relative affine Weyl group of S_e. It is the semidirect product of the relative Weyl group W_{S_e} and the free abelian group S_e/S_e^0. There is a natural action of $W_{S_e}^{\text{aff}}$ on $\mathcal{A}(S_e, r)$: W_{S_e} acts by conjugation in the usual way while S_e/S_e^0 acts by translations on $\mathfrak{s}_e(0)$. It can be shown that the moduli space of S_e-toral connections of slope r/e is given by $\mathcal{A}(S_e, r)/W_{S_e}^{\text{aff}}$ **[BS13]**.

We can now defined the truncated orbit of a toral formal type. Fix $\mathcal{F} \in \mathcal{A}(S_e, r)$. By compatibility of $\mathfrak{p}_e$ and $\mathfrak{s}_e$, there are natural injections $\mathfrak{s}_e^{-r}/\mathfrak{s}_e^1 \frac{dz}{z} \hookrightarrow \mathfrak{p}_e^{-r}/\mathfrak{p}_e^1 \frac{dz}{z} \subset \mathfrak{p}_e^\vee$. The parahoric subgroup P_e thus acts on $\mathfrak{p}_e^{-r}/\mathfrak{p}_e^1 \frac{dz}{z}$ by the coadjoint action, and we define the truncated orbit of $\mathcal{F}$ to be $\mathscr{O}_{\mathcal{F}} = \operatorname{Ad}^*(P_e)(\mathcal{F})$. Equivalently, one can view $\mathcal{F}$ as an element of $\mathfrak{p}_e^{-r}$, and then $\mathscr{O}_{\mathcal{F}}$ is obtained by conjugation by P_e followed by truncation. Note that this action factors through the finite-dimensional group P_e/P_e^{r+1}.

4.5. Moduli spaces of toral framable connections. We are now ready to discuss moduli spaces of meromorphic connections with toral and unramified formal types. First, we need to understand what "framability" means in the context of toral singularities.

DEFINITION 4.7. Let $\nabla = d + [\nabla]_\phi$ be a global connection on $\mathbb{P}^1$ with a singular point at 0, and let $\mathcal{F}$ be an S_e-formal type of depth r. We say that ∇ is *framable at* 0 *with respect to* $\mathcal{F}$ if

(1) there exists $g \in \mathrm{GL}_n(\mathbb{C})$ such that $[\nabla]_{g\phi} = g \cdot [\nabla]_\phi \in \mathfrak{p}_e^{-r}\frac{dz}{z}$ and $[\nabla] - \mathcal{F} \in \mathfrak{p}_e^{1-r}\frac{dz}{z}$; and
(2) there exists an element $p \in P_e^1$ such that $\mathrm{Ad}^*(p)([\nabla]_{g\phi}\frac{dz}{z})|_{\mathfrak{p}_e} = \mathcal{F}$.

Intuitively, this means that there is a global trivialization in which the leading terms of the matrix of ∇ and $\mathcal{F}$ agree and that there is a further formal gauge change by an element of P_e^1 in which the matrix of ∇ and $\mathcal{F}$ are equal, viewed as functionals on $\mathfrak{p}_e$.

Fix three disjoint (possibly empty) subsets $\{a_1, \dots, a_m\}$, $\{b_1, \dots, b_\ell\}$, and $\{c_1, \dots, c_p\}$ of $\mathbb{P}^1$ with $m \geq 1$. Let $\boldsymbol{\mathcal{F}} = (\mathcal{F}_1, \dots, \mathcal{F}_m)$ be a collection of toral formal types at the a_i's, $\boldsymbol{\mathcal{D}} = (\mathcal{D}_1, \dots, \mathcal{D}_\ell)$ a collection of nontoral unramified irregular formal types at the b_j's, and $\boldsymbol{\mathcal{O}} = (\mathcal{O}_1, \dots, \mathcal{O}_p)$ a collection of nonresonant adjoint orbits at the c_k's. One can now consider the category $\mathscr{C}(\boldsymbol{\mathcal{F}}, \boldsymbol{\mathcal{D}}, \boldsymbol{\mathcal{O}})$ of meromorphic connections ∇ satisfying the following properties:

(1) ∇ has irregular singularities at the a_i's and b_j's, regular singularities at the c_k's, and no other singular points;
(2) for each i and j, ∇ is framable at a_i (resp. b_j) with respect to the formal type $\mathcal{F}_i$ (resp. $\mathcal{D}_j$); and
(3) for each k, ∇ has residue at c_k in $\mathcal{O}_k$.

As in the Fuchsian and unramified cases, one can construct the corresponding moduli space $\mathcal{M}(\boldsymbol{\mathcal{F}}, \boldsymbol{\mathcal{D}}, \boldsymbol{\mathcal{O}})$ as the Hamiltonian reduction of a product of symplectic $\mathrm{GL}_n(\mathbb{C})$-manifolds, one for each singular point. It remains to define the symplectic manifold $\mathcal{M}_{\mathcal{F}}$ encoding the local data at a toral singularity. It no longer suffices to take the truncated orbit $\mathcal{O}_{\mathcal{F}} = \mathrm{Ad}^*(P_e)(\mathcal{F})$ because it no longer contains full information about the principal part. Indeed, $\mathfrak{s}_e(i)\frac{dz}{z}$ contributes to the residue for positive i as long as $i < e$.

Recall that P_e is the pullback of a certain standard parabolic subgroup $Q_e \subset \mathrm{GL}_n(\mathbb{C})$ under the map $\mathrm{GL}_n(\mathfrak{o}) \to \mathrm{GL}_n(\mathbb{C})$ induced by $z \mapsto 0$. More intrinsically, $Q_e \cong P_e/(1 + \mathrm{GL}_n(\mathfrak{o}))$. The "extended orbit" $\mathcal{M}_{\mathcal{F}} \subset (Q_e \backslash \mathrm{GL}_n(\mathbb{C})) \times \mathfrak{gl}_n(\mathfrak{o})^\vee$ is now defined by

$$\mathcal{M}_{\mathcal{F}} = \{(Q_e g, \alpha) \mid (\mathrm{Ad}^*(g)\alpha)|_{\mathfrak{p}_e} \in \mathrm{Ad}^*(P_e)(\mathcal{F})\}. \tag{4.5}$$

Thus, α is the principal part, and the global trivialization g is needed to move the principal part into $\mathfrak{p}_e^{-r}\frac{dz}{z}$. One only considers Q_e-orbits of such g since Q_e preserves $\mathfrak{p}_e^{-r}\frac{dz}{z}$. It is easy to see that $\mathcal{M}_{\mathcal{F}}$ is a symplectic manifold [**BS13**]. The group $\mathrm{GL}_n(\mathbb{C})$ acts on $\mathcal{M}_{\mathcal{F}}$ via $h \cdot (Q^b g, \alpha) = (Q^b g h^{-1}, \mathrm{Ad}^*(h)\alpha)$; this action is Hamiltonian with moment map $(Q^b g, \alpha) \mapsto \alpha|_{\mathfrak{gl}_n(\mathbb{C})} = \mathrm{Res}(\alpha)\frac{dz}{z}$.

THEOREM 4.8. The moduli space $\mathcal{M}(\boldsymbol{\mathcal{F}}, \boldsymbol{\mathcal{D}}, \boldsymbol{\mathcal{O}})$ is given by

$$\mathcal{M}(\boldsymbol{\mathcal{F}}, \boldsymbol{\mathcal{D}}, \boldsymbol{\mathcal{O}}) \cong \left[\left(\prod_i \mathcal{M}_{\mathcal{F}_i}\right) \times \left(\prod_j \mathscr{O}_{\mathcal{D}_j}\right) \times \left(\prod_k \mathscr{O}_k\right)\right] /\!\!/_0 \, \mathrm{GL}_n(\mathbb{C}).$$

This theorem was proved by Bremer and Sage in [**BS13**] with the hypothesis that all irregular singular points are uniform toral. The proof in this general case follows by combining the methods of [**BS13**] and [**HY14**]. Note that as in the previous cases, there is no dependence on the specific points.

REMARK 4.9. When all irregular singular points are toral, there are two other variants of this moduli space. One can consider the moduli space $\widetilde{\mathcal{M}}(\boldsymbol{\mathcal{F}}, \boldsymbol{\mathcal{O}})$ of *framed connections*, which classifies framable connections together with the additional data of "compatible framings". This was constructed by Boalch in the unramified case [**Boa01**] and by Bremer and Sage when ramified uniform toral singularities are allowed [**BS13**]. This moduli space is a smooth symplectic manifold, and $\mathcal{M}(\boldsymbol{\mathcal{F}}, \boldsymbol{\mathcal{O}})$ is obtained from it by Hamiltonian reduction by a certain torus. There is also a larger Poisson moduli space classifying framed connections "with fixed combinatorics". This means that at each a_i, the formal type is allowed to be arbitrary in $\mathcal{A}(S_{e_i}, r_i)$ for fixed e_i and r_i. Again, the construction is due to Boalch in the unramified case [**Boa01**] and to Bremer and Sage in the uniform toral case [**BS12**].

As usual, one can also define the stable moduli space $\mathcal{M}^{\mathrm{st}}(\mathbf{F}, \mathbf{D}, \mathbf{O})$ consisting of the irreducible framable connections with the given formal types. The *toral Deligne–Simpson problem* is then the determination of when the stable moduli space is nonempty.

We now compute these moduli spaces in the simplest case that involves a ramified formal type: rank 2 framable meromorphic connections with a Coxeter toral singularity of slope 1/2 and (possibly) a regular singular point. As we will see in the next section, meromorphic connections with a Coxeter toral singular point are always irreducible, so here, the stable moduli space coincides with the whole moduli space.

EXAMPLE 4.10. Let $\mathcal{F} = a\omega_2^{-1}\frac{dz}{z}$ with $a \in \mathbb{C}^*$, and let $\mathscr{O} \subset \mathfrak{gl}_2(\mathbb{C})$ be a nonresonant adjoint orbit. The moduli space $\mathcal{M}(\mathcal{F}, \mathscr{O})$ is the space of $\mathrm{GL}_2(\mathbb{C})$-orbits of triples (Bg, α, Y) with $(Bg, \alpha) \in \mathscr{M}_{\mathcal{F}}$, $Y \in \mathscr{O}$, and $\mathrm{Res}(\alpha) + Y = 0$. (Here, B is the group of upper triangular matrices.) This is the same as the space of B-orbits of triples (B, α, Y).

We first compute the truncated orbit of $\mathcal{F}$, using the fact that $I = T \ltimes I^1$ where $I^1 = 1 + \mathfrak{t}\omega + \mathfrak{i}^2$. Since

$$(1 + \mathrm{diag}(b, c)\omega)a\omega^{-1}(1 + \mathrm{diag}(b, c)\omega)^{-1} = a\omega^{-1} + \mathrm{diag}(b - c, c - b) \pmod{\mathfrak{i}^1},$$

we get

$$\mathrm{Ad}^*(I)(\mathcal{F}) = \mathrm{Ad}^*(T)\left\{\begin{pmatrix} u & az^{-1} \\ a & -u \end{pmatrix}\frac{dz}{z} \mid u \in \mathbb{C}\right\}. \tag{4.6}$$

Thus, in order to satisfy the condition $\alpha\frac{dz}{z}|_{\mathfrak{i}} \in \mathscr{O}_{\mathcal{F}}$, we must have

$$\alpha = \begin{pmatrix} u & vz^{-1} + w \\ v^{-1} & -u \end{pmatrix}\frac{dz}{z} \tag{4.7}$$

for some $u, v, w \in \mathbb{C}$ with $v \neq 0$. In fact, each B-orbit has a unique representative with $v = 1$ and $u = 0$. This means that the only adjoint orbits $\mathscr{O}$ that give nonempty moduli space are those containing $\left(\begin{smallmatrix} 0 & -w \\ -1 & 0 \end{smallmatrix}\right)$ for some w. Thus, $\mathcal{M}(\mathcal{F}, \mathscr{O})$ is a singleton if $\mathscr{O}$ is regular nilpotent or regular semisimple with trace zero; otherwise, it is empty. We remark that in the regular nilpotent case, the unique such connection is the GL_2-version of the Frenkel–Gross rigid connection, and this argument shows that it is indeed uniquely determined by its local behavior.

4.6. The ramified Deligne–Simpson problem for Coxeter connections. Little is known about the ramified Deligne–Simpson problem in general. We now restrict attention to a simple special case in which the problem has recently been solved by Kulkarni, Livesay, Matherne, Nguyen, and Sage [**KLM+22, LSN23**]: connections with a *maximally ramified* irregular singularity (i.e., an irregular singularity of slope r/n with $\gcd(r, n) = 1$) and (possibly) an additional regular singular point. Without loss of generality, we can view such connections as connections on $\mathbb{G}_m$ with the irregular singularity at 0. Following [**KS21**], we refer to such connections as *Coxeter connections.* As we have seen, classical examples include the Frenkel–Gross and Airy connections. Coxeter connections have played a significant role in recent work on the irregular geometric Langlands program [**FG09, HNY13, Zhu17, KS21, JKY21**]. They have also arisen in Lam and Templier's proof of mirror symmetry for minuscule flag varieties [**LT17**].

It is shown in Theorem 4.4 of [**KLM+22**] (c.f. [**KS23**]) that maximally ramified formal connections are precisely the $\mathcal{C}$-toral connections. Moreover, $\mathcal{C}$-toral connections are irreducible [**KS21**], so meromorphic connections with a $\mathcal{C}$-toral singularity are also irreducible. Thus, the Deligne–Simpson problem for Coxeter connections is the determination of when the moduli space $\mathcal{M}(\mathcal{F}, \mathscr{O})$ is nonempty for $\mathcal{F}$ a $\mathcal{C}$-formal type and $\mathscr{O}$ a nonresonant adjoint orbit.

THEOREM 4.11. [**KLM+22**, Theorem 5.4] *Let $\mathcal{F}$ be a maximally ramified formal type of slope r/n with $\gcd(r, n) = 1$. Then $\mathcal{M}(\mathcal{F}, \mathscr{O})$ is nonempty if*

(1) $\mathrm{Tr}(\mathscr{O}) = -\operatorname{Res}\mathrm{Tr}(\mathcal{F})$, *and*
(2) *there are at most r Jordan blocks for each eigenvalue of $\mathscr{O}$.*

We can reformulate this result in a more geometric way. The set Orb^n of adjoint orbits in $\mathfrak{gl}_n(\mathbb{C})$ is a partially ordered set under the Zariski closure relation: $\mathscr{O} \preceq \mathscr{O}'$ if and only if $\mathscr{O} \subset \overline{\mathscr{O}'}$. The set of matrices with fixed characteristic polynomial q is a closed subset of $\mathfrak{gl}_n(\mathbb{C})$ which is stable under conjugation, so as posets $\mathrm{Orb}^n = \bigsqcup \mathrm{Orb}^n_q$, where the disjoint union runs over all monic polynomials of degree n.

Let $q = \prod_{i=1}^s (x - c_i)^{m_i}$ be a polynomial of degree n with the c_i's distinct complex numbers. Let $\mathrm{Part}(m)$ be the set of partitions of m; it is a poset under the dominance order. By the theory of the Jordan canonical form, there is a bijection $\mathrm{Orb}^n_q \xrightarrow{\sim} \prod_{i=1}^s \mathrm{Part}(m_i)$ sending an orbit to the partitions determined by the sizes of the Jordan blocks for each eigenvalue. Under this identification, the closure ordering corresponds to the product of the dominance orderings.

We will show that the subset

$$\mathrm{DS}(\mathcal{F}, q) = \{\mathscr{O} \in \mathrm{Orb}^n_q \mid \mathcal{M}(\mathcal{F}, \mathscr{O}) \neq \varnothing\} \tag{4.8}$$

is a principal filter in Orb^n_q. (This means that it is an upward-closed subset with a unique minimum element.) Given positive integers r and m, there exists a unique smallest partition $\lambda^{m,r} \in \mathrm{Part}(m)$ with at most r parts. Indeed, if we write $m = kr + r'$ with $k, r \in \mathbb{Z}$ and $0 \leq r' < r$, then this partition is given by $\lambda^{n,r} = \{(k+1)^{r'}, k^{r-r'}\}$. We now let $\mathcal{O}_q^r$ be the orbit in Orb_q^n corresponding to the element

$$(\lambda^{m_1,r}, \lambda^{m_2,r}, \dots, \lambda^{m_s,r}) \in \prod_{i=1}^{s} \mathrm{Part}(m_i).$$

This is the unique smallest orbit for which each eigenvalue has at most r eigenvalues.

EXAMPLE 4.12. For nilpotent orbits (i.e., $q = x^n$), a representative of $\mathcal{O}_{x^n}^r$ is given by the matrix with 1's on the rth subdiagonal if $1 \leq r \leq n-1$ and 0 if $r \geq n$. Note that these matrices are precisely the residues of $\omega_n^{-r}\frac{dz}{z}$.

COROLLARY 4.13. Let $\mathcal{F} = p(\omega_n^{-1})\frac{dz}{z}$ be a maximally ramified formal type of slope r/n, and let $q = \prod_{i=1}^{s}(x - c_i)^{m_i}$ be a degree n polynomial with the c_i's distinct modulo $\mathbb{Z}$. Then,

(1) $\mathrm{DS}(\mathcal{F}, q)$ is empty if and only if $np(0) \neq -\sum_{i=1}^{s} m_i c_i$.
(2) If $\mathrm{DS}(\mathcal{F}, q)$ is nonempty, it is the principal filter generated by $\mathcal{O}_q^r$.
(3) For fixed q, $\mathrm{DS}(\mathcal{F}, q)$ only depends on $\deg p$ and $p(0)$.

EXAMPLE 4.14.

(a) If $\mathrm{slope}(\mathcal{F}) = 1/n$, then $\mathcal{M}(\mathcal{F}, \mathcal{O})$ is empty unless the trace condition holds and $\mathcal{O}$ is regular, i.e., maximal in Orb^n.
(b) If $\mathrm{slope}(\mathcal{F}) = (n-1)/n$, then $\mathcal{M}(\mathcal{F}, \mathcal{O})$ is nonempty unless the trace condition fails or $\mathcal{O} = \{aI\}$ for some $a \in \mathbb{C}$.
(c) If $\mathrm{slope}(\mathcal{F}) > 1$, then $\mathcal{M}(\mathcal{F}, \mathcal{O})$ is nonempty unless the trace condition fails. More generally, this is the case if $\mathrm{slope}(\mathcal{F}) \geq m_i/n$ for all i, with $q = \prod_{i=1}^{s}(x - c_i)^{m_i}$ as above.

REMARK 4.15. In general, it is not easy to find explicit connections in $\mathcal{M}(\mathcal{F}, \mathcal{O})$. However, it can be done if $\mathcal{F}$ is *homogeneous*, meaning that $\mathcal{F} = a\omega_n^{-r}\frac{dz}{z}$ and $\mathcal{O}$ is nilpotent. Indeed, an algorithm is described in [**LSN23**] which constructs a binary, strictly upper triangular matrix $b_{\mathcal{O}}$ for any $\mathcal{O} \succeq \mathcal{O}_{x^n}^r$ such that $d + (a\omega_n^{-r} + b_{\mathcal{O}})\frac{dz}{z}$ gives an element of $\mathcal{M}(a\omega_n^{-r}\frac{dz}{z}, \mathcal{O})$. In particular, $b_{\mathcal{O}_{x^n}^r} = 0$.

4.7. Rigidity for Coxeter connections. We now discuss rigidity for Coxeter connections. As in the case of Fuchsian connections, there are several possible notions of rigidity for connections with formal types as in Theorem 4.8: physical rigidity, "moduli space rigidity" (where the moduli space is a singleton), and "stable moduli space rigidity". In general, it is very difficult to determine directly whether a connection is rigid in any of these senses. However, it turns out that there is a more accessible version called *cohomological rigidity*, which coincides with physical rigidity for irreducible connections.

Let $j : U \hookrightarrow \mathbb{P}^1$ be the inclusion of a nonempty Zariski-open set. A connection ∇ on U is called cohomologically rigid if $H^1(\mathbb{P}^1, j_{!*}\mathrm{ad}_\nabla) = 0$. Here, ad_∇ is the associated adjoint vector bundle. For irreducible connections, this cohomology vanishing condition implies that the connection has no infinitesimal deformations. It is a theorem of Bloch and Esnault that cohomological and physical rigidity coincide for irreducible connections [**BE04**]. We will simply call such connections rigid.

We can now give a complete classification of rigid framable Coxeter connections under the additional hypothesis that the connection has unipotent monodromy at ∞.

THEOREM 4.16. [**KLM+22**, Theorem 6.1] Let $\mathcal{F} = p(\omega_n^{-1})\frac{dz}{z}$ be a rank n maximally ramified formal type of slope r/n with $p(0) = 0$, and let $\mathcal{O}$ be any nilpotent orbit with at most r Jordan blocks. Then there exists a rigid connection with formal type $\mathcal{F}$ and unipotent monodromy determined by $\mathcal{O}$ if and only if $\mathcal{O} = \mathcal{O}^r_{x^n}$ and $r|(n \pm 1)$.

In particular, the Frenkel–Gross and Airy connections are rigid. Note that there are no rigid Coxeter connections of this type with the slope of the irregular singularity greater than $(n+1)/n$.

REMARK 4.17. It was originally proved in [**KS21**] that $d + \omega_n^{-r}\frac{dz}{z}$ is rigid if and only if $r|(n \pm 1)$.

We conclude this section by giving a brief sketch of the proof. Let ∇ be a Coxeter connection. There is an explicit formula for the cohomology group $\dim(H^1(\mathbb{P}^1, j_{!*}\mathrm{ad}_\nabla))$ in terms of the global and local differential Galois groups of ∇ and of the irregularity of the adjoint connection at 0. Let $\mathcal{I}$ denote the global differential Galois group, and let $\mathcal{I}_0$ and $\mathcal{I}_\infty$ denote the local differential Galois groups at 0 and ∞. These are all closed algebraic subgroups of $\mathrm{GL}_n(\mathbb{C})$. Also, let $\mathrm{Irr}(\mathrm{ad}_{\widehat{\nabla}_0})$ denote the irregularity of the formal connection $\mathrm{ad}_{\widehat{\nabla}_0}$. It is shown in [**FG09**, Proposition 11] that

$$\dim(H^1(\mathbb{P}^1, j_{!*}\mathrm{ad}_\nabla)) = \mathrm{Irr}(\mathrm{ad}_{\widehat{\nabla}_0}) - \dim(\mathfrak{gl}_n(\mathbb{C})^{\mathcal{I}_0}) - \dim(\mathfrak{gl}_n(\mathbb{C})^{\mathcal{I}_\infty}) + 2\dim(\mathfrak{gl}_n(\mathbb{C})^{\mathcal{I}}). \tag{4.9}$$

It is a general fact that if $\widehat{\nabla}$ is a toral connection of slope s, then $\mathrm{Irr}(\mathrm{ad}_{\widehat{\nabla}_0}) = sn(n-1)$ [**KS19**, Lemma 19]. Since $s = r/n$ in this case, we have $\mathrm{Irr}(\mathrm{ad}_{\widehat{\nabla}_0}) = r(n-1)$.

The differential Galois group of a $\mathcal{C}$-toral connection is computed in [**KS23**], following [**FG09**] and [**KS21**]. It has the form $\mathcal{I}_0 \cong H \ltimes \langle\theta\rangle$, where H is a certain torus containing a regular semisimple element and θ is an order n element of $N(H)$. One can then show as in [**KS21**] that $\mathfrak{gl}_n(\mathbb{C})^{\mathcal{I}_0} = \mathbb{C}$. Since $\mathcal{I}_0 \subset \mathcal{I}$, we also get $\mathfrak{gl}_n(\mathbb{C})^{\mathcal{I}} = \mathbb{C}$.

Finally, if $N \in \mathcal{O}$, then $\mathcal{I}_\infty$ is the one-parameter subgroup generated by N, i.e., the unipotent group $\{\exp(aN) \mid a \in \mathbb{C}\}$. It follows that $\mathfrak{gl}_n(\mathbb{C})^{\mathcal{I}_\infty}$ is the centralizer of N.

A short calculation shows that

$$\dim(H^1(\mathbb{P}^1, j_{!*}\mathrm{ad}_\nabla)) = \dim(\mathcal{O}) + (r-n-1)(n-1) \geq 0. \tag{4.10}$$

Recall that by Corollary 4.13, $\mathrm{DS}(\mathcal{F}, x^n)$ is the principal filter generated by $\mathcal{O}^r_{x^n}$, so in particular, there exists a connection with formal types $\mathcal{F}$ and $\mathcal{O}$ if $\mathcal{O} \succeq \mathcal{O}^r_{x^n}$. If $\mathcal{O} \succ \mathcal{O}^r_{x^n}$, so $\dim(\mathcal{O}) > \dim(\mathcal{O}^r_{x^n})$, then the dimension of the cohomology group is positive. Thus, rigidity is only possible if $\mathcal{O} = \mathcal{O}^r_{x^n}$. It remains to check when $\dim(\mathcal{O}^r_{x^n}) = (n+1-r)(n-1)$. It was shown in [**KS21**] that this occurs precisely when $r|(n \pm 1)$.

5. Meromorphic G-connections

5.1. ***G*-connections.** A rank n connection may be viewed as a $\mathrm{GL}_n(\mathbb{C})$-*connection*: a principal $\mathrm{GL}_n(\mathbb{C})$-bundle endowed with a meromorphic connection. This perspective allows us to generalize our set-up to consider G-connections where G is an complex reductive group. We can now ask the same fundamental question:

to what extent does specified local behavior determine a global G-connection? In this section, we briefly describe what is needed to address this question as well as what is known.

Let G be a connected reductive group over $\mathbb{C}$ with Lie algebra $\mathfrak{g}$, and let $\mathrm{Rep}(G)$ be the category of finite-dimensional complex representations of G. Let $\mathcal{E}$ be a principal G-bundle over $\mathbb{P}^1$. By Tannakian formalism, $\mathcal{E}$ may be viewed as a tensor functor from $\mathrm{Rep}(G)$ to the category of (finite rank) vector bundles via $V \mapsto V_{\mathcal{E}} = \mathcal{E} \times_G V$. A *meromorphic G-connection* is a tensor functor $(\mathcal{E}, \nabla)$ from $\mathrm{Rep}(G)$ to the category of vector bundles endowed with a meromorphic connection. We will always assume that $\mathcal{E}$ is trivializable. Fix a trivialization ϕ. This induces corresponding trivializations ϕ_V for each $V_{\mathcal{E}}$. The tensor functor is then determined by $[\nabla]_\phi \in \mathfrak{g}(\mathbb{C}(z))\frac{dz}{z}$; if (V, ρ) is a representation, then the induced connection on $V_{\mathcal{E}}$ is given by $d + [\rho([\nabla]_\phi)]$. We may thus view ∇ (in this trivialization) as the abstract connection $d + [\nabla]_\phi$. Formal G-connections are defined similarly.

Changing the trivialization changes the matrix of the G-connection by gauge change:

$$[\widehat{\nabla}]_{g\phi} = g \cdot [\widehat{\nabla}]_\phi = \mathrm{Ad}^*(g)([\widehat{\nabla}]_\phi) - (dg)g^{-1}. \tag{5.1}$$

Here, we are considering formal connections, and $g \in G(F)$; the right invariant Mauer–Cartan form $(dg)g^{-1}$ lies in $\mathfrak{g}(F)$, so the formula makes sense. For meromorphic connections, one takes $g \in G$, so the nonlinear gauge term is of course 0.

We remark that there is an equivalence of categories between GL_n-connections and rank n connections given by $(\mathcal{E}, \nabla) \mapsto (V_{\mathcal{E}}, \nabla_V)$, where V is the standard representation of $\mathrm{GL}_n(\mathbb{C})$.

For each $p \in \mathbb{P}^1$, the local behavior of a meromorphic G-connection ∇ at p is governed by the corresponding formal connection $\widehat{\nabla}_p$. Our original three questions about existence, uniqueness, and moduli spaces of meromorphic connections with specified local behavior now make sense for G-connections. As we will see, for Fuchsian and unramified G-connections, the construction of moduli spaces may be generalized to arbitrary reductive G with essentially no changes. However, very little is known about the analogues of the Deligne–Simpson and rigidity problems. It is more challenging to make sense of the three problems in the ramified case, and we will describe how to do so. We will conclude the paper with a discussion of Coxeter G-connections for G simple.

Throughout this section, we fix a Borel subgroup and maximal torus $G \supset B \supset T$, and let $\mathfrak{b} \supset \mathfrak{t}$ be the corresponding Borel subalgebra and Cartan subalgebra. Let U be the unipotent radical of B, and let $\mathfrak{u}$ be its Lie algebra. We also fix a nondegenerate invariant symmetric bilinear form $\langle , \rangle$ on $\mathfrak{g}$. We use $\mathrm{Res}\,\langle , \rangle$ to identify $\mathfrak{g}(F)\frac{dz}{z}$ with the space of smooth $\mathbb{C}$-linear functionals $\mathfrak{g}(F)^\vee$.

5.2. Fuchsian and unramified framable G-connections. Just as for rank n connections, in order to consider moduli spaces of meromorphic connections with only regular singular points, we restrict to *Fuchsian G-connections.* These are defined just as before: one simply takes G-connections of the form $\nabla = d + \sum_{i=0}^k \frac{A_i}{z-a_i} dz$, with $A_i \in \mathfrak{g}$. The isomorphism classes of a regular singularity are parameterized by the (local) monodromy conjugacy class. As long as each residue of the Fuchsian connection ∇ satisfies an appropriate notion of "nonresonance"[3], the local monodromies are determined by the residue. In other words, we can view nonresonant adjoint orbits in $\mathfrak{g}$ as formal types for Fuchsian connections. We immediately get the analogue of Proposition 2.3:

PROPOSITION 5.1. Given nonresonant adjoint orbits $\mathscr{O}_1, \ldots, \mathscr{O}_k \subset \mathfrak{g}$, the corresponding moduli space of Fuchsian G-connections is given by

$$\mathscr{M}(\mathscr{O}_1, \ldots \mathscr{O}_k) = \{(A_1, \ldots, A_k) \in \mathscr{O}_1 \times \cdots \times \mathscr{O}_k \mid \sum_{i=1}^{k} A_i = 0\}/G. \tag{5.2}$$

The generalization of unramified framable connections to G-connections is also straightforward. There is a Jordan form for formal G-connections due to Babbitt and Varadarajan that is entirely analogous to the LT form for rank n formal connections.

THEOREM 5.2 ([**BV83**]). Any formal G-connection is isomorphic to a connection of the form

$$d + (D_r z^{-r/b} + D_{r-1} z^{-(r-1)/b} + \cdots + D_1 z^{-1} + D_0 + N)\frac{dz}{z}, \tag{5.3}$$

where $D_i \in \mathfrak{t}$, $D_r \neq 0$, $N \in \mathfrak{u}$ commutes with each D_i, and $b \in \mathbb{N}$.

The rational number r/b is an invariant called the slope.

The formal connection $\widehat{\nabla}$ is called unramified if one may take $b = 1$ in the Jordan form, and we view the matrix of this Jordan form as an unramified formal type. Thus, the unramified formal types are subsets of the space of principal parts $\mathfrak{g}(\mathfrak{o})^\vee \cong \mathfrak{g}(\mathbb{C}(z^{-1}))\frac{dz}{z}$. The truncated orbits of an unramified formal type $\mathcal{D}$ is just the coadjoint orbit $\mathrm{Ad}^*(G(\mathfrak{o})(\mathcal{D}) \subset \mathfrak{g}(\mathfrak{o})^\vee$.

We again only consider *framable G-connections*; if we assume that ∞ is a regular point, these are connections of the form

$$d + \left(\sum_{i=1}^{k} \sum_{\nu=0}^{r_i} \frac{A_\nu^{(i)}}{(z - a_i)^\nu} \right) \frac{dz}{z} \tag{5.4}$$

with $A_\nu^{(i)} \in \mathfrak{g}$ and $A_{r_i}^{(i)}$ nonnilpotent if $r_i > 0$.

We can construct the moduli space of framable connections with unramified formal types $\mathcal{D}_i$ at the singular points a_i just as for $\mathrm{GL}_n(\mathbb{C})$. Indeed, one need only replace $\mathrm{GL}_n(\mathbb{C})$ by G in (3.5).

We can now formulate the Fuchsian and unramified framable Deligne–Simpson problem. A meromorphic connection is called *Lie-irreducible* if it does not admit a reduction of structure to a proper parabolic subgroup. For framable (and Fuchsian) connections, this simply means that there is no proper parabolic subalgebra of $\mathfrak{g}$ containing all of the coefficients $A_\nu^{(i)}$ from (5.4). The stable moduli space is the subspace consisting of the Lie-irreducible connections with the given formal types.

The Deligne–Simpson and unramified Deligne–Simpson problems for G-connections are just the determination of when the relevant stable moduli spaces are nonempty. However, in spite of the obvious parallelism with the situation for GL_n, I am not aware of any significant results on these problems outside of type A.

---

[3] A formal G-connection $d + (B_0 + B_1 z + \ldots)\frac{dz}{z}$ is called *nonresonant* if $\mathrm{ad}(B_0) \in \mathrm{End}(\mathfrak{g})$ has no eigenvalues in $\mathbb{Z} \setminus \{0\}$.

5.3. Fundamental G-strata and toral G-connections. As we have seen for $G = \mathrm{GL}_n(\mathbb{C})$, if $\widehat{\nabla}$ is a formal G-connection of nonintegral slope, then the naive leading term of $\widehat{\nabla}$ with respect to any trivialization will be nilpotent. One can generalize the theory of fundamental strata discussed in §4.3 to G-connections to obtain a more useful notion of the leading term [**BS18**]. It is also possible to define regular G-strata and toral G-connections [**BS20**]. In this context, lattice chain filtrations are replaced by a class of Lie-theoretically defined filtrations on $\mathfrak{g}(F)$ called *Moy–Prasad filtrations*, which are parameterized by points in the *Bruhat–Tits building* $\mathcal{B}$ for $G(F)$.

To minimize technicalities, from now on we assume that G is simple and simply connected. The standard Iwahori subgroup $I \subset G(F)$ is the pullback of B under the "evaluation at 0" map $G(\mathfrak{o}) \to G$. A proper subgroup $P \subset G(F)$ is called a parahoric subgroup if it contains a conjugate of I. For example, parahoric subgroups in $\mathrm{SL}_n(F)$ are stabilizers of lattice chains just as for GL_n, but here, the conjugacy classes of parahoric subgroups are determined by subsets of $\mathbb{Z}/n\mathbb{Z}$ instead of classes of such subsets under the translation action. The Bruhat–Tits building is a simplicial complex whose facets are in bijective correspondence with the parahoric subgroups of $G(F)$. The chambers (i.e., the maximal simplices) correspond to Iwahori subgroups while vertices correspond to maximal parahoric subgroups. We call the closure of the chamber corresponding to I the *fundamental alcove.* The building is divided into *apartments*–real affine spaces of dimension equal to the rank of G–which are parameterized by the split maximal tori of $G(F)$. The apartment labeled by $T(F)$ is called the standard apartment; it contains the fundamental alcove.

Given $x \in \mathcal{B}$, let $G(F)_x$ (resp. $\mathfrak{g}(F)_x$) denote the parahoric subgroup (resp. subalgebra) corresponding to the facet containing x. The Moy–Prasad filtration $\{\mathfrak{g}(F)_{x,r}\}_{r\in\mathbb{R}}$ associated to x is a decreasing $\mathbb{R}$-filtration of $\mathfrak{g}(F)$ by $\mathfrak{o}$-lattices. It satisfies $\mathfrak{g}(F)_{x,0} = \mathfrak{g}(F)_x$ and is 1-periodic, in the sense that $\mathfrak{g}(F)_{x,r+1} = z\,\mathfrak{g}(F)_{x,r}$. If we set $\mathfrak{g}(F)_{x,r+} = \bigcup_{s>r} \mathfrak{g}(F)_{x,s}$, then the set of r where $\mathfrak{g}(F)_{x,r} \neq \mathfrak{g}(F)_{x,r+}$ is discrete. If x is in the standard apartment, then this filtration is induced by a grading $\mathfrak{g}(\mathbb{C}[z,z^{-1}]) = \bigoplus_{r\in\mathbb{R}} \mathfrak{g}(F)(r)$. We remark that for SL_n or GL_n, a lattice chain filtration with stabilizer P is just a rescaling of the Moy–Prasad filtration at x_P, the barycenter of the facet labeled by P [**BS18**]. For example, $\mathfrak{i}^s = (\mathfrak{gl}_n(F))_{x_I,s/n}$ and $\mathfrak{i}(s) = (\mathfrak{gl}_n(F))_{x_I}(s/n)$.

A G-stratum is a triple (x,r,β) with $x \in \mathcal{B}$, $r \geq 0$, and

$$\beta \in (\mathfrak{g}(F)_{x,r}/\,\mathfrak{g}(F)_{x,r+})^\vee \cong (\mathfrak{g}(F)_{x,-r}/\,\mathfrak{g}(F)_{x,-r+})\,\frac{dz}{z}.$$

It is called fundamental if every representative in this coset is nonnilpotent. The definition of a formal G-connection containing a stratum is just the same as for GL_n.

We can now state the generalization of Theorem 4.3.

THEOREM 5.3 ([**BS18**, Theorem 2.14]). Every formal G-connection $\widehat{\nabla}$ contains a fundamental stratum (x,r,β), where x is in the fundamental alcove and $r \in \mathbb{Q}$; the depth r is positive if and only if $\widehat{\nabla}$ is irregular singular. Moreover,

(1) If $\widehat{\nabla}$ contains the stratum (y,r',β'), then $r' \geq r$.
(2) If $\widehat{\nabla}$ is irregular singular, a stratum (y,r',β') contained in $\widehat{\nabla}$ is fundamental if and only if $r' = r$.

The common depth of the fundamental strata contained in $\widehat{\nabla}$ is one possible definition of the slope of $\widehat{\nabla}$.

To define toral G-connections, we need the notion of a regular G-stratum. For convenience, we will restrict to points in the standard apartment, so that the functionals β with always have a homogeneous representative $\beta^\flat$.

Let $S \subset G(F)$ be a maximal torus. The associated Cartan subalgebra $\mathfrak{s}$ is endowed with a unique Moy–Prasad filtration $\{\mathfrak{s}_r\}$ which comes from a grading with graded pieces $\mathfrak{s}(r)$. We say that x is compatible (resp. graded compatible) with $\mathfrak{s}$ if $\mathfrak{s}_r = \mathfrak{g}_{x,r} \cap \mathfrak{s}$ (resp. $\mathfrak{s}(r) = \mathfrak{g}_x(r) \cap \mathfrak{s}$) for all $r \in \mathbb{R}$. The fundamental stratum (x, r, β) with x in the standard apartment and $r > 0$ is called *S-regular* if x is compatible with $\mathfrak{s}$ and S is the connected centralizer of $\beta^\flat$. If $\widehat{\nabla}$ contains an S-regular stratum, then we say that $\widehat{\nabla}$ is *S-toral*.

Just as for GL_n, there is a bijection between conjugacy classes of maximal tori in $G(F)$ and conjugacy classes in the Weyl group [**KL88**]. A Weyl group element is called *regular* if its action on $\mathfrak{t}$ has a regular semisimple eigenvector; the corresponding eigenvalue is also called regular. If γ is a conjugacy class in W, we say that the maximal torus has type γ if it is in the corresponding class of tori.

THEOREM 5.4. [**BS20**] Let γ be a conjugacy class in W. Then the following are equivalent.

(1) γ is regular and $e^{2\pi i r}$ is a regular eigenvalue for γ.
(2) If S is of type γ, then there exists a regular semisimple element in $\mathfrak{s}(r)$.
(3) There exists an S-toral connection with slope r for some S of type γ.

For GL_n, a maximal torus if of regular type if and only if it is uniform or almost uniform.

Finally, we generalize Theorem 4.5. Let $\mathcal{A}(S, r)$ be the open subset of $\bigoplus_{j \in [-r,0]} \mathfrak{s}(j)$ whose leading component (i.e., the component in $\mathfrak{s}(-r)$) is regular semisimple. This is called the set of *S-formal types* of depth r.

THEOREM 5.5. [**BS20**] If $\widehat{\nabla}$ contains the S-regular stratum (x, r, β), then $\widehat{\nabla}$ is $\mathfrak{g}_{x,0+}$-gauge equivalent to a connection with matrix in $\mathcal{A}(S, r)\frac{dz}{z}$.

Upon choosing nice representatives for the maximal tori admitting toral connections, this theorem can be viewed as giving rational canonical forms for formal connections.

REMARK 5.6. As in Remark 4.6, there is an action of the relative affine Weyl group W_S^{aff} on $\mathcal{A}(S, r)$, and the moduli space of S-toral connections of slope e is given by $\mathcal{A}(S, r)/W_S^{\mathrm{aff}}$ [**BS20**].

5.4. Moduli spaces of toral G-connections. Let us return to our checklist of ingredients necessary to construct moduli spaces. We assume that S is compatible with a point x in the standard apartment, and take $\mathcal{F} \in \mathcal{A}(S, r)$. In fact, we can (and will) choose S so that it is graded compatible with a point x in the fundamental alcove. The truncated orbit is then defined as $\mathrm{Ad}^*(G(F)_x)(\mathcal{F})$. However, a difficulty now arises: the parahoric subgroup $G(F)_x$ is not necessarily a subgroup of $G(\mathfrak{o})$. This is problematic: principal parts are elements of $\mathfrak{g}(\mathfrak{o})^\vee$, and in the construction of moduli spaces, one needs to be able to restrict these functionals to $\mathfrak{g}(F)_x$ to check whether they give the desired formal type. We thus need to make the additional hypothesis that one can choose x so that the closure of the facet containing x contains the vertex corresponding to $\mathfrak{g}(\mathfrak{o})$. Note that this condition is always satisfied for GL_n, since S_e is graded compatible to the barycenter x_{P_e}.

If x satisfies this hypothesis, then $G(F)_x$ is the pullback to $G(\mathfrak{o})$ of a standard parabolic subgroup $Q_x \subset G$. We define the extended orbit $\mathcal{M}_{\mathcal{F}} \subset (Q_x\backslash G) \times \mathfrak{g}(\mathfrak{o})^\vee$ just as before:

$$\mathcal{M}_{\mathcal{F}} = \{(Q_x g, \alpha) \mid (\mathrm{Ad}^*(g)\alpha)|_{\mathfrak{g}(F)_x} \in \mathrm{Ad}^*(G(F)_x)(\mathcal{F})\}.$$

The construction of moduli spaces with toral singularities now goes through as in Theorem 4.8. (See, for example, [**Liv17**], where this is worked out for $G = \mathrm{Sp}_{2n}(\mathbb{C})$.) The ramified Deligne–Simpson problem for G-connections is the determination of when the stable moduli space, consisting of the Lie-irreducible connections with the given formal types, is nonempty.

5.5. Coxeter G-connections. Very little is known about the Deligne–Simpson problem for Fuchsian or unramified G-connections. Indeed, I do not even know of any conjectural statements. A major issue is that the conditions appearing in the results of Crawley-Boevey and Hiroe are not Lie-theoretic in nature. Instead, they involve a translation to an analogous problem for quiver representations, and this translation has no known analogue outside of type A. On the other hand, the solution of the Deligne–Simpson (and rigidity) problem for Coxeter connections is Lie-theoretic, involving the structure of the poset of adjoint orbits, and thereby admits the possibility of a generalization to other types. Moreover, Coxeter G-connections can be defined for any simple group (or for a reductive group with connected Dynkin diagram) [**KS21**]. In this final section, we discuss precise conjectures (and some partial results) for the Deligne–Simpson and rigidity problems for Coxeter G-connections.

Let G be a simple group with fixed Borel subgroup and maximal torus $B \supset T$. Let Φ denote the set of roots and $\Delta \subset \Phi$ the set of simple roots. We let $\mathfrak{g}_\alpha \subset \mathfrak{g}$ denote the root space corresponding to the root α; by convention, we set $\mathfrak{g}_0 = \mathfrak{t}$. We fix an "affine pinning" of $\mathfrak{g}$, namely a generator $X_\alpha \in \mathfrak{g}_\alpha$ for each negative simple root $\alpha \in -\Delta$ and a generator $E \in \mathfrak{g}_\theta$, where θ is the highest root. We now set $\omega_{-1} = z^{-1}E + \sum_{\alpha\in-\Delta} X_\alpha$. Note that for SL_n, an appropriate choice of the affine pinning gives $\omega_{-1} = \omega_n^{-1}$ as defined in §4.1.

It can be shown that ω_{-1} is regular semisimple [**KS21**], and we denote its connected centralizer by $\mathcal{C}$. We call $\mathcal{C}$ the Coxeter maximal torus and its Lie algebra $\mathfrak{c}$ the Coxeter Cartan subalgebra. The terminology is justified by the fact that the conjugacy class of $\mathcal{C}$ corresponds to the Coxeter class in the Weyl group W.

It turns out that $\mathcal{C}$ is graded compatible with x_I, the barycenter of the fundamental alcove. Let h be the Coxeter number of G, i.e., the order of the Coxeter elements in W. The Moy–Prasad grading associated to x_I has nonzero components only in degrees $\frac{1}{h}\mathbb{Z}$. It may be described explicitly in terms of the heights of the roots of $\mathfrak{g}$:

$$\mathfrak{g}_{x_I}(m + \tfrac{s}{h}) = z^m \left[\left(\bigoplus_{\mathrm{ht}(\alpha)=s} \mathfrak{g}_\alpha\right) \oplus \left(z \bigoplus_{\mathrm{ht}(\alpha)=s-h} \mathfrak{g}_\alpha\right)\right] \quad \text{for } 0 \le s < h. \tag{5.5}$$

The graded piece $\mathfrak{c}(r/h) = \mathfrak{g}_{x_I}(r/h) \cap \mathfrak{c}$ is one-dimensional when $\gcd(r, h) = 1$. If $0 < -r < h$, we choose a generator $\omega_r \in \mathfrak{c}(r/h)$; in general, $\omega_{m+r/h} \stackrel{\text{def}}{=} z^m\omega_r$ is a generator. In type A_{n-1}, we simply set $\omega_r = \omega_n^r$.

If $r>0$ is relatively prime to h, then the set $\mathcal{A}(\mathcal{C})$ of $\mathcal{C}$-formal types of depth r/h is the subset of $\left(\bigoplus_{i=0}^{r}\mathfrak{c}(-i/h)\right)\frac{dz}{z}$ with nonzero component in degree $-r/h$. We call the formal type *homogeneous* if it has zero components outside of this degree.

A meromorphic G-connection on $\mathbb{G}_m$ is called a *Coxeter G-connection* if it has a $\mathcal{C}$-toral irregular singularity at 0 and possibly an additional regular singular point at ∞. If $\mathcal{F}$ is a $\mathcal{C}$-formal type and $\mathscr{O}\subset\mathfrak{g}$ is a nonresonant adjoint orbit, we can define the moduli space of framable Coxeter G-connections $\mathcal{M}(\mathcal{F},\mathscr{O})$ with these formal types. Explicitly,

$$\mathcal{M}(\mathcal{F},\mathscr{O})\cong\{(\alpha,Y)\mid\alpha\in\mathfrak{g}(\mathbb{C}[z^{-1}])\tfrac{dz}{z},Y\in\mathscr{O},\alpha|_{\mathfrak{i}}\in\mathrm{Ad}^*(I)(\mathcal{F}),\\ \text{and }\mathrm{Res}(\alpha)+Y=0\}/B.$$

Coxeter connections are automatically Lie-irreducible because the formal connection at 0 is already Lie-irreducible.

We can now state our conjectural solution of the Deligne–Simpson problem for Coxeter G-connections. Let Orb^G be the poset of adjoint orbits under the partial order determined by containment of orbit closures. Let q be an element in the adjoint quotient $\mathfrak{g}/G\cong\mathfrak{t}/W$. The elements in $\mathfrak{g}$ mapping to q form a G-stable closed subset; we let Orb_q^G be the collection of adjoint orbits contained in this subset. There is a poset isomorphism $\mathrm{Orb}^G\cong\bigsqcup_q\mathrm{Orb}_q^G$. Let $\mathrm{DS}(\mathcal{F},\mathscr{O})$ be the set of $\mathscr{O}\in\mathrm{Orb}_q^G$ for which $\mathcal{M}(\mathcal{F},\mathscr{O})$ is nonempty.

CONJECTURE 5.7. Let G be a simple complex group with Lie algebra $\mathfrak{g}$. Fix a $\mathcal{C}$-formal type $\mathcal{F}$ of slope r/h with $\gcd(r,h)=1$. Then for each $q\in\mathfrak{t}/W$, $\mathrm{DS}(\mathcal{F},\mathscr{O})$ is either empty or is a principal filter generated by a (unique) orbit $\mathscr{O}_q^r\in\mathrm{Orb}_q^G$ which is independent of the choice of $\mathcal{F}$ with the given slope. Moreover,

(1) If $r>h$, then $\mathscr{O}_q^r=0$, so $\mathcal{M}(\mathcal{F},\mathscr{O})$ is always nonempty.
(2) If $q=0$ (so Orb_0^G is the set of nilpotent orbits), then $\mathrm{Res}(\omega_{-r}\frac{dz}{z})$ is a representative of $\mathscr{O}_0^r$.

For G simple, $\mathfrak{c}(0)=0$; this explains the absence of the trace condition which appears in Theorem 4.11.

Finally, we discuss rigidity for Coxeter G-connections. As in §4.7, we will actually investigate cohomological rigidity, i.e., when $H^1(\mathbb{P}^1,j_{!*}\mathrm{ad}_\nabla)=0$. For Lie-irreducible connections, the vanishing of this cohomology group implies that ∇ has no infinitesimal deformations. Unlike in type A, it is not known whether physical rigidity and cohomological rigidity coincide. From now on, we will simply call cohomologically rigid connections rigid.

We first consider *homogeneous Coxeter connections*: meromorphic G-connections of the form $\widehat{\nabla}_r^a=d+a\omega_{-r}\frac{dz}{z}$ with $r>0$ relatively prime to h and $a\in\mathbb{C}^*$. When $r=1$, we get the Frenkel–Gross rigid G-connection. In [**KS21**], Kamgarpour and Sage showed that a homogeneous Coxeter G-connection with $r=h+1$ is always rigid. We call these connections *Airy G-connections.* Note that they are nonsingular at ∞.

It turns out that the Frenkel–Gross and Airy connections are the only homogeneous Coxeter G-connections which are rigid for all simple types. The complete classification of rigid homogeneous Coxeter G-connections is given in [**KS21**].

THEOREM 5.8. The Coxeter connection $\nabla=\nabla_r^a$ is rigid if and only if

(i) $r = h + 1$ in which case ∇ is regular at infinity;
(ii) $r = 1$ in which case ∇ is regular singular at infinity with principal unipotent monodromy;
(iii) $1 < r < h$ satisfies the following conditions

Root system	Conditions on r
A_{n-1}	$r \mid n \pm 1$
B_n	$r \mid n+1,\ r \mid 2n+1$
C_n	$r \mid 2n \pm 1$
D_n	$r \mid 2n,\ r \mid 2n-1$
E_7	$r = 7$

in which case ∇ is regular singular at infinity with unipotent monodromy $\exp(\operatorname{Res}(\omega_{-r}\frac{dz}{z}))$.

We expect that the restriction to homogeneous formal types is unnecessary. Indeed, the following statement would follow from the $q = 0$ case of Conjecture 5.7.

CONJECTURE 5.9. Let $\mathcal{F}$ be a $\mathcal{C}$-formal type of slope r/h, and let $\mathcal{O} \subset \mathfrak{g}$ be any nilpotent orbit with $\mathcal{O} \succeq \mathcal{O}_0^r$ (as defined in the previous conjecture). Then, there exists a rigid connection with the given formal type and unipotent monodromy determined by $\mathcal{O}$ if and only if r satisfies the conditions in Theorem 5.8.

References

[BE04] S. Bloch and H. Esnault, *Local Fourier transforms and rigidity for $\mathcal{D}$-modules*, Asian J. Math. **8** (2004), no. 4, 587–605. MR2127940

[BK93] C. J. Bushnell and P. C. Kutzko, *The admissible dual of* GL(N) *via compact open subgroups*, Annals of Mathematics Studies, vol. 129, Princeton University Press, Princeton, NJ, 1993, DOI 10.1515/9781400882496. MR1204652

[Boa01] P. Boalch, *Symplectic manifolds and isomonodromic deformations*, Adv. Math. **163** (2001), no. 2, 137–205, DOI 10.1006/aima.2001.1998. MR1864833

[Boa08] P. Boalch, *Irregular connections and Kac–Moody root systems*, arXiv:0806.1050, 2008.

[BS12] C. L. Bremer and D. S. Sage, *Isomonodromic deformations of connections with singularities of parahoric formal type*, Comm. Math. Phys. **313** (2012), no. 1, 175–208, DOI 10.1007/s00220-012-1493-0. MR2928222

[BS13] C. Bremer and D. S. Sage, *Moduli spaces of irregular singular connections*, Int. Math. Res. Not. IMRN (2013), 1800–1872.

[BS18] C. Bremer and D. S. Sage, *A theory of minimal K-types for flat G-bundles*, Int. Math. Res. Not. IMRN (2018), 3507–3555.

[BS20] C. Bremer and D. S. Sage, *Flat G-bundles and regular strata for reductive groups*, arXiv:1309.6060, 2020.

[Bus87] C. J. Bushnell, *Hereditary orders, Gauss sums and supercuspidal representations of* GL_N, J. Reine Angew. Math. **375/376** (1987), 184–210, DOI 10.1515/crll.1987.375-376.184. MR882297

[BV83] D. G. Babbitt and V. S. Varadarajan, *Formal reduction theory of meromorphic differential equations: a group theoretic view*, Pacific J. Math. **109** (1983), no. 1, 1–80. MR716289

[CB03] W. Crawley-Boevey, *On matrices in prescribed conjugacy classes with no common invariant subspace and sum zero*, Duke Math. J. **118** (2003), no. 2, 339–352, DOI 10.1215/S0012-7094-03-11825-6. MR1980997

[Del77] P. Deligne, *Applications de la formule des traces aux sommes trigonométriques* (French), Cohomologie étale, Lecture Notes in Math., vol. 569, Springer, Berlin, 1977, pp. 168–232. MR3727438

[FG09] E. Frenkel and B. Gross, *A rigid irregular connection on the projective line*, Ann. of Math. (2) **170** (2009), no. 3, 1469–1512, DOI 10.4007/annals.2009.170.1469. MR2600880

[Hir17] K. Hiroe, *Linear differential equations on the Riemann sphere and representations of quivers*, Duke Math. J. **166** (2017), no. 5, 855–935, DOI 10.1215/00127094-3769640. MR3626566

[HNY13] J. Heinloth, B.-C. Ngô, and Z. Yun, *Kloosterman sheaves for reductive groups*, Ann. of Math. (2) **177** (2013), no. 1, 241–310, DOI 10.4007/annals.2013.177.1.5. MR2999041

[HY14] K. Hiroe and D. Yamakawa, *Moduli spaces of meromorphic connections and quiver varieties*, Adv. Math. **266** (2014), 120–151, DOI 10.1016/j.aim.2014.05.024. MR3262356

[JKY21] K. Jakob, M. Kamgarpour, and L. Yi, *Airy sheaves for reductive groups*, Proc. Lond. Math. Soc. (3) **126** (2023), no. 1, 390–428, DOI 10.1112/plms.12494. MR4535023

[Kat87] N. M. Katz, *On the calculation of some differential Galois groups*, Invent. Math. **87** (1987), no. 1, 13–61, DOI 10.1007/BF01389152. MR862711

[Kat88] N. M. Katz, *Gauss sums, Kloosterman sums, and monodromy groups*, Annals of Mathematics Studies, vol. 116, Princeton University Press, Princeton, NJ, 1988, DOI 10.1515/9781400882120. MR955052

[Kat90] N. M. Katz, *Exponential sums and differential equations*, Annals of Mathematics Studies, vol. 124, Princeton University Press, Princeton, NJ, 1990, DOI 10.1515/9781400882434. MR1081536

[KL88] D. Kazhdan and G. Lusztig, *Fixed point varieties on affine flag manifolds*, Israel J. Math. **62** (1988), no. 2, 129–168, DOI 10.1007/BF02787119. MR947819

[KLM+22] M. C. Kulkarni, N. Livesay, J. P. Matherne, B. Nguyen, and D. S. Sage, *The Deligne-Simpson problem for connections on $\mathbb{G}_m$ with a maximally ramified singularity*. part B, Adv. Math. **408** (2022), no. part B, Paper No. 108596, 28, DOI 10.1016/j.aim.2022.108596. MR4458913

[Kos02] V. P. Kostov, *On the Deligne-Simpson problem*, Tr. Mat. Inst. Steklova **238** (2002), no. Monodromiya v Zadachakh Algebr. Geom. i Differ. Uravn., 158–195; English transl., Proc. Steklov Inst. Math. **3(238)** (2002), 148–185. MR1969311

[Kos03] V. P. Kostov, *On some aspects of the Deligne-Simpson problem*, J. Dynam. Control Systems **9** (2003), no. 3, 393–436, DOI 10.1023/A:1024673222003. MR1990242

[Kos04] V. P. Kostov, *The Deligne-Simpson problem—a survey*, J. Algebra **281** (2004), no. 1, 83–108, DOI 10.1016/j.jalgebra.2004.07.013. MR2091962

[KS19] M. Kamgarpour and D. S. Sage, *A geometric analogue of a conjecture of Gross and Reeder*, Amer. J. Math. **141** (2019), no. 5, 1457–1476, DOI 10.1353/ajm.2019.0038. MR4011806

[KS21] M. Kamgarpour and D. S. Sage, *Rigid connections on $\mathbb{P}^1$ via the Bruhat-Tits building*, Proc. Lond. Math. Soc. (3) **122** (2021), no. 3, 359–376, DOI 10.1112/plms.12346. MR4230058

[KS23] M. Kamgarpour and D. S. Sage, *Connections whose differential Galois groups are reductive of maximal degree*, arXiv:2309.11742, 2023.

[Lev75] A. H. M. Levelt, *Jordan decomposition for a class of singular differential operators*, Ark. Mat. **13** (1975), 1–27, DOI 10.1007/BF02386195. MR500294

[Liv17] N. D. Livesay, *Moduli Spaces of Flat GSp-Bundles*, ProQuest LLC, Ann Arbor, MI, 2017. Thesis (Ph.D.)–Louisiana State University and Agricultural & Mechanical College. MR4463452

[LSN23] N. Livesay, D. S. Sage, and B Nguyen, *Explicityit constructions of connections on the projective line with a maximally ramified irregular singularity*, arXiv:2303.06581, 2023.

[LT17] T. Lam and N. Templier, *Mirror symmetry for minuscule flag varieties*, `arXiv:1705.00758`, 2017.

[Sab08] C. Sabbah, *An explicit stationary phase formula for the local formal Fourier-Laplace transform*, Singularities I, Contemp. Math., vol. 474, Amer. Math. Soc., Providence, RI, 2008, pp. 309–330, DOI 10.1090/conm/474/09262. MR2454354

[Sag00] D. S. Sage, *The geometry of fixed point varieties on affine flag manifolds*, Trans. Amer. Math. Soc. **352** (2000), no. 5, 2087–2119, DOI 10.1090/S0002-9947-99-02295-3. MR1491876

[Sag17] D. S. Sage, *Regular strata and moduli spaces of irregular singular connections*, New trends in analysis and interdisciplinary applications, Trends Math. Res. Perspect., Birkhäuser/Springer, Cham, 2017, pp. 69–75, DOI 10.1007/978-3-319-48812-7_10. MR3695627

[Sim91] C. T. Simpson, *Products of matrices*, Differential geometry, global analysis, and topology (Halifax, NS, 1990), CMS Conf. Proc., vol. 12, Amer. Math. Soc., Providence, RI, 1991, pp. 157–185. MR1158474

[Tur55] H. L. Turrittin, *Convergent solutions of ordinary linear homogeneous differential equations in the neighborhood of an irregular singular point*, Acta Math. **93** (1955), 27–66, DOI 10.1007/BF02392519. MR68689

[Was76] W. Wasow, *Asymptotic expansions for ordinary differential equations*, Robert E. Krieger Publishing Co., Huntington, NY, 1976. Reprint of the 1965 edition. MR460820

[Zhu17] X. Zhu, *Frenkel-Gross' irregular connection and Heinloth-Ngô-Yun's are the same*, Selecta Math. (N.S.) **23** (2017), no. 1, 245–274, DOI 10.1007/s00029-016-0238-x. MR3595893

Department of Mathematics, Louisiana State University, Baton Rouge, LA 70803

Current address: Department of Mathematics, University at Buffalo, Buffalo, New York 14260

Email address: dsage@buffalo.edu

SELECTED PUBLISHED TITLES IN THIS SERIES

804 **Mahir Bilen Can and Jörg Feldvoss, Editors,** A Glimpse into Geometric Representation Theory, 2024

802 **Ralph M. Kaufmann, Martin Markl, and Alexander A. Voronov, Editors,** Higher Structures in Topology, Geometry, and Physics, 2024

801 **K. A. Brown, T. J. Hodges, M. Vancliff, and J. J. Zhang, Editors,** Recent Advances in Noncommutative Algebra and Geometry, 2024

800 **Avinoam Mann, Louis H. Rowen, David J. Saltman, Aner Shalev, Lance W. Small, and Uzi Vishne, Editors,** Amitsur Centennial Symposium, 2024

799 **Alberto A. Condori, Elodie Pozzi, William T. Ross, and Alan A. Sola, Editors,** Recent Progress in Function Theory and Operator Theory, 2024

798 **Marat V. Markin, Igor V. Nikolaev, and Carsten Trunk, Editors,** Advances in Functional Analysis and Operator Theory, 2024

797 **Sangita Jha, Mrinal Kanti Roychowdhury, and Saurabh Verma, Editors,** Recent Developments in Fractal Geometry and Dynamical Systems, 2024

796 **John Cremona, John Jones, Jennifer Paulhus, Andrew V. Sutherland, and John Voight, Editors,** LuCaNT: LMFDB, Computation, and Number Theory, 2024

795 **Michael A. Jones, David McCune, and Jennifer M. Wilson, Editors,** Mathematical Analyses of Decisions, Voting and Games, 2024

794 **Jean-Pierre Magnot, Editor,** Recent Advances in Diffeologies and Their Applications, 2024

793 **Abba Gumel, Editor,** Mathematical and Computational Modeling of Phenomena Arising in Population Biology and Nonlinear Oscillations, 2024

792 **Shaoming Guo, Zane Kun Li, and Brian Street, Editors,** Recent Developments in Harmonic Analysis and its Applications, 2024

791 **Mee Seong Im, Bach Nguyen, and Arik Wilbert, Editors,** Algebraic and Topological Aspects of Representation Theory, 2024

790 **Peter Crooks and Alexandru I. Suciu, Editors,** Compactifications, Configurations, and Cohomology, 2023

789 **I. S. Krasil′shchik, A. B. Sossinsky, and A. M. Verbovetsky, Editors,** The Diverse World of PDEs, 2023

788 **I. S. Krasil′shchik, A. B. Sossinsky, and A. M. Verbovetsky, Editors,** The Diverse World of PDEs, 2023

787 **Hemen Dutta, Editor,** Mathematical Modelling, 2023

786 **Hemen Dutta, Editor,** Mathematical Modelling, 2023

785 **A. Leroy and S. K. Jain, Editors,** Algebra and Coding Theory, 2023

784 **Dinh-Liem Nguyen, Loc Hoang Nguyen, and Thi-Phong Nguyen, Editors,** Advances in Inverse Problems for Partial Differential Equations, 2023

783 **Alla Detinko, Michael Kapovich, Alex Kontorovich, Peter Sarnak, and Richard Schwartz, Editors,** Computational Aspects of Discrete Subgroups of Lie Groups, 2023

782 **Galina Filipuk, Alberto Lastra, and Sławomir Michalik, Editors,** Recent Trends in Formal and Analytic Solutions of Diff. Equations, 2023

781 **Emanuel Indrei, Diego Marcon, and Levon Nurbekyan, Editors,** Geometric and Functional Inequalities and Recent Topics in Nonlinear PDEs, 2023

780 **Erik Koelink, Stefan Kolb, Nicolai Reshetikhin, and Bart Vlaar, Editors,** Hypergeometry, Integrability and Lie Theory, 2022

779 **Samuele Anni, Valentijn Karemaker, and Elisa Lorenzo García, Editors,** Arithmetic, Geometry, Cryptography, and Coding Theory 2021, 2022